NEUROTOX '91
Molecular Basis of Drug & Pesticide Action

Papers presented at: Neurotox '91, an SCI Meeting held at The University of Southampton, UK, 7–11 April 1991.

NEUROTOX '91

Molecular Basis of Drug & Pesticide Action

Editor

I. R. DUCE

Department of Life Science,
University of Nottingham,
University Park,
Nottingham NG7 2RD, UK

Published for SCI

by

ELSEVIER APPLIED SCIENCE

LONDON and NEW YORK

ELSEVIER SCIENCE PUBLISHERS LTD
Crown House, Linton Road, Barking, Essex IG11 8JU, England

WITH 33 TABLES AND 117 ILLUSTRATIONS

© 1992 SCI

British Library Cataloguing in Publication Data

Neurotox '91 (University of Southampton)
Neurotox '91: molecular basis of drug &
pesticide action.
I. Title   II. Duce, I. R.
574.192
ISBN 1851667466

Library of Congress Cataloguing-in-Publication Data

Neurotox '91 (1991 : University of Southampton)
NEUROTOX '91 ; molecular basis of drug & pesticide action /
editor, I. R. Duce.
    p. cm.
"Papers presented at: Neurotox '91, an SCI meeting held at the
University of Southampton, UK, 7–11 April 1991"—Half t.p.
Includes bibliographical references and indexes.
ISBN 1-85166-746-6
1. Molecular neurobiology—Congresses. 2. Neurotoxicology—
Congresses. 3. Neuropharmacology—Congresses. I. Duce, I. R.
II. Society of Chemical Industry (Great Britain) III. Title.
IV. Title: Molecular basis of drug & pesticide action. V. Title:
Molecular basis of drug and pesticide action.
QP356.2.N486   1992   91–39933
591.2′4—dc20   CIP

Preface

NEUROTOX '91 was the fourth meeting in a series which started in 1979. The '91 meeting, like its predecessors, was held under the patronage of the Society of Chemical Industry, and despite the unfortunate proximity of hostilities in the Arabian Gulf attracted a truly international mix of industrial and academic pesticide scientists.

This volume contains the text of invited papers read at the meeting and presents the dramatic developments which so excited those who attended. The potential of molecular neurobiology for gaining knowledge of target sites for neurotoxicants is now starting to be realised. These studies, in conjunction with developments in molecular imaging and modelling, provide new opportunities for chemists and biologists to gain insights into molecular interactions underlying intoxication. Molecular techniques have also enabled rapid advances on a second front, where the cloning of genes controlling pesticide resistance should have a profound impact on our understanding of this commercially important problem.

The understanding of molecular events will undoubtedly be vital in future developments in chemical control of pests; however, the value of understanding the way in which the nervous system controls behaviour and how behaviour can be modified by chemicals of bath synthetic and natural origins was highlighted. Natural products and their synthetic analogues have continued to provide new and interesting molecules which are already proving their worth as tools for the neuroscientist and may offer leads for commercial synthesis.

Lest we are carried away on a wave of enthusiasm by the exciting developments described in this book, a number of industrial colleagues point out in their papers the relatively small impact such research has had on the discovery of the existing commercial pesticides. This caveat must of course be considered seriously but the optimism and enthusiasm which pervaded NEUROTOX '91 leads me to anticipate that in future a balance between the pragmatism of industry and the speculative approach of academia offers the best chance of success for both. Conferences such as NEUROTOX have a role to play in fostering this relationship and it is my hope that this volume provides a valuable account of current developments in pesticide science and some of the challenges facing us.

I. R. Duce
Nottingham, UK
July 1991

Prologue

Presented at NEUROTOX '91, Southampton, April 1991
International Symposium on the Molecular Basis of Drug and Pesticide Action

Invertebrate Neuroscience and its Potential Contribution to Insect Control

GÜNTHER VOSS & RAINER NEUMANN

Agro Division, Ciba-Geigy Ltd, 4002 Basle, Switzerland

When we attended the first NEUROTOX Meeting in York 12 years ago, we were simply quiet consumers of lectures, posters and group discussions on insect neurobiology and pesticide action. We went back to our laboratory in Basle with expectations that at least some of the newest findings and academic recommendations provided at the meeting would prove beneficial for the work of industrial chemists and biologists alike. This was particularly true with regard to facilitating a more rational discovery of toxicants as expressed by Dr Graham-Bryce in his foreword to the Proceedings of the 1979 Symposium.

In contrast to York, however, NEUROTOX '91 in Southampton gives us a less comfortable feeling. Firstly, the former lab bench scientists have turned into managers who now realise that speaking to those who know better is more difficult than listening to them. Secondly, the fun of science, its progress, prospects and promises as being communicated among the participants of this meeting puts us at a clear disadvantage: they are all much more attractive than the well-known complaints of and constraints to the insect control business: (1) a stagnant world market, (2) the pressure on product prices, (3) the lack of hard currency in Third World countries, (4) cheap product imitations, (5) insecticide resistance and pest acceleration, (6) smaller and smaller innovation steps versus higher costs of R&D, and (7) the media-driven public belief that most plant protection chemicals can be replaced by 'soft' technologies including biological products, even though most of them are still unreliable, uneconomic and of a very limited practical value.

We are not neurobiologists ourselves, we have never inserted a microelectrode into an insect nerve or muscle, and our past experimental background is limited to using vertebrate and invertebrate enzymes as tools in comparative and analytic biochemistry, for screening purposes, in insecticide resistance monitoring programmes, mode of action studies and other practical applications of interest to industrial R&D. When we agreed to present our thoughts at this prestigious symposium on 'Invertebrate neuroscience and its potential contribution to insect control' we did not do it because we felt most qualified, but we accepted the challenge as managers who are often said to make quick decisions on problems they are unable to grasp themselves. We hope, however, that many years of experience in industry and many working relations with chemists, screening biologists, biochemists, field entomologists, production chemists and colleagues from marketing will partly compensate for our deficiencies in the area of neurobiology.

Perhaps some of you would have preferred to listen to an introductory lecture providing 'neuroscientific visions' instead of being bothered with the realities and contradictions of industrial R&D. Nevertheless, it is these realities and contradictions that we wish to discuss. Among them are:

— the undeniable economic importance and value of neuroactive insecticides versus their negative public image;
— the very small number of useful modes of neurotoxic action as compared to hopes and claims that additional and more selective neurotargets exist;
— the discrepancy between our continuing reliance on random *in vivo* screening and often heard promises that the rational *in vitro* design of insecticides is just around the corner.

We have tried to summarize our personal attitude towards and judgement of neuroscience and its value for R&D insect control in four theses. The first one reads as follows:

Thesis 1—Real or perceived safety problems associated with neuroactive insecticides force industry to divert more R&D resources to new and alternative pest control technologies.

Pesticides are no exception to the experience of mankind that beneficial inventions are often not free of risks. Since organophosphates and carbamates inhibit acetylcholinesterases of both invertebrates and mammals, their misuse can cause accidental intoxications in unprotected and uneducated users, especially in developing countries. Some of these and other neuroactive, but also non-neuroactive insecticides can also affect aquatic and terrestrial wildlife, or kill beneficial insects such as bees, parasites and predators. In spite of an increasingly broad industrial and regulatory risk assessment for the benefit of users, consumers and the environment (Table 1), and in spite of all the progress made in minimizing adverse effects through product enhancement or replacement, our mass media continue to generalize and to inflate the actual risks of plant protection chemicals in general, and neuroactive insecticides in particular. This is especially true in affluent countries where safe and well-studied chemical specialities are regarded more as consumer or environmental poisons than as valuable productivity tools for agriculture. Insufficient information about the benefits of agrochemical products coupled with public campaigns capitalizing on the 'fear of the unknown' can have quite dramatic

TABLE 1
Toxicological and Environmental Studies

Toxicology		Environment
Acute:	oral, dermal, inhalation	Physico–chemical parameters
Subacute:	dermal, (inhalation)	Hydrolysis
	4–weeks feeding	Photolysis
Subchronic:	3–months feeding	Leaching in soil (laboratory, field)
Chronic:	2–years feeding	Evaporation from water and soil
Special:	mutagenicity	Soil metabolism (aerobic, anaerobic)
	carcinogenicity	Plant metabolism
	2–gen. reproduction	Ecotoxicity (laboratory, field)
	neurotoxicity	Behaviour in the food chain
	teratogenicity	
	skin, eye irritation	
	allergy potential	
	animal metabolism	

effects as exemplified by a survey published in *Scientific American* in 1982 (Table 2). When college students, members of the league of women voters and business men in the US were asked to rank 30 types of activities or products with a known risk potential, pesticides were placed in positions 4, 9 and 15, respectively, as compared to their real place in rank 28.

The plant protection industry and its customers in agriculture have to accept the challenge that their own benefit/risk assessments often differ from those of the public, and that actions may be taken by Government bodies that would change the pest control picture and make chemical control less attractive. As a consequence, industry has already shifted considerable R&D, production and

TABLE 2
Risks in the USA[a]

1	Smoking	150 000
2	Alcohol	100 000
3	Motor vehicles	50 000
4	Handguns	17 000
5	Electric power	14 000
6	Motorcycles	3 000
7	Swimming	3 000
8	Surgery	2 800
9	X-rays	2 300
11	General aviation	1 300
14	Hunting	800
15	Home appliances	200
25	Vaccinations	10
28	**PESTICIDES**	**<10**
30	Spray cans	<10

[a] Reproduced from *Scientific American*.

Prologue

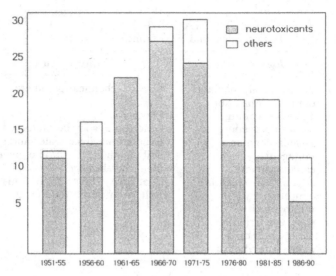

Fɪɢ. 1. Numbers and types of insecticides/acarides introduced between 1950 and 1990. Data from *Pesticide Manual, 1990.*

marketing efforts to new and 'softer' technologies, which include safer formulations, packaging and application systems, more selective chemicals with better fitness for integrated pest and resistance management, and various biological products. The impact of this development on the relative importance of neurotoxicants is evident (Fig. 1): ingredients introduced into the market during the 1960s were almost exclusively neuroactive, whereas the 1970s and 1980s have seen a growing proportion of products with other modes of action, but also a significant decrease of the total number of insecticides and acaricides introduced. The obvious trend to non-neurotoxicants which we expect to continue is probably a reflection of both the growing need for and acceptance of novel types of products capable of solving new problems, and the difficulty of identifying and utilizing additional neuroactive principles that provide new effects on insect nerves and sufficient biological activity at the same time.

The ongoing shift to 'soft' insecticides, however, sharply contrasts with the present situation in the market (Fig. 2). According to our own estimate these low toxicity non-neurotoxic insecticides (synthetic analogs of juvenilhormones, chitin biosynthesis inhibitors and a few others) have not even gained a 10% share of the world-wide insect control market. The large number of neuroactive insecticides, depicted as circles sized in proportion to sales (primarily organophosphates, carbamates, and pyrethroids) continue to dominate the market, although several of them are quite toxic. There is no doubt that they still provide the most effective, reliable and economic solutions to pest problems in agriculture, stored products, animal health and public hygiene, and we do not believe that quick and adequate replacements are at hand.

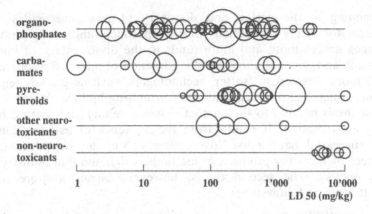

FIG. 2 Insecticides and their acute mammalian toxicities (circles: proportional to sales). Data from *Pesticide Manual, 1990* and CIBA-GEIGY Marketing.

According to published figures (Table 3) the total sales value of insecticides and acaricides will reach approximately 7 billion US$ by the year 1995. DDT and the cyclodienes will continue to lose market share, whereas the pyrethroids are still expected to grow. Organophosphates will grow slightly and remain by far the most important single group of insecticides. Carbamates will hold their position, whereas the remaining products—other neuroactive compounds as well as all non-neurotoxicants—are expected to almost double their sales volume between 1988 and 1995. Disregarding the GABA activated ion channels and a few other neurotargets of minor importance, we must conclude that only two modes of action represent the base for 80% of the insecticide market: the voltage dependent sodium channels and the acetylcholinesterases of the insect nervous system. Our reliance on these two targets appears to be tremendous; their protection from becoming ineffective through resistance is an absolute must.

After many years of unfulfilled hopes and unkept promises we sense a growing reluctance on the part of industry to fund speculative neuroscience

TABLE 3
Neuroactive Insecticides and Their Economic Importance

Types of successful insecticides/acaricides	Mio US$ world sales[a]			Interference with or inhibition of
	1972	1988	1995	
Cyclodienes DDT	1 580	500	350	GABA activated ion channels
Pyrethroids	none	1 150	1 500	Voltage–sensitive sodium channels
Organophosphates	1 375	2 325	2 520	Acetylcholinesterase
Carbamates	885	1 400	1 400	Acetylcholinesterase
All others	175	700	1 330	

[a] Source: County NatWest Woodmac 1990.

projects aiming at the design and discovery of new insecticides. Firstly, requirements for safety evaluations of new products and re-registrations of existing ones absorb more and more funds to the disadvantage of innovative basic research and, secondly, the shrinking fraction of financial resources left is more and more directed to 'softer' technologies, such as growth regulators, pheromones, microbial products and other promising biological agents.

Our first thesis pointed to the conflict between reality and wishes, between rationality and emotion. It results from the existence of and the continuing need for successful neuroactive insecticides versus the public demand, to replace them by less risky products or methods, a demand that industry cannot afford to neglect. The second thesis, however, carries a more technical message. It reads as follows:

Thesis 2—Insecticides have to comply with the needs of the market and society. The multitude of positive product features required for safety, performance, selectivity, manufacturing, costs and patenting cannot be 'designed' by neuroscience.

Industry is expected to discover and develop 'ideal insecticides' designed to satisfy the needs of several 'user groups' in our society: farmers and workers in agriculture, employees of the plant protection industry, consumers of agricultural produce, environmentalists, legislators and others. Thus, the ideal insecticides should be

—highly researched but low-priced;
—superior to existing products but not more expensive;
—highly selective but big in market size;
—long-lived in the market but should not create resistance problems;
—broadly active on pests but inactive on non-target species;
—long-lasting on crops and in insect pests but not causing residues in crops and the environment;
—fast-acting on pests but preferably not as a neurotoxicant;
—mobile in plants but immobile in soil.

In order to cope with these contradictions and to 'optimize the compromise' between them, the entire product creation process for new insect control agents has become much more sophisticated and problem oriented than it has been in the past (Fig. 3), when markets were wide open to absorb all the many broad-spectrum insecticides discovered and developed by the plant protection industry. Today the insect control market is saturated and a rigid, insufficiently harmonized legislation, strong competition, increasing costs as well as high performance standards difficult to exceed determine the environment in which industrial R&D has to discover and develop products with acceptable modes of action, stabilities and physicochemical properties. Customers, society and industrial marketing request and expect R&D to solve new or still existing practical problems, such as multi-resistant diamond-back moths in South-East Asia, whiteflies and aphids in cotton, planthoppers in rice, scale insects in

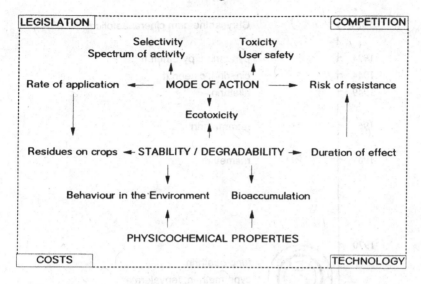

FIG. 3. R&D insect control—framework and topics.

citrus, spider mites and psyllids in deciduous fruits, or the corn rootworm in soils of the American corn belt. Only a more and more sophisticated random screening on live target species in the appropriate development stage can provide the right answers; in-vitro and in-situ test systems based on neurophysiological models prepared from locusts, cockroaches, houseflies, *Manduca* and others are inappropriate tools for identifying the best products for the envisaged market at the lowest cost in the shortest possible time. What we need are safer and more selective insecticides designed for markets and problems large and important enough to justify development, not inhibitors of enzymes or receptor ligands designed for or discovered in neurobiological model systems. Although 'biorational' in-vitro concepts are often perceived to be scientifically and intellectually more demanding than 'simple' random screening on target insects followed by chemical, physicochemical, biological, toxicological and economic optimization, industrial R&D must continue to rely on its traditional but highly successful 'try and see' approach.

A hypothetical question can easily clarify our point of view: could neuroscience have contributed to the discovery of today's pyrethroids, or—to phrase it in a different way—what could neurobiology contribute to product optimization and development if pyrethrins were to be discovered now instead of 70 years ago? An answer to this question may well provide information where knowledge could have led to faster success in the past and where it could assist in making better decisions in the future.

The history of pyrethroids (Fig. 4) can be divided into three phases: (1) the discovery of useful insecticidal properties within the Pyrethrum plant some 170 years ago; (2) the elucidation of the underlying chemical structures about 100

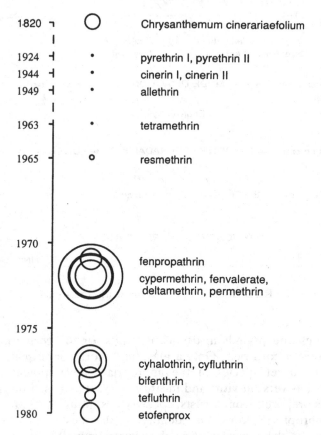

FIG. 4. The history of pyrethroids (circle area proportional to sales).

years later; and (3) the modification of these structures aiming at very respectable sales products that fulfill the needs of the market.

Let us look at the first phase: can neurophysiology help to find natural insecticides? Our answer is, 'No!' We would not like to spend limited R&D funds on such attempts, because tests on target insects are faster, cheaper and more relevant. Moreover, they provide a broader information base, since the whole range of potential targets can be tested at the same time on a live insect.

Also the second phase, structural elucidation of natural products, does not need neuroscience; perhaps an in-vitro bioassay could support in-vivo data. The third phase, however, lead optimization, appears to be an area for using neurobiological expertise. Has its application really happened, could it have happened and if not, can it happen in the future? We would be surprised if those chemists who engaged themselves in the stepwise optimization of pyrethroids that even led to pyrethroid ethers like ethofenprox would give any credit to neuroscience when writing their success stories. Primarily it was not the intrinsic activity that needed to be improved when moving from the pyrethrins to the pyrethroids, but the products' hydrolytic and oxidative stability. Therefore, we conclude that neurophysiology did not help synthesis chemists in the discovery and optimization process of pyrethroids, whose sales, however, have helped to finance a lot of basic neurophysiological research.

What about organophosphates and carbamates? Farmers do not necessarily prefer to use the most potent cholinesterase inhibitors. They want products that are both effective and safe, properties mainly based on differential metabolism. Also for another class of insecticides, the non-neurotoxic chitin synthesis inhibitors, intrinsic activity at the site of action was found to be of secondary importance. Here a compromise had to be achieved between stability in the insect gut and acceptable environmental degradation.

We do not hesitate to state that so far neuroscience has not been able to design new types of insecticides which would have had the potential to compete with existing products in the market. Obviously, the aim of industry to search for even more selective compounds targeted to pre-defined uses will even further minimize the chances for neurobiological success in product design and discovery.

Also our third thesis does not leave room for much optimism. It reads:

Thesis 3—The number of neurotargets providing acceptable and economic pest control opportunities is much smaller than anticipated by academic research.

Our colleague Dr Jack Benson has compiled about two dozen neuronal target sites for insecticides, that are either known to exist in invertebrates or have been proposed as hypothetical targets (Table 4)

—The first part refers to the potential of utilizing receptor ligand recognition sites. In the area of 'classic' neurotransmission, only the nicotinic acetylcholine and octopamine receptors as affected by nicotine, cartap and the new nitromethylenes on the one hand, and the formamidines on the other hand, are of practical importance. Commercial insecticides working on the muscarinic acetylcholine receptor site or acting as glutamate, serotonin, dopamine and histamine mimics are unknown, as are commercial insecticides based on neuropeptides.

—Only one commercial target has so far been identified among the receptor activated ion channel types: the GABA activated chloride channel appears to be the target of the aging cyclodiene insecticides. The non-receptor activated voltage dependent sodium channel, however, represents one of the most important targets for insecticides: it is affected by DDT and the pyrethroids. The natural product derivatives milbemycin and avermectin are said to work on the passive chloride channel, whereas the voltage dependent calcium and potassium channels as well as the metabolic sodium pump have never been shown to be targets for insecticides.

—There has to be at least one system for the removal of each transmitter

TABLE 4
Neuronal Targets and Insecticides

Receptor Ligand Recognition Sites

    cholinergic (nicotinic)—nicotine, cartap, nitromethylenes
    cholinergic (muscarinic)—none
    GABAergic—see Cl channel
    octopaminergic—formamidines
    glutamatergic—none
    serotonergic—none
    dopaminergic—none
    histaminergic—none
    neuropeptides—none

Ion Channels

    Cl channel—GABA—cyclodienes
    Na—voltage—DDT, pyrethroids
    Cl channel—passive—milbemycin, avermectin
    Ca channels—voltage—none
    K channels—voltage—none

Metabolic Ion Pumps

    Na pump—Na exclusion—none

Transmitter Re-uptake and Breakdown

    AChE—ACh—organophosphates, carbamates
    MAO—monoamines—none

Transmitter Synthesis

    Choline acetyl transferase—none
    Glutamate and other decarboxylases—none
    GABA transaminase—none
    Neuropeptide post-translational processing—none

Second Messenger Systems

    Cyclic AMP—e.g. octopamine—none
    Cyclic GMP—e.g. eclosion hormone—none
    Phosphotidylinositol—e.g. salivary glands—none

after it has served its purpose at the site of action, be it by metabolism or direct re-uptake. Two metabolic systems are well known in insects: acetylcholinesterase and monoamine oxidase. Inhibitors of cholinesterases, the organophosphates and carbamates, have become the most successful commercial insecticides, whereas monoamine oxidases have so far never shown potential as an insecticide target. The same is true for several transmitter synthesis and second messenger systems.

We cannot hide feelings of irritation and disappointment that 50 and 40 years after the introduction of DDT and the first organophosphates, respectively, insect control continues to rely on only two major neurotargets that really work: the voltage dependent sodium channels and acetylcholinesterases. Three targets are of minor economic importance: the nicotinic acetylcholine receptor which may become more attractive as the nitromethylene insecticides develop, the octopamine receptor and the GABA activated chloride channels. The remaining 20 targets listed, however, have so far failed to fulfill their promise for the insect control business.

In spite of the fact that at least five million chemicals have now been screened for insecticidal activity, only two major and three minor neurotargets of practical value have emerged. This unpleasant situation will have to be discussed and thoroughly analysed by those who know much more about the insect nervous system than we do; our own contribution can only be to ask a few strategic and philosophical questions:

—What are the chances of success for discovering novel types of chemical structures that interfere with sodium channels and acetylcholinesterases in insects? Do we really need such compounds? Would their development not even increase the risk of losing the two most valuable modes of action through target site related cross-resistance? What are the chances for new products designed to cope with modified sodium channels and acetylcholinesterases to assist in resistance management?

—Why have the cyclodienes remained the only GABAergic insecticides, and the formamidines the only ones that interfere with the octopamine receptor? Why have all conventional and biorational attempts failed to identify new chemistry with dieldrin- and chlordimeform-like activity? Are the octopamine and the nicotinic acetylcholine receptors promising targets to improve insect–insect selectivities, and if yes, why?

—What do we do with the apparently existing, but unutilized hypothetical invertebrate neurotargets? Do we continue to speculate that the day will come when at least a few of them will represent more than exciting subjects for academic research? How realistic are these hopes, how feasible are these targets as a base for safer, more selective, and economic commercial products? If these hypothetical targets would really work for test compounds applied orally or by contact, would we not have identified them in our large-scale industrial screening operations coupled to subsequent mode of action studies? And, last but not least, should a cost and priority minded manager in industrial R&D allow vague speculations to draw scarce R&D funds from more realistic projects that also wait to be implemented?

Real and perceived safety problems, the growing need to give the market what it requests, and the apparently very small number of useful targets limit the opportunities of neuroscience to contribute to the design, discovery and

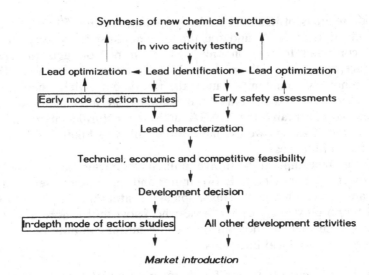

FIG. 5. R&D insect control—contribution of neuroscience.

development of better and safer insecticides that are wanted by everybody. In fact, we may claim that so far neuroscience has obtained more benefits from insecticides than the insect control business has from neuroscience.

We hope that the provocative statements made while presenting the first three theses can provide a base for some in-depth discussions. Thesis 4, however, carries a more optimistic and conciliatory message:

Thesis 4—Assessing and explaining the modes of action, resistance and selectivity of toxic and behaviour modifying chemicals represents the main value of invertebrate neuroscience.

Neurophysiology and biochemistry start to play their main role as soon as a new active ingredient with proven in-vivo insecticidal activity has been identified (Fig. 5). We are convinced that early—though admittedly superficial—mode of action studies done in parallel to chemical and biological optimization can tell us something about a new lead's value and assist in making better decisions faster. With progressing product development, however, in-depth mode of action studies become an essential part of the whole information package required to satisfy both regulatory authorities as well as the high quality standards of research-minded companies who want to know how their products act on the physiological, biochemical or molecular level in target and non-target organisms. It is primarily in this area where neuroscience will continue to produce its most valuable achievements, and where industry is prepared to cooperate with academic research.

Again, the group of pyrethroids can be used as a good example for the progress made in the mode of action area: Today we know, that

—pyrethroids act on axonal sodium channels, a finding confirmed by single channel recordings;

—only a small number of channels have to be affected to produce the well-known high potency of these insecticides;

—temperature-dependent toxicities can be explained at the neuronal level;

—and cross-resistance against DDT and pyrethroids have a common site.

All this is important and useful information for which neuroscientists in both academic and industrial laboratories deserve a lot of credit, although many questions are still unanswered: for the pyrethroids, we lack an explanation for the differences between type I and type II compounds at the molecular level, and for anticholinesterases and compounds acting on the acetylcholine receptor a closer insight into the relative importance of synapses located at different sites within the nervous system, and of isozymes would be welcome. We also need to know more about the lethal modes of action of those insecticides that do not directly interfere with the insect's nervous system, such as respiration inhibitors. What is their effect on the proper functioning of the nervous system as compared to other insect tissues.

In order to further improve the safe use of neuroactive insecticides, we also plead for more research efforts in comparative neurobiology and biochemistry. Knowledge in these areas will lead to a better understanding of pre-existing and acquired insensitivities towards present and future insect control agents; to a better understanding of the neurological mechanisms causing selectivities between larger and smaller taxonomic units or among insect growth stages, and the mechanisms leading to resistance. The modified acetylcholinesterases of several insect species are very good examples for a practical application of neuroscience: these enzymes have become effective biochemical monitoring tools for organophosphate and carbamate resistance in pest populations.

In order to obtain a broader view of the neurobiological 'wants' of industry, we have asked a few fellow researchers of our own department to list areas where they feel neuroscience should and can contribute. Here are their main answers:

—Without underestimating the technical difficulties, more insect species of economic importance should be included in basic and applied neuroscience.

—Once a chemical lead has been identified in vivo, neurophysiological or biochemical test systems can predict, confirm or exclude a particular mode of action. This is valuable information. Depending on the type of knowledge gained, chemists may or may not profit, when optimizing new structures for biological efficacy and safety.

—Although the chances for discovering competitive neuroactive insecticides with novel modes of action appear to be small, much research will have to be done by industry and academia as soon as such products move to the market.

—Differential metabolism of insecticides is a well-established base for selectivities among representatives of different taxonomic units. What, however, is the role of the nervous system of different species in providing selectivities?

—Insecticide resistance is a great challenge for industry, agriculture and society. Research that does not just describe findings, but assists in early detection, prevention or successful management of the problem, will find the support of industrial companies.

—Natural or synthetic chemicals that modify insect behaviour, are becoming part of the R&D project portfolios of probably all major pesticide companies. The development, refinement and use of efficient and meaningful screening systems measuring behavioural responses in target insects are prerequisites for success.

—Where public concern against neuroactive insecticides appears to be unfairly exaggerated or unjustified, both academic and industrial neuroscientists must speak up in order to defend some of our most valuable, though admittedly not perfect, agrochemical specialities, for which we do not yet have satisfactory alternatives. The forum for getting better public acceptance, however, will not be scientific journals and conferences, but the media and some of our educational institutions. Should we not share the objectives, the methods and the results of a fascinating research area with its impressive progress but also its obvious shortcomings with those who worry more about the risks of neuroactive insecticides than they appreciate their benefits?

Obituary

CHARLES POTTER, 1907–1989

Charles Potter exerted a strong influence in inaugurating this series of conferences on Insect Neurobiology and Pesticide Action; the concept of this unique forum for communication between scientists concerned with improved pest control but from diverse disciplines was quite typical of the breadth and depth of his vision combined with his flair for perception of opportune, appropriate actions. His long, active career spanning some six decades was distinguished by a series of imaginative decisions which profoundly benefitted the progress of research on crop protection and on pesticides and their action. Yet Potter's contributions have never perhaps been adequately recognized, for he was an *éminence grise* deriving fulfilment from knowledge of the advances made under his influence rather than from the immediate approbation of the scientific establishment whose ill-judged decisions he was at intervals motivated to contest.

Potter's first research (1929–1938) at Imperial College, London greatly advanced knowledge of the control of stored products pests and enunciated for the first time the principle of using active residual films of contact insecticides, specifically, in this instance, pyrethrum in white oil. This technique, applied throughout World War II to protect stored foodstuffs from insect infestation, established a precedent for the effective use of DDT in films, when that insecticide became readily available at the end of the war. During 1944 and 1945 Potter, now entomologist at Rothamsted Experimental Station, made some of the first trials of the control of UK plant pests with DDT. He was appointed Head of the Insecticides and Fungicides Department in 1947, at a time when the introduction of organophosphorus and organochlorine insecticides, combined with the need for improved crop protection, had greatly stimulated new research on insecticides. With characteristic foresight, and drawing on his personal experience with pyrethrum, Potter chose to concentrate on the relationship between insecticidal activity and the structure of pyrethroids rather than on the more fashionable organochlorine and organophosphorus compounds. This was an outstandingly important decision, because in the early 1970s the need for new insecticides, not unduly persistent and with improved activity, became urgent; Rothamsted was then able, before any other academic or industrial research group, to provide, via patents assigned to the National Research Development Corporation, appropriate compounds—the more photostable synthetic pyrethroids.

Potter also achieved worldwide recognition for his elaboration of equipment (for example the Potter Spraying Tower) and techniques to assay insecticidal action and study the factors that influence toxicity; he initiated chemical and biological research on the field control of agricultural pests including the aphid transmitter of virus in potatoes and of the problems associated with seed dressing. Under his direction studies were started on the isolation and chemical identification of behaviour controlling compounds such as food and oviposition attractants and of the effect of insecticides on beneficial insects. Long before the potential severity of the problem was recognized, he directed a strong team of biologists and biochemists in research on the genetics and biochemical basis of the mechanisms by which insects acquire tolerance to insecticides; the global significances of this project was later to be recognized.

Potter gave advice without stint or consideration of personal recognition, on the wide range of topics on which he was expert; he was much in demand on official committees and by organizations both at home and abroad. When he retired from Rothamsted in 1972, aged 65, he embarked on a new phase of his career which he found especially stimulating and exciting. As a consultant to the Wellcome Foundation Research Station at Berkhamsted, he headed a multidisciplinary research group which increased knowledge of insecticide action in two major new series of compounds; he even returned to the bench to make his own practical contribution.

Charles Potter was a truly great scientist who exerted an immensely beneficial influence on pesticide research in the 20th century both directly by his work and by the inspiration he imparted to all who came under his guidance. To have known him and worked under his direction was indeed an exceptional privilege.

Michael Elliott

Acknowledgements

The organizing committee of NEUROTOX '91 wish to thank the following organizations for their contributions to the bursary fund:

American Cyanamid Co. (USA)
Bayer AG
Ciba Geigy Ag (Switzerland)
EI du Pont de Nemours & Co. (USA)
FMC Corporation (USA)
Hoechst UK Limited (UK)
ICI Agrochemicals (UK)
Merck & Co. Inc. (USA)
Merck Sharp and Dohme (UK)
Nippon Chemiphar (Japan)
Rhône–Poulenc Ag Co. (USA)
Rohm and Haas Co. (USA)
Roussel Uclaf (France)
Schering Agrochemicals Limited (UK)
Shell Research Limited (UK)
Society of Chemical Industry
US Department of Army

Contents

SECTION 1

Natural Products

1

Philanthotoxin-433 and Analogs, Noncompetitive Antagonists of Glutamate and Nicotinic Acetylcholine Receptors

Gary Chiles,[a] Seok-Ki Choi,[a] Amira Eldefrawi,[b] Mohyee Eldefrawi,[b] Shinji Fushiya,[a] Robert Goodnow, Jr.,[a] Aris Kalivretenos,[a] Koji Nakanishi[a] & Peter Usherwood[c]

[a] Department of Chemistry, Columbia University, New York, New York 10027, USA
[b] School of Medicine, University of Maryland, Baltimore, Maryland 21201, USA
[c] Department of Life Science, University of Nottingham, Nottingham NG7 2RD, UK

INTRODUCTION

Piek and co-workers reported that the solitary digger wasp *Philanthus triangulum* F., a wasp found in the Sahara Desert that preys on honeybees, makes a venom which blocks glutamate receptors (GluR) on locust muscles.[1] A toxin, δ-philanthotoxin (δ-PhTX), which is an antagonist of GluR, was isolated from the venom sacs of the female wasp and assigned structure **1** (Fig. 1) from spectroscopic data and the synthesis of its DL-form.[2] Using a honeybee worker paralysis assay, we independently isolated the toxin PhTX-433 (numerals denote the number of methylenes in the polyamine moiety) and established its structure as **1**, with the S configuration, by synthesis and circular dichroic spectroscopy. Natural PhTX-433 and two other isomers, -343 and -334, were also synthesized.[3] PhTX-433 and its analogs are efficient noncompetitive reversible inhibitors of the quisqualate-sensitive glutamate receptor (QUISR) in locust skeletal muscle.[3] A number of similar but structurally more complex toxins have been isolated from various spider venoms[4-8] (Fig. 1). The isolation and characterization of polyamine toxins from spiders and wasps has stimulated considerable interest in view of the important involvement of GluR in mammalian central nervous systems and in insect neurotransmission.[9,10]

GLUTAMATE RECEPTOR SUBTYPES

As is the case with other receptors, the definition of GluR subtypes based on agonists and antagonists is undergoing constant change with the improvement

3

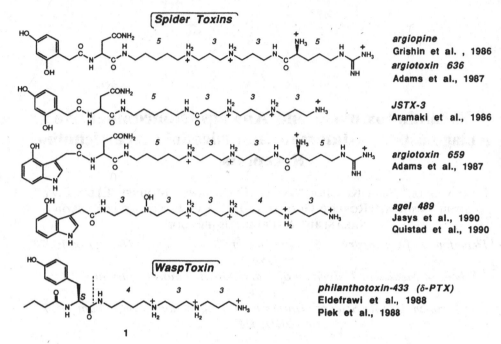

FIG. 1. Insect polyamine neurotoxins.

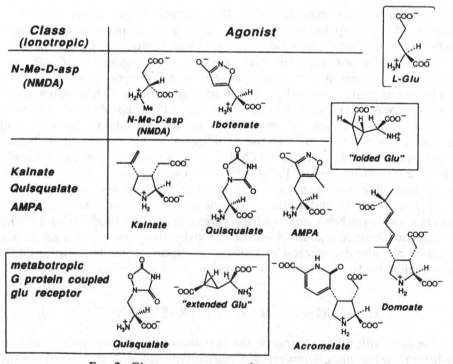

FIG. 2. Glutamate receptor subtypes—vertebrates.

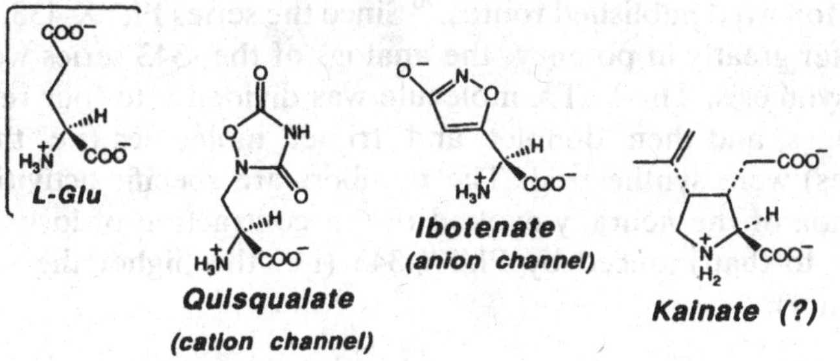

FIG. 3. Glutamate receptor subtypes—insect.

of pharmacological tools.[11] The current trends are summarized in Figs 2 and 3, respectively, for vertebrates and insects. Vertebrate GluR are divided into two major subtypes of ionotropic receptors: NMDAR and kainate/quisqualate/AMPAR. Recently, a metabotropic GluR has been sequenced and expressed. Somewhat surprisingly, this receptor has little sequence similarity to conventional G protein coupled receptors; it has seven putative transmembrane domains, but is characterized by large hydrophilic sequences on both sides of the membrane.[12] The excitatory amino acid L-glutamic acid (L-Glu) shown in Fig. 2 can adopt various conformers, depending on the environment. Recent electrophysiological studies with rat spinal cord and *Xenopus* oocytes injected with rat brain RNA have shown that, of the four possible stereoisomers of L-Glu where C-2 and C-3 have a bridging cyclopropane ring (synthesized[13] by Ohfune and co-workers), the 'folded' and 'extended' forms depicted in Fig. 2 act as potent agonists for the ionotropic[14] and metabotropic[15] GluRs, respectively. Previously, Lea and Usherwood[34] had shown that a partially folded form of L-Glu binds to locust muscle QUISR (see also Ref. 35). These are important findings, since the conformationally restricted molecules define in a stereospecific manner the shape of the glutamate molecule when it binds to the respective GluRs.

STRUCTURE ACTIVITY STUDIES OF ANALOGS WITH INVERTEBRATE QUISQUALATE RECEPTOR

Blagbrough,[16] Piek,[17] Nakajima[18] and co-workers have synthesized several analogs related to the spider and wasp toxins, which are more active than the natural compounds. In the following, we summarize our results of structure activity studies, published[19] and unpublished, as monitored by blocking of the cation-selective channel gated by QUISR of locust (*Schistocerca gregaria*) muscle. Most of the compounds have been submitted to extensive electrophysiological and pharmacological studies with vertebrate NMDA, kainate and nicotinic acetylcholine receptors (nAChR) as well, and interestingly show the same general trends (discussed later). The synthesis of most analogs, with a few exceptions, followed published routes.[20] Since the series PhTX-433, -343 and -334 did not differ greatly in potency, the analogs of the -343 series were used, due to ease of synthesis. The PhTX molecule was divided into four regions for systematic changes, and then 'doublet' and 'triplet' molecules (i.e. those with multiple changes) were synthesized. The numbers are specific activities based on their inhibition of the neurally-evoked twitch contraction of locust skeletal muscle, relative to that induced by PhTX-343 (i.e. the higher the value, the higher the potency).

Region I (Fig. 4)

(a) Polyamine is necessary. Shortening to less than -43 (number of CH_2's) reduces activity.

(b) Subtle differences exist between natural PhTX-433 (1·3) and the unnatural -343 (1·0) and -334 (1·5) isomers. This may reflect distances of anionic charges within the GluR.

(c) Hydrophobic side chain, e.g. Bu, enhances potency. Further analogs have been synthesized.[21]

(d) Quaternization of nitrogen eliminates potency (due to steric hindrance?).

Region II (Fig. 4)

(a) Activity is reduced when the molecule is elongated by an amide (GABA), but is potentiated when COOH is that of Lys or Arg.

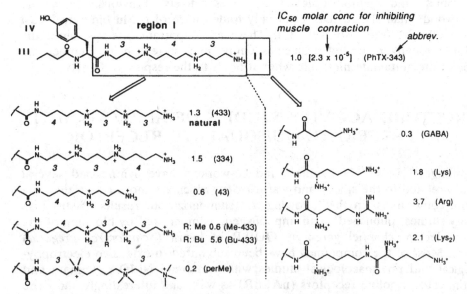

FIG. 4. Activity summary for regions I and II.

b) Higher potency of Arg than Lys may arise from the capability of guanidinium to delocalize its charge to better accommodate GluR binding requirements or to form stronger H-bond to carboxylate.

Region III (Fig. 5)

a) Hydrophobic moiety is necessary. High potency of C10 should be noted. However, see Fig. 9.
b) High activity of azidobenzamide analog N3Ph enables one to use this as a photoaffinity label.
c) DNP12 with longer linker than DNP4 is active; this type is candidate for affinity columns.

Region III-cont. (Fig. 6)

Attachment of Asp leads to dramatic loss in activity. The spin labeled analog is also inactive.

Region IV (Fig. 7)

a) OH of Tyr is not important: cf. Phe, OAc and OBn.
b) A bulky anchoring group with moderate hydrophobicity appears to be necessary: Leu versus Gly. Trp is highly active; intramolecular H-bond of NO_2 diminishes hydrophilicity.
c) The series $F \rightarrow Cl_2 \rightarrow Br_2 \rightarrow I_2$ suggests polarizability affects binding.

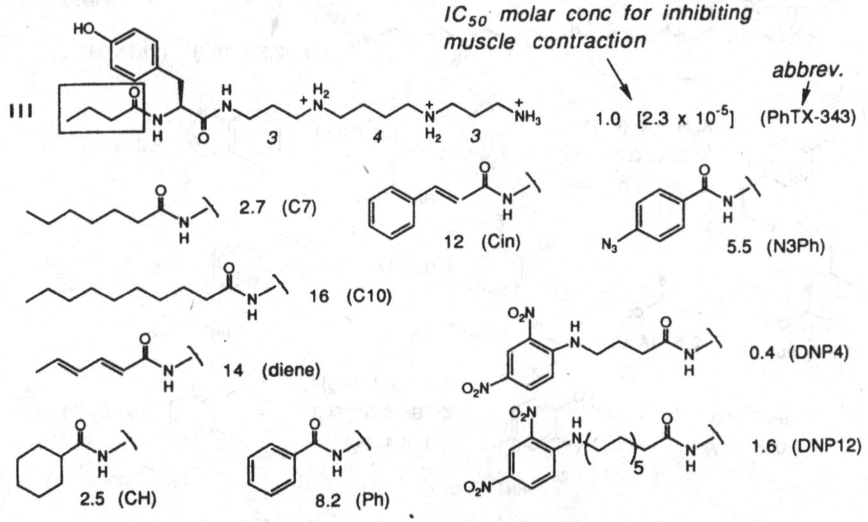

FIG. 5. Activity summary for region III.

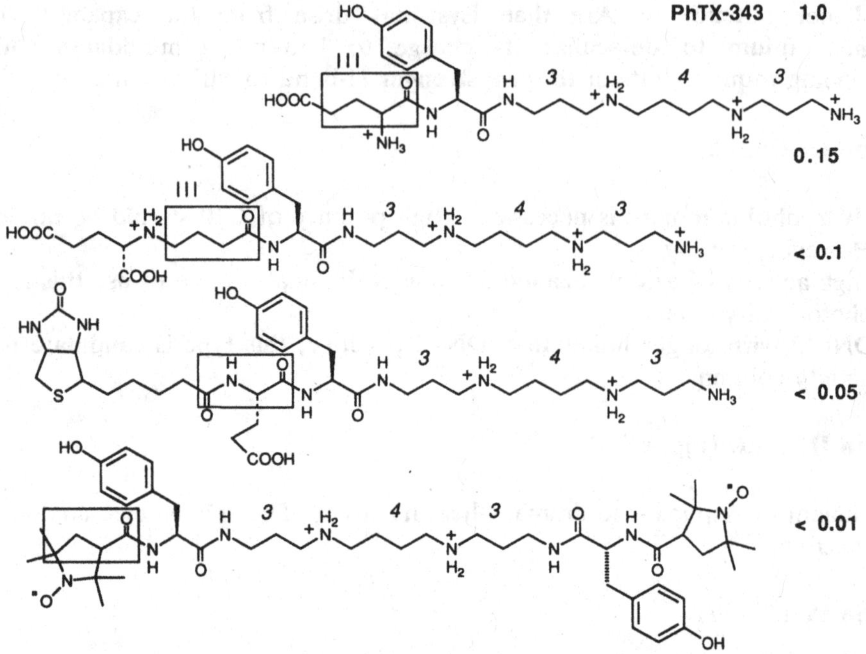

G. Chiles et al.

FIG. 6. Activity summary for region III for hydrophilic analogs.

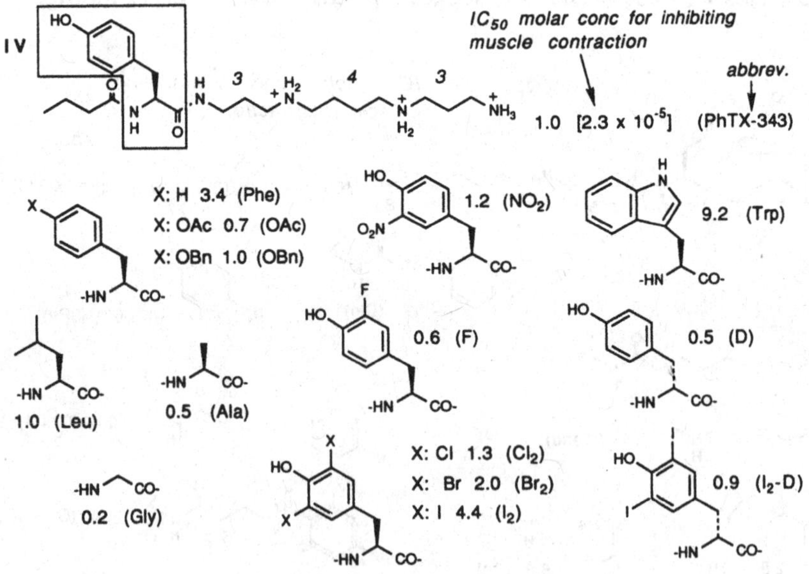

FIG. 7. Activity summary for region IV.

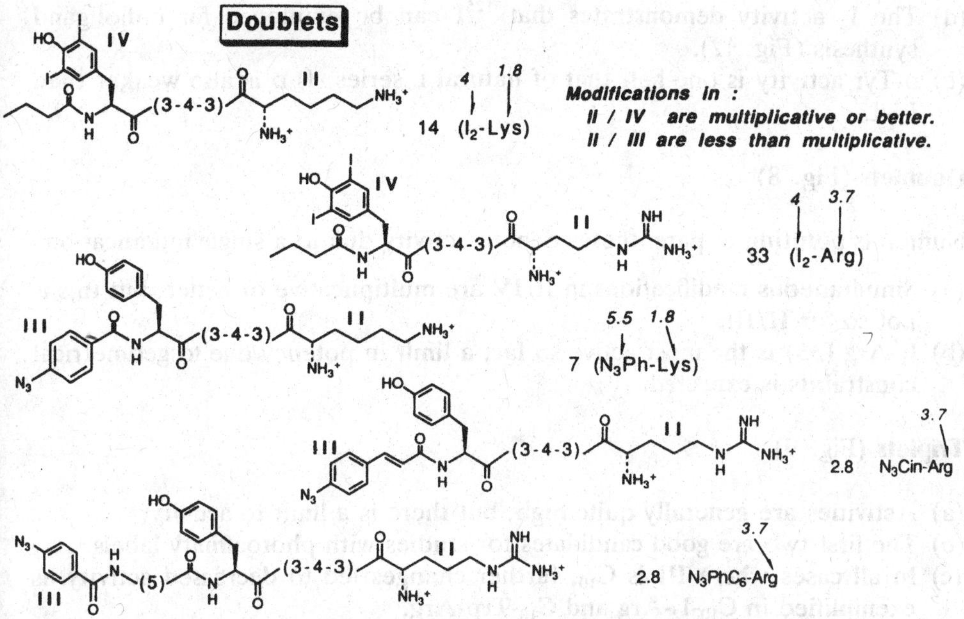

Doublets

Modifications in :
II / IV are multiplicative or better.
II / III are less than multiplicative.

14 (I$_2$-Lys)

33 (I$_2$-Arg)

7 (N$_3$Ph-Lys)

2.8 N$_3$Cin-Arg

2.8 N$_3$PhC$_6$-Arg

FIG. 8. Activity summary for doublet PhTX analogs.

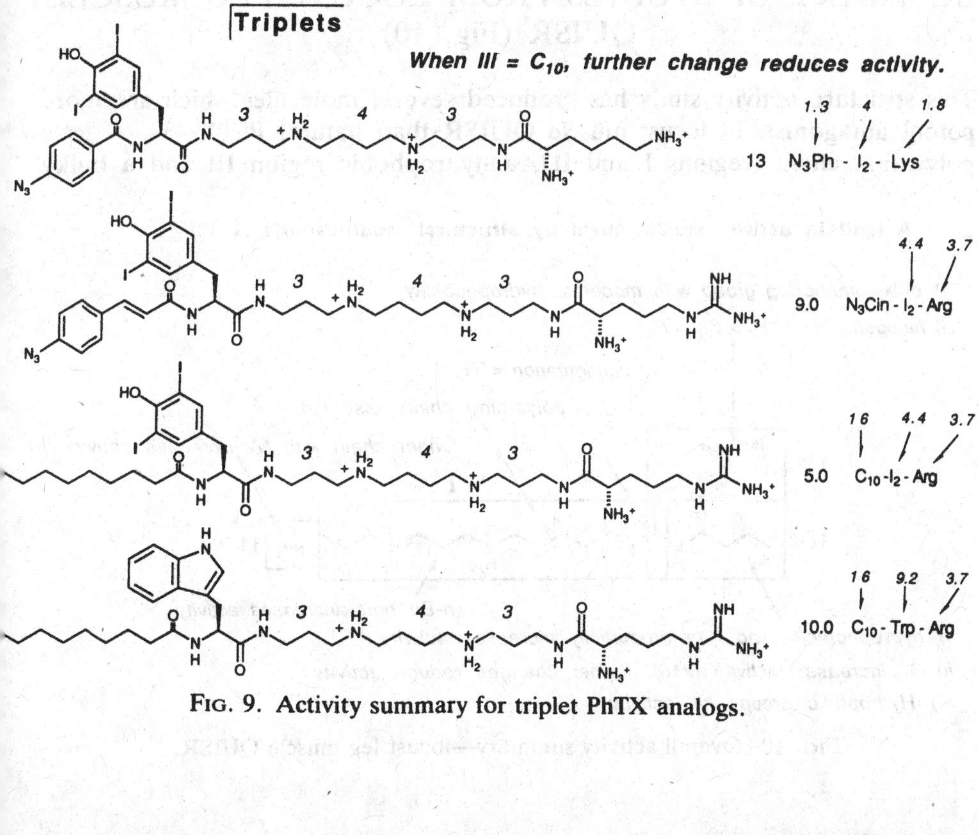

Triplets

When III = C$_{10}$, further change reduces activity.

13 N$_3$Ph - I$_2$ - Lys

9.0 N$_3$Cin - I$_2$ - Arg

5.0 C$_{10}$-I$_2$-Arg

10.0 C$_{10}$ - Trp - Arg

FIG. 9. Activity summary for triplet PhTX analogs.

(d) The I_2 activity demonstrates that ^{125}I can be employed for radioligand synthesis (Fig. 12).

(e) D-Tyr activity is one-half that of natural L series. I_2-D is also weaker than I_2-L.

Doublets (Fig. 8)

Numerals pointing to parentheses denote activity due to a single modification.

(a) Simultaneous modifications in II/IV are multiplicative or better, but this is not so for II/III.

(b) I_2-Arg (33) is the most active so far; a limit in potency due to geometrical constraints is expected.

Triplets (Fig. 9)

(a) Activities are generally quite high, but there is a limit to activity.

(b) The first two are good candidates for studies with photoaffinity labels.

(c) In all cases where III is C_{10}, further changes led to decreased activity as exemplified in C_{10}-I_2-Arg and C_{10}-Trp-Arg.

SUMMARY OF STUDIES FROM LOCUST LEG MUSCLE QUISR (Fig. 10)

This structure activity study has produced several molecules which are more potent antagonists of locust muscle QUISR than natural PhTX-433. A long polycation chain (regions I and II), a hydrophobic region III and a bulky

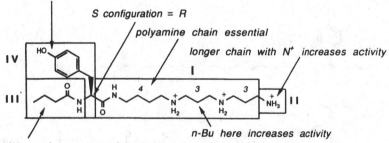

A limit in activity enhancement by structural modifications (x 33)

i) bulky anchoring group with moderate hydrophobicity

ii) halogens: $I_2 > Br_2 > Cl_2 > F$

S configuration = R

polyamine chain essential

longer chain with N^+ increases activity

n-Bu here increases activity

i) hydrophobicity and / or aromaticity increases activity

ii) C_{10} increases activity (x16); further changes reduce activity

iii) Hydrophilic groups kill activity

FIG. 10. Overall activity summary—locust leg muscle QUISR.

anchoring group with moderate hydrophobicity (region IV) are important potency determinants. It appears that the role of the butyryl side chain is to anchor the toxin in a hydrophobic pocket in or around the receptor channel to assist the binding of the polyamine moiety,[22] but this will not be clear until assays of similar analogs[21] with alkyl side chains protruding from other sections of the polyamine are completed. Doublet and triplet molecules which embrace structural modifications leading to increased potency have been prepared, and the results of the muscle assay showed that these are not necessarily multiplicative. The maximal potency obtained thus far is that of I_2-Arg (Fig. 8). The studies have led to several potent analogs with photoaffinity and radioactive labels.

ACTION OF ANALOGS ON OTHER RECEPTORS
(Fig. 11)

Extensive electrophysiological and biochemical displacement studies performed on in-vitro excitable tissues and *Xenopus* oocytes have demonstrated that contrary to initial expectations, the PhTX analogs are reversible noncompetitive antagonists of vertebrate and invertebrate NMDAR, KAINR, QUISR and nAChR. PhTX-433, when administered at high concentrations ($c. > 10^{-6}$ M), reversibly inhibited the responses of *Xenopus* oocytes injected with rat brain[23,24] or chick[24] brain mRNA encoding QUISR, NMDAR and KAINR; but with low concentrations of PhTX (10^{-11}–10^{-7} M), the effect on NMDAR and KAINR was potentiation.[23,24] Similarly, whether measured electrophysiologically (patch clamp) on frog muscles and cockroach ganglia or biochemically on honeybee brain and *Torpedo* electric organ, PhTX-433 clearly interacts with

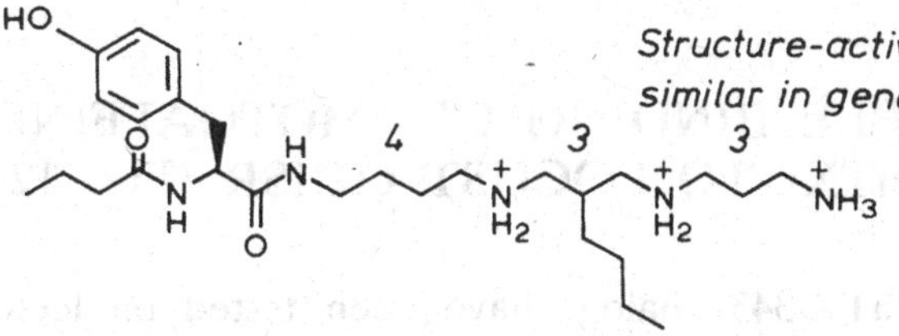

Structure-activity relations: similar in general

Bu-433 potent in all systems

Effect of structural variations: NMDA R<nACh R

Relative affinities, IC_{50} $(x10^{-6}M)$

	QUIS R (locust)	NMDA R (rat brain)	nACh R (electric ray)
PhTX-433	23	43·8	1·09
Bu-433	4·1 (x5·6)	4·7 (9·3)	0·14 (x7·8)

FIG. 11. PhTX action on QUISR, NMDAR, nAChR.

nAChRs at between 10^{-7} and 10^{-3} M.[25] Electrophysiological results suggest that PhTX interacts with the activated open conformation of nAChR to produce a blocked state. The binding of [^{3}H]acetylcholine to nAChR was potentiated by low concentrations of PhTX but inhibited by high concentrations, thus suggesting the presence of at least two binding sites.[24] The inhibitory actions of 37 PhTX analogs were tested on rat brain NMDAR and *Torpedo* nAChR by biochemical displacement assays (Fig. 11).[26] The structure activity relationships were similar to those found in the studies on locust muscle, but antagonism of nAChR varied more with toxin structure than antagonism of NMDAR. As with the locust muscle QUISR, the analog with the *n*-butyl side-chain was more potent than native PhTX-433 by several times at both the NMDAR and nAChR (see table in Fig. 11; the smaller the value, the greater the potency. Note the high affinity of Bu-433 to the nAChR of *Torpedo*). The complex nature of PhTX interactions with NMDAR and nAChR observed during these studies suggests that PhTX may bind to two sites, an external polyamine binding site and a channel binding site.[26] Figure 11 also summarizes the effect of structural modifications of PhTX on *Torpedo* nAChR, which are again similar to the results obtained with the locust QUISR. The binding of ^{125}I$_2$-Lys and ^{125}I$_2$-Arg (Fig. 8) to rat brain tissues was not significantly inhibited by 10^{-4} M of the excitatory amino acids QUIS, KAIN, NMDA and their noncompetitive blockers[24]. These results show that PhTX analogs are inappropriate ligands for labelling or purifying GluR. However, the increased activities of many analogs seen with the extended structure of PhTXs make them suitable for attaching photoaffinity labels at various sites in the molecule for tertiary structure studies of GluR, nAChR and other receptors. The radioactive ^{125}I is introduced conveniently at the last stage of synthesis by treating the unprotected analog (200 nmol scale) with Na^{125}I and chloramine-T. Although the yield is 23–30%, the monoiodo- and diiodo-products can be readily separated by HPLC.[27]

IRREVERSIBLE BINDING OF PHOTOAFFINITY LABELED PhTXs TO LOCUST QUISR (Fig. 12)[24,28]

Five photosensitive PhTX-343 analogs have been tested on locust muscle QUISR. At 10^{-9} M they reversibly inhibited the postjunctional currents recorded from muscle fibers during motor nerve stimulation, primarily by noncompetitively antagonizing postjunctional QUISR, probably through open channel block. When the preparation was irradiated with 270 nm light during toxin application and nerve stimulation, the currents were irreversibly inhibited. These results seemingly demonstrate that the toxins bind covalently to the receptor in its open channel form during irradiation, thereby irreversibly blocking the channel gated by QUISR. It is interesting to note that the monoiodo azidocinnamide analog (activity relative to PhTX-343,–10) is more

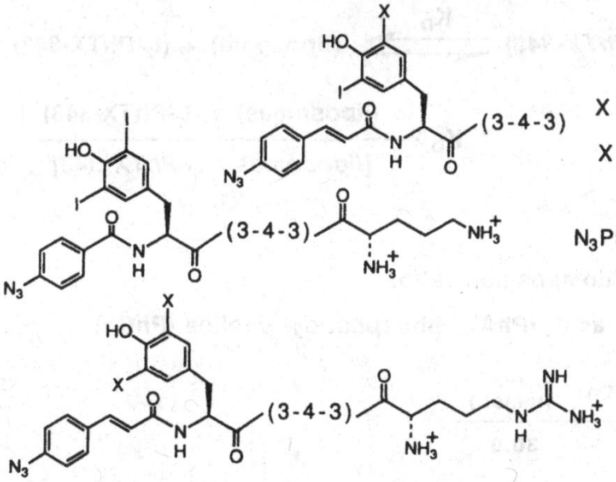

X : H N_3Cin-I 10

X : I N_3Cin-I_2 4

N_3Ph-I_2-Lys 13

X : H N_3Cin-Arg 2.8

X : I N_3Cin-I_2-Arg 9.0

***locust leg nerve-muscle*, 10^{-9} M conc:**

***reversible* block of excitatory postsynaptic currents.**

***irreversible* block after irr 270 nm.**

FIG. 12. Action of photosensitive PhTX analogs on excitatory (glutamatergic) post-synaptic potential locust leg muscle.

active than the diiodo analog (activity 4). The use of ligands labeled with ^{125}I will be tested for the isolation of receptors from appropriate sources. The effects of these analogs on the GluR of rat brain are also being investigated using a *Xenopus* oocyte mRNA expression system.

BINDING OF PhTX ANALOG TO LIPID VESICLES (Fig. 13)

At this stage, nothing is known regarding the mode of binding of PhTX analogs to the various EAA receptors. The tertiary structures of the receptors comprising several transmembrane domains are unknown and, moreover, the complex nature of PhTX interactions, e.g. potentiation at low concentrations and antagonism at high concentrations, suggests that there may be at least two binding sites for this toxin. PhTX and its analogs could be electrostatically bound to the lipid bilayer surface, or the polyamine moiety could be inserted into the channel. In the case of the kainate class of GluR, (e.g. GluR-K1,[29] GluR-K2 and -K3,[30] chick kainate binding protein (KBP),[31] and frog brain (KBP),[32] the cytoplasmic segment between the two transmembrane subunits M1 and M2 has a high population of glutamate and aspartate residues (up to nine in K1, K2, K3) while the transmembrane segment has no charged amino acid residues. Therefore, it is likely that the polyamine chain of PhTX could bind to the cytoplasmic domains of these subunits, provided the toxin first permeates the lipid layer.

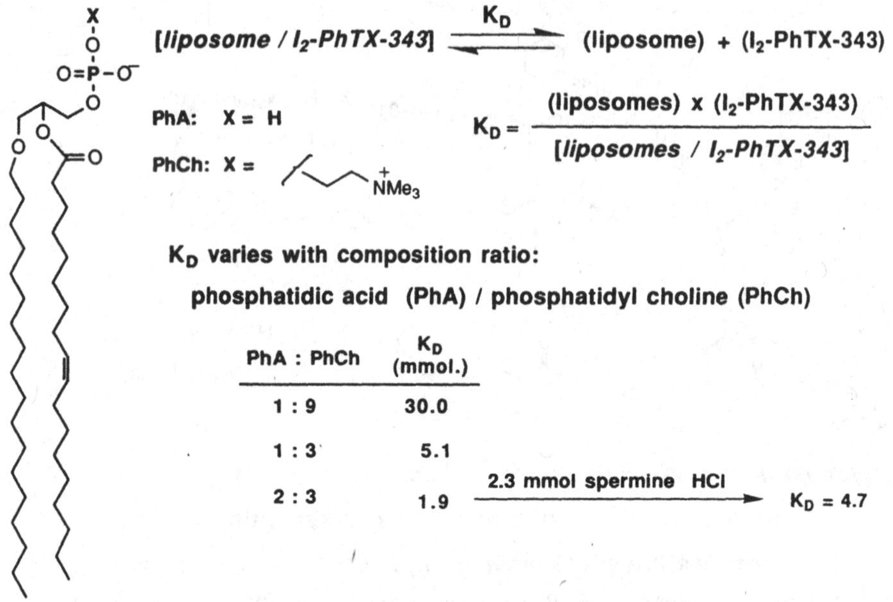

$$[liposome / I_2\text{-}PhTX\text{-}343] \underset{}{\overset{K_D}{\rightleftharpoons}} (liposome) + (I_2\text{-}PhTX\text{-}343)$$

PhA: X = H

PhCh: X =

$$K_D = \frac{(liposomes) \times (I_2\text{-}PhTX\text{-}343)}{[liposomes / I_2\text{-}PhTX\text{-}343]}$$

K_D varies with composition ratio:

phosphatidic acid (PhA) / phosphatidyl choline (PhCh)

PhA : PhCh	K_D (mmol.)
1 : 9	30.0
1 : 3	5.1
2 : 3	1.9

2.3 mmol spermine HCl $\longrightarrow$ $K_D = 4.7$

Fig. 13. I_2-PhTX-343 binding to liposomes: NMR T_2.

Since the planned photoaffinity mapping experiments using the photosensitive radioligands of PhTX are likely to be performed in liposomes, the binding of the iodinated analog, I_2-PhTX-343, to lipid vesicles has been investigated by measurements of NMR transverse relaxation as a function of the liposome composition and the analog concentration (Fig. 13) and nuclear Overhauser effects.[33] The dissociation constant is dependent on the ratio of phosphatidic acid and phosphatidyl choline. As expected, the higher the proportion of the former (which bears a negative charge), the stronger the binding. Also, the addition of spermine loosens the binding of the analog as seen in the increase of K_D from 1·9 to 4·7.

SUMMARY (Fig. 14)

The results of the present structure activity studies are summarized in Fig. 14. We plan to carry out tertiary structure studies of the binding site of an appropriate receptor using the photoaffinity labeled radioligands. Several potent analogs of this type have been made, and preliminary photolysis studies demonstrate that they do indeed bind to the receptor active site. Photoaffinity labeled PhTX analogs carrying the affinity label in other regions of the molecule have been prepared for further photoaffinity mapping studies, which will be coupled with fragmentation of the labeled receptor and sequencing of the peptidic fragment.

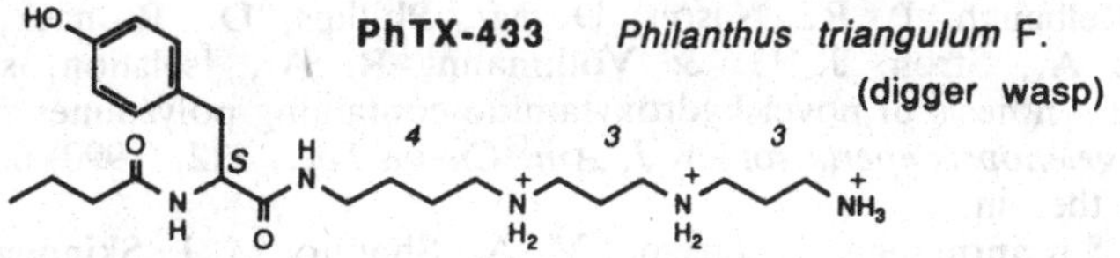

PhTX-433 *Philanthus triangulum* F.
(digger wasp)

PhTX and analogs

noncompetitive antagonist of glu & nACh receptors

1) Binding assay, electrophysiology, *Xenopus* oocyte patch clamp assay

Low concentrations : Agonistic responses of NMDA, kainic acid and quisqualic acid are potentiated. Binding of acetylcholine to nAChR potentiated.

High concentrations : Inhibition of actions.

2) Complex nature of PhTX interactions suggests at least 2 binding sites, *external polyamine binding site* and a *channel binding site*.

FIG. 14. Summary of structure-activity studies.

ACKNOWLEDGMENTS

These studies were supported by NIH grant AI 10187 to (K.N.), British Agriculture and Food Research Council (to P.N.R.U.) and NIH ES02594 (to A.T.E. and M.E.).

REFERENCES

1. Piek, T., Mantel, P. & Engels, E., Neuromuscular block in insects caused by the venom of the digger wasp *Philanthus triangulum*. F. *Comp. Gen. Pharmacol.*, **2** (1971) 317.
2. Piek, T., Fokkens, R. H., Karst, H., Kruk, C., Lind, A., Van Marle, J., Nakajima, T., Nibbering, N. M. M., Shinozaki, H., Spanjer, W. & Tong, Y. C., Polyamine like toxins—a new class of pesticides? In *Neurotox '88: Molecular Basis of Drug & Pesticide Action*, ed. G. G. Lunt. Elsevier, Amsterdam, 1988, pp. 61–76.
3. Eldefrawi, A. T., Eldefrawi, M. E., Konno, K., Mansour, N. A., Nakanishi, K., Oltz, E. & Usherwood, P. N. R., Structure and synthesis of a potent glutamate receptor antagonist in wasp venom. *Proc. Natl. Acad. Sci.*, **85** (1988) 4910.
4. (a) Grishin, E. V., Volkova, T. M., Arseniev, A. S., Reshetova, O. S., Onoprienko, V. V., Magazanik, L. G., Antonov, S. M. & Federova, I. M., Structure-functional characterization of argiopin—an ion channel blocker from the venom of the spider *Argiope Iobata*. *Bioorg. Khim.*, **12** (1986) 1121. (b) Grishin, E. V., Volkova, T. M. & Arseniev, A. S., Isolation and structure analysis of components from venom of the spider *Argiope lobata*. *Toxicon*, **27** (1989) 541.
5. Aramaki, Y., Yasuhara, T., Higashijima, T., Yoshioka, M., Miwa, A., Kawai, N. & Nakajima, T., Chemical characterization of spider toxins, JSTX and NSTX. *Proc. Japan. Acad. (Ser. B)*, **62** (1986) 359.
6. Adams, M. E., Carney, R. L., Enderlin, F. E., Fu, E. W., Jarema, M. A., Li, J. P., Miller, C. A., Schooley, D. A., Shapiro, M. J. & Venema, V. J., Structures and biological activities of three synaptic antagonists from orb web spider venom. *Biochem. Biophys. Res. Commun.*, **148** (1987) 678.

7. Jasys, V. J., Kelbaugh, P. R., Nason, D. M., Phillips, D., Rosnack, K. J., Saccomano, N. A., Stroh, J. G. & Volkmann, R. A., Isolation, structure elucidation, and synthesis of novel hydroxylamine-containing polyamines from the venom of the *Agelenopsis aperta* spider. *J. Am. Chem. Soc.*, **112,** (1990) 6696, and references cited therein.

8. Quistad, G. B., Suwanrumpha, S., Jarema, M. A., Shapiro, M. J., Skinner, W. S., Jamieson, G. C., Lui, A. & Fu, E. W., Structure of paralytic acylpolyamines from the spider *Agelenopsis aperta*. *Biochem. Biophys. Res. Commun.*, **169** (1990) 51.

9. Jackson, H. & Usherwood, P. N. R., Spider toxins as tools for dissecting elements of excitatory amino acid transmission. *Trends Neurosci.*, **11** (1988) 278.

10. Usherwood, P. N. R., Sudan, H., Standley, C., Blagbrough, I. S., Bycroft, B. W. & Mather, A. J., Mechanisms of neurotoxicity of low molecular weight spider toxins. In *Basic Science in Toxicology*, ed. G. N. Volans, J. Sims, F. M. Sullivan & P. Turner. Taylor & Francis, London, 1990, 569.

11. Watkins, J. C., Krogsgaard-Larsen, P. & Honoré, T., Structure-activity relationships in the development of excitatory amino acid receptor agonists and competitive antagonists. *Trends Pharm., Sciences*, **11** (1990) 25.

12. Masu, M., Tanabe, Y., Tsuchida, K., Shigemoto, R. & Nakanishi, S., Sequence and expression of a metabotropic glutamate receptor. *Nature*, **349** (1991) 760.

13. Yamanoi, K., Ohfune, Y., Watanabe, K., Li, P. N. & Takeuchi, H., Syntheses of trans- and cis-α-(carboxycyclopropyl)glycines. Novel neuroinhibitory amino acids as L-glutamate analogues. *Tetrahedron Lett.*, **29** (1988) 1181.

14. Shinozaki, H., Ishida, M., Shimamoto, K. & Ohfune, Y., A conformationally restricted analogue of L-glutamate, the (2S, 3R, 4S) isomer of L-α-(carboxycyclopropyl)glycine, activates the NMDA-type receptor more markedly than NMDA in the isolated rat spinal cord. *Brain Research*, **480** (1989) 355.

15. (a) Ishida, M., Akagi, H., Shimamoto, K., Ohfune, Y. & Shinozaki, H., A potent metabotropic glutamate receptor antagonist: electrophysiological action of a conformationally restricted glutamate analogue in the rat spinal cord and *Xenopus* oocytes. *Brain Research*, **537** (1990) 311. (b) Nakagawa, Y., Saitoh, K., Ishihara, T., Ishida, M. & Shinozaki, H., (2S, 3S, 4S)α-(Carboxycyclopropyl)glycine is a novel agonist of metabotropic glutamate receptors. *Eur. J. Pharm.*, **184** (1990) 205.

16. (a) Blagbrough, I. S., Bruce, M., Bycroft, B. W., Mather, A. J. & Usherwood, P. N. R., Invertebrate pharmacological assay of novel, potent glutamate antagonists: acylated spermines. *Pestic. Sci.* **30** (1990) 397. (b) Blagbrough I. S., et al. *Neurotox '91, Abstracts,* 1991. pp. 81–90.

17. Karst, H., Piek, T., Van Marle, J., Lind, A. & Van Weeren-Kramer, J. *Comp. Biochem. Physiol.* (In press).

18. Asami, T., Kageuchika, H., Hashimoto, Y., Shudo, K., Miwa, A., Kawai, N. & Nakajima, T., Acylpolyamines mimic the action of joro spider toxin (JSTX) on crustacean muscle glutamate receptors. *Biomed. Res.*, **10** (1989) 185.

19. (a) Bruce, M., Bukownik, R., Eldefrawi, A. T., Eldefrawi, M. E., Goodnow, R., Kallimopoulos, T., Konno, K., Nakanishi, K., Niwa, M. & Usherwood, P. N. R., Structure-activity relationships of analogues of the wasp toxin philanthotoxin: noncompetitive antagonists of quisqualate receptors. *Toxicon*, **28** (1990) 1323. (b) Nakanishi, K., Goodnow, R., Konno, K., Niwa, M., Bukownik, R., Kallimopoulos, T. A., Usherwood, P. N. R., Eldefrawi, A. T. & Eldefrawi, M. E., Philanthotoxin-433 (PhTX-433), a non-competitive glutamate receptor inhibitor. *Pure & Appl. Chem.*, **62** (1990) 1223.

20. Goodnow, R., Konno, K., Niwa, M., Kallimopoulos, T., Bukownik, R., Lenares, D. & Nakanishi, K., Synthesis of glutamate receptor antagonist philanthotoxin-433 (PhTX-433) and its analogs. *Tetrahedron*, **46** (1990) 3267.

21. Kalivretenos, A. & Nakanishi, K., Synthesis of Alkyl-branched spermine type polyamines. Application to the Synthesis of Novel Philanthotoxin-433 analogs. *J. Org. Chem.* (In preparation).

22. Usherwood, P. N. R. & Blagbrough, I. S., Antagonism of insect muscle glutamate receptors—with particular reference to arthropod toxins. In *Insecticide Action: From Molecule to Organism*, ed. T. Narahashi & J. E. Chambers. Plenum Press, New York, 1990, pp. 13–31.
23. Ragsdale, D., Gant, D. B., Anis, N. A., Eldefrawi, A. T., Eldefrawi, M. E., Konno, K. & Miledi, R., Inhibition of rat glutamate receptors by philanthotoxin. *J. Pharmacol. Exp. Therap.*, **251** (1989) 156–163.
24. Brackley, P. Goodnow, R., Nakanishi, K., Sudan, H. L. & Usherwood, P. N. R., Spermine and philanthotoxin potentiate excitatory amino acid responses of *Xenopus* oocytes injected with rat and chick brain RNA. *Neuroscience Lett.*, **114** (1990) 51.
25. Rozenthal, R., Scoble, G. T., Alburquerque, E. X., Idriss, M., Sherby, S., Sattelle, D. B., Nakanishi, K., Konno, K., Eldefrawi, A. T. & Eldefrawi, M. E., Allosteric inhibition of nicotinic acetylcholine receptors of vertebrates and insects by philanthotoxin. *J. Pharmacol. Exp. Therap.*, **249** (1989) 123.
26. Anis, N., Sherby, S., Goodnow, R., Niwa, M., Konno, K., Kallimopoulos, T., Bukownik, R., Nakanishi, K., Usherwood, P. N. R., Eldefrawi, A. T. & Eldefrawi, M. E., Structure-activity relationships of philanthotoxin analogs and polyamines on *N*-methyl-D-aspartate and nicotinic acetylcholine receptors. *J. Pharmacol. Exp. Ther.*, **254** (1990) 764.
27. Goodnow, R. A., Bukownik, R., Nakanishi, K., Usherwood, P. N. R., Eldefrawi, A. T., Anis, N. A. & Eldefrawi, M. E., Synthesis and binding of $[^{125}I_2]$philanthotoxin-343, $[^{125}I_2]$philanthotoxin-343-lysine, and $[^{125}I_2]$philanthotoxin-343-arginine to rat brain membranes. *J. Med. Chem.* **34** (1991) 2389–94.
28. Chiles, G. & Usherwood, P. N. R., unpublished results (1991).
29. Hollmann, M., O'Shea-Greenfield, A., Rogers, S. W. & Heinemann, S., Cloning by functional expression of a member of the glutamate receptor family. *Nature*, **342** (1989) 643.
30. Nakanishi, N., Schneider, N. A. & Axel, R., A family of glutamate receptor genes: evidence for the formation of heteromultimeric receptors with distinct channel properties, *Neuron*, **5** (1990) 569.
31. Gregor, P., Mano, I., Maoz, I., McKeown, M. & Teichberg, V. I., Molecular structure of the chick cerebellar kainate-binding subunit of a putative glutamate receptor. *Nature*, **342** (1989) 689.
32. Wada, K., Dechesne, C. J., Shimasaki, S., King, R. G., Kusano, K., Buonanno, A., Hampson, D., Banner, C., Wenthold, R. J. & Nakatani, Y., Sequence and expression of a frog brain complementary DNA encoding a kainate-binding protein. *Nature*, **342** (1989) 684.
33. Goodnow, R. A., Turner, C. J. & Nakanishi, K., unpublished results (1991).
34. Lea, T. J. & Usherwood, P. N. R., The size of action of ibotenic acid and the identification of two populations of glutamate receptors on insect muscle fibres. *Comp. Gen. Pharmacol.*, **4** (1973) 333–50.
35. Bycroft, B. W. & Jackson, D. E. The putative binding conformations of glutamate at a well defined invertebrate glutamatergic synapse. In '*Neurotox '88: Molecular Basis of Drug and Pesticide Action*', ed. G. G. Lunt. Elsevier, Amsterdam, pp. 461–8.

2

Spider Neurotoxins as Tools for the Investigation of Glutamatergic Synaptic Transmission

L. G. Magazanik, S. M. Antonov, V. Yu. Bolshakov, I. M. Fedorova & S. A. Gapon

Sechenov Institute of Evolutionary Physiology and Biochemistry, Russian Academy of Sciences, 194223 St. Petersburg, Russia

INTRODUCTION

There is considerable evidence for the involvement of glutamate and probably other excitatory amino acids in synaptic transmission. Recent rapid progress in the investigation of glutamatergic mechanisms is due mainly to the discovery of sufficiently specific chemical tools affecting the postsynaptic responses of glutamate.[1–3] A new step in answer to this problem is connected with the introduction of neurotoxins isolated from spider venoms into pharmacological studies of glutamatergic transmission (see review[4]). Neurotoxins of low molecular weight (<1 kDa), such as JSTX isolated from Joro spider venom,[5,6] argiotoxin-636[4,7,8,20] from *Argiope trifasciata* venom, a family of neurotoxins including argiopine (ARG) isolated from *Argiope lobata* venom[9–17] and α-agatoxin from *Agelenopsis aperta* venom[18] are highly potent and selective antagonists of glutamate responses. Spider venoms are also rich in presynaptic toxins which specifically affect different stages of neurotransmitter secretion.[10,18,19]

MODE OF POSTSYNAPTIC TOXIN ACTION

The mechanism of blocking by neurotoxin was investigated in detail in experiments on glutamate sensitive muscle fibres of insects (locust and blowfly) and crustacea.[5–11] Use-dependent inhibition of locust muscle twitch,[7] appearance of biphasic decay of EPSP and analysis of noise induced by glutamate in neurotoxin-treated blowfly muscle[9–11] provides evidence for its ability to block the open state of glutamate-activated channels. This effect was also demonstrated by the recording of single channels in experiments on crayfish and locust muscles.[13,20] However, the interaction of neurotoxin with open glutamate channel is not the only postsynaptic mode of action, ARG elicits

19

either a closed or open channel block. In the blowfly preparation both types of block were manifested in the same concentration range and differed only in the direction of their potential dependence: hyperpolarization enhanced the blocking effect of ARG on the open channels but reduced that on closed ones.[10,11] In crayfish preparations, the block induced by 1–100 nM ARG only reduced the number of Glu-activated channels and did not influence the channel kinetics. In the presence of 1 μM ARG evidence of potential independent open channel block was demonstrated.[13] The direct competition between ARG-like neurotoxins and glutamate for the same binding site on the receptor is doubtful.

The specificity of neurotoxin blocking seems to be high. In the experiments on cholinergic frog neuromuscular junctions, ARG affected end-plate currents too. In control experiments the decay of postsynaptic currents was exponential but in both preparations in the presence of ARG it became biphasic (Fig. 1). Under the same conditions (resting potential and room temperature) dissociation constant (K_D) of ARG interaction with open glutamate channels was found to be 36 times lower, compared to that of open acetylcholine channels, due to higher association and lower dissociation rate constant of the ARG–channel complex.[10,11] This means that ARG has preferential affinity for open glutamate channels.

The distinct qualitative difference in effects of ARG on glutamate- and cholinergic postsynaptic membranes was revealed by the analysis of the current–voltage relationship. It has been shown that ARG failed to interact with closed acetylcholine channels, although the interaction of ARG with closed glutamate channels is very important and in some preparations probably is the leading mechanism for its specific antiglutamate effect.[10,11,13,21]

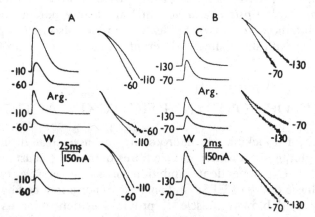

Fig. 1. The effect of argiopine on excitatory postsynaptic currents (EPSCs) recorded in (A) blowfly larvae muscle and (B) frog muscle.[10] EPSC (averaged 16 responses) recorded at different holding potential level (indicated) and corresponding semi-logarithmic plots of EPSC decay. C, control; Arg, after 30 min treatment with 0·44 μM (A) and 47 μM (B) argiopine; W, after 60 min washing.

Reversibility of neurotoxin action differs greatly, being dependent on the chemical structure of the neurotoxin and on the kind of preparation. JSTX seems to possess the least reversibility of action on any preparation.[5,6,22] The blocking effect of ARG could be more easily removed by long-lasting washing in the experiments on blowfly muscle[10,11] or molluscan neurones[21] than on crayfish muscle[13] or isolated frog spinal cord.[12,16,17]

Thus the postsynaptically active neurotoxins from spider venoms are highly potent and specific glutamate antagonists. They were used in our experiments as tools for the identification and classification of glutamate receptors in molluscan neurones and frog spinal cord.

EFFECTS ON MOLLUSCAN NEURONES

A variety of glutamate (Glu) responses differing in ionic mechanism (K^+, Cl_-, cationic) are typical for molluscan neurones.[23,24] The application of 100 μM Glu by microperfusion to identified neurone Ped-9 isolated from the pedal ganglia of a freshwater mollusc *Planorbarius corneus* induced two kinds of responses.[21] A slowly developing outward current appeared after a pronounced latency (about 1 s) in Ped-9 clamped at a holding potential of -40 mV, which was close to or more positive than the resting level (Fig. 2A). The amplitude of this Glu-induced current decreased with hyperpolarization and increased with depolarization of the neurone. When the neurone was clamped at -90 mV a fast inward current with a very short latency could be recorded. Depolarization of the neurone induced a decrease of inward current amplitude.

The receptors studied appear to belong to the non-NMDA group of excitatory amino acid receptors, since all our attempts to reveal NMDA receptors in molluscan neurones were to no avail. Quisqualic acid (QA) reproduced both kinds of response, being 6–10 times less potent than Glu, the threshold QA concentration was 1–5 μM. On the contrary, application of kainic acid (KA) and ibotenate evoked only outward and inward current, respectively (Fig. 2B and C).

The slow outward current elicited (at resting potential level) by Glu is due to an increase of potassium permeability:

(1) its E_r is very close to E_k estimated earlier;[25]
(2) the effect of $[K^+]_0$ change on E_r of the Glu response is in accordance with the Nernst equation predictions;
(3) this current can be inhibited by the potassium channel blocker tetraethylammonium.

The fast initial component of Glu-induced current is due to an increase in chloride permeability:

(1) its E_r is very close to E_{Cl};
(2) injection of Cl^- ions from a microelectrode induces a predicted shift of E_r of the Glu response;
(3) furosemide inhibits selectively the fast Glu-induced chloride current.

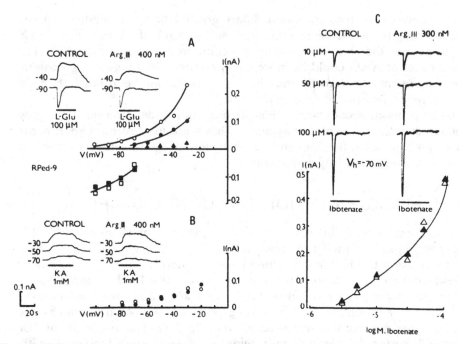

FIG. 2. Effect of argiopinine III on glutamate (A), kainate (B) and ibotenate (C) responses on a Ped-9 neurone.[21] (A) and (B): I–V plots for responses. Circles, amplitude of potassium currents; squares, amplitude of chloride currents. Open symbols, control responses. Filled circles and squares, after 40 min treatment with 400 nM ARG. III; filled triangles, after an additional 30 min treatment with increased concentration of ARG. III up to 1 μM. Insets: examples of currents recorded at different holding potential level. (C): dose response curves of ibotenate induced currents before (open triangles) and after 40 min treatment with 300 nM ARG. III (filled triangles). Insets: responses to different doses of ibotenate before and after ARG. III. Holding potential, -70 mV.

Argiopine (ARG) and structurally related argiopinine III (ARG. III) isolated from the venom of *Argiope lobata*[14] were used for further detailed investigation of glutamate receptors in molluscan neurones. ARG. III (50 nM–1 μM) strongly, but not completely, inhibited the slow potassium currents. On the other hand, ARG. III failed to affect the chloride Glu-induced current (Fig. 2). The inhibitory effect of ARG. III developed slowly over a 30–40-min period and could be partially reversed during 60 min of washing. The inhibitory effect was potential dependent (Fig. 2A): at -70 mV the current was reduced to $35 \pm 5\%$, ($n = 5$) of initial level, but at -30 mV it fell to $64 \pm 3\%$, ($n = 11$). Evidently the depolarization of the neurone weakens the ARG. III blocking action. The inhibitory effects of ARG and ARG. III were qualitatively quite similar: only the late potassium Glu-induced current was affected, the potential dependence of ARG action was the same (the shift of holding potential from -30 mV to -70 mV decreased the ARG blocking effect by 1·8 times), and the

ARG effect was only partially reversible. The only difference demonstrated was in the potency of these two neurotoxins. Apparent dissociation constants obtained from corresponding dose–response curves prove that ARG is 3 times less potent than ARG. III: $K_D = 230 \pm 30$ nM $(n = 5)$ and $K_D = 76 \pm 3$ nM $(n = 6)$, respectively.

The ability of ARG. III to block the responses induced by superfusion of several agonists was compared. QA-induced potassium current was inhibited by ARG. III just like the Glu-induced one: K_D values differed slightly $(K_D = 52 \pm 12$ nM, $n = 3$ in the case of QA responses). ARG. III failed to block QA-induced chloride current as well (Fig. 2C). The most interesting difference was found in the experiments with KA application. In contrast to Glu- and QA-induced responses, ARG. III and ARG had no influence on KA-induced potassium currents (Fig. 2B). Thus the selective effect of these antagonists indicated the existence in molluscan neurones of two distinct Glu receptors (QA- and KA-type) controlling the potassium conductivity. This existence was also confirmed by cross-desensitization experiments, where a lack of interaction was observed between quisqualate and kainate.[21]

What is the mechanism whereby ARG blocks the quisqualate-type receptor on molluscan neurone? In the present experiments the blocking effect of ARG and ARG. III on potassium current induced by Glu and QA was more pronounced at -70 mV than at -30 mV, but it cannot serve as a basis for believing that the antagonists preferentially interacted with open potassium channels. On the contrary, the inability of ARG to inhibit the KA effect speaks against this possibility. There are good reasons to believe that QA and KA may open the same population of potassium channels by activation of GTP-binding protein:

1. Pretreatment of neurones with pertussis toxin (PTX) selectively inhibited the slow potassium currents evoked by both QA and KA application but failed to affect the fast chloride current.
2. These responses to QA or KA were mimicked by intracellular injection of GTP-γ-S or GppNHp. Long-lasting (45–50 min) treatment of neurones with 0·5 μM ARG did not inhibit the current induced by GTP-γ-S.
3. The average maximal amplitudes of currents induced by QA application and GTP-γ-S injections were practically equal.
4. Glu, QA or KA application at the height of GTP-γ-S induced potassium current could not evoke an additional response.
5. KA application on the height of maximal QA response and vice versa did not evoke an additional current.

These results and the above mentioned data suggest that two distinct Glu receptors (of QA- and KA-type) can share a common potassium channel by means of G-protein activation.[21] Only one of these receptors, namely the QA-type, is antagonized by ARG and ARG. III. The absence of an ARG antagonist effect towards GTP-γ-S- and KA-evoked potassium currents indicates that ARG fails to block the open potassium channels.

Thus ARG and related neurotoxins allow one to distinguish two pharmacologically separate but functionally common glutamate receptors on molluscan neurone.

EFFECTS ON GLUTAMATE RECEPTORS OF ISOLATED FROG SPINAL CORD

Pharmacological studies indicate that the glutamate receptors of non-NMDA types play a critical role in synaptic function in the brain. They mediate preferentially the excitatory postsynaptic potential (EPSP) resulting from unitary synaptic activation.[3,26] This means that non-NMDA receptors must be involved in generation of EPSPs in monosynaptic inputs to spinal cord motoneurones. It is very important to use potent and selectively acting antagonists of non-NMDA receptors for their identification and investigation of pharmacological and functional properties. ARG and related neurotoxins appear to be a good tool.

It was shown in the experiments performed on the isolated spinal cord of the frog *Rana ridibunda* that ARG is a selective antagonist of non-NMDA receptors.[12,16,17] Short-lasting (30 s) application of different agonists (Glu, QA, KA, NMDA or aspartate) which were added to the perfusate evoked depolarization of motoneurones. The dose–response curves were based on measurements of the amplitudes before and after treatment with ARG. ARG (0·1–0·5 μM) effectively inhibited the responses to Glu and QA (to about 50% of initial amplitude), while depolarizations induced by aspartate and KA were unaffected (Fig. 3). Thus ARG revealed a high specificity for its blocking effect. It is really a valuable tool for distinguishing different subtypes of non-NMDA receptors.

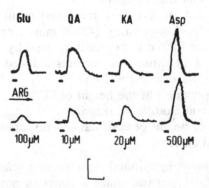

FIG. 3. Effect of argiopine on responses of frog motoneurones on the application of glutamate agonists. Glu, glutamate; QA, quisqualate; KA, kainate; ASP, aspartate. Bars, time of agonist application, concentration indicated below. Upper row, before, and lower row, after 30 min treatment with 1 μM argiopine. Calibration: 10 mV and 2 min.

ARG was used for investigation of the involvement of non-NMDA receptors in synaptic transmission from dorsal roots (DR) and supraspinal structures (reticular formation, RF). (EPSPs) evoked by stimulation of (1) the 9th DR, (2) the small groups of DR fibers, and (3) the RF at the bottom of the IV ventricle were recorded intracellulary in the motoneurones with resting potentials exceeding −60 mV.[16,17]

In every experiment ($n = 28$) 10 min after ARG treatment (0·3–10 μM) EPSPs were transiently enhanced (Fig. 4A, B). By the 30th minute the amplitude of the early (monosynaptic) component of DR EPSP recovered to the initial level (Fig. 4A and C), whereas the amplitudes of early RF and late (polysynaptic) components of both EPSPs were decreased (Fig. 4A, B, D). At 40–45 min the amplitudes of the polysynaptic DR and RF EPSPs and monosynaptic RF EPSP greatly decreased and were completely blocked after 60 min of ARG exposure (Fig. 4B), while the early component of DR EPSP was little affected, being close to control amplitude (Fig. 4A, C). Neither an increase of ARG concentration, nor a prolongation of its treatment could significantly change such different behaviour of the DR and the RF monosynaptic responses.

It seemed reasonable to compare antagonist effects on mono- and polysynaptic DR and RF responses of ARG and the newly developed kainate/quisqualate antagonist CNQX. It has been shown that CNQX preferentially blocked QA-induced depolarizations (in the concentration range

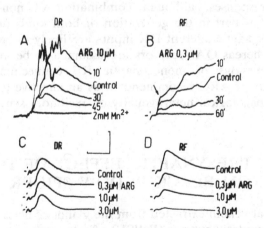

Fig. 4. The effect of argiopine (ARG) on EPSP recorded in frog motoneurones.[17] (A), (B): EPSPs evoked by stimulation of the whole DR (16 responses averaged) and of the RF (10 responses averaged). The time after ARG application is shown for each recording. Bottom line of part (A): 2 mM Mn^{2+}, electrotonic component of EPSP. Arrows indicate the monosynaptic components of EPSPs. (C), (D): the effect of increasing ARG concentration on monosynaptic DR- and RF-evoked EPSPs in the same motoneurone. Ten single EPSPs have been averaged. Callibration: part (A), 10 ms and 5 mV; parts (B), (C) and (D), 10 ms and 1 mV.

1–3 μM).[27] Indeed, in our experiments CNQX in the low concentrations (1–3 μM) diminished monosynaptic RF EPSP more effectively than monosynaptic DR EPSP, the increase of the dose up to 10 μM was required for a full blockade of both monosynaptic responses.[17] It means that the specificity of ARG as an antagonist of QA-type of glutamate receptors is much greater than of the best drug which is available now.

The results obtained indicate that the monosynaptic inputs to motoneurones from DR and RF differ. The same neurotransmitter appears to be involved in these inputs, since kynurenate (non-selective antagonist of glutamate receptors) in concentrations 1–2 mM was found to block both excitatory responses, including the monosynaptic DR EPSP. D,L-2-Amino-5-phosphonovaleric acid, AP5 (selective antagonist of NMDA receptors) in a concentration of 0·1 mM could slightly diminish only the late, but not the early monosynaptic component of the DR responses ($n = 5$). No obvious effect of AP5 on monosynaptic RF response was observed ($n = 5$). Evidently the receptors of NMDA type do not play a prominent role in the generation of both monosynaptic responses under study.

The use of ARG revealed certain differences in pharmacological properties of the receptors involved in the sensorimotor (DR) and supraspinal (RF) inputs in frog motoneurones. The resistance of monosynaptic DR EPSP to ARG blocking effect correlates with inability of ARG to block KA response of motoneurons at low concentrations and with inability of ARG to block KA response in molluscan neurons.

Evidence suggests that only non-NMDA receptors are involved in DR and RF monosynaptic responses, although combination of non-NMDA and NMDA receptors takes part in the generation of both kinds of polysynaptic response. The monosynaptic afferent DR inputs are likely to be mediated by the KA receptors, whereas QA receptors are assumed to be involved in the interneuronal transmissions of monosynaptic RF connections. The high potency and specificity of ARG recommends it as an effective tool for further analysis of the pharmacological heterogeneity of excitatory synaptic inputs in brain neurones.

SELECTIVE PRESYNAPTIC EFFECT OF TOXINS ISOLATED FROM BLACK WIDOW SPIDER VENOM

It has been shown that venom extracted from the glands of black widow spider *Latrodectus mactans tredecimguttatus* (BWSV) affects transmitter release from the nerve endings of various animals: vertebrates, insects, crustacea.[19,28–31] BWSV greatly enhances the rate of spontaneous quantal release of many transmitters such as acetylcholine, GABA, serotonin and excitatory amino acids. Fractionation of BWSV revealed several protein factors which differed in their toxicity for mice and in their ability to act on houseflies, the crayfish stretch receptor and the cockroach heart.[28] The first factor has been identified as a single protein of mol. wt 130 kDa and referred to as α-latrotoxin (p-LT).[28,31] α-LT affects the release of various kinds of transmitters, but unlike BWSV only from the nerve endings of vertebrate animals. Even high doses of α-LT were ineffective in experiments on invertebrate preparations.[31,32]

Recently, as a result of chromatographic separation procedures, an homogeneous weakly acidic protein of mol. wt 120 kDa was isolated from BWSV.[33,34] This protein, referred to as α-latroinsectotoxin (α-LIT) failed to affect transmitter release from frog nerve endings, but displayed toxicity to *Musca domestica* larvae and was highly effective in experiments on isolated neuromuscular preparation of blowfly *Calliphora vicina* larvae.

The presynaptic action of α-LT and α-LIT was manifested by consistent changes in the frequency of spontaneous quantal transmitter release from the motor nerve endings of frog and blowfly (MEPPs and MEPSPs respectively). Following bath application of 2–5 nM α-LT on frog preparations, MEPP frequency soon increased, climbing steeply from 2 to 100–150 per s after 20 min, although similar application of 50 nM α-LIT did not affect the MEPP frequency. Opposite results were obtained on fly preparations where application of 1–4 nM α-LIT induced after a short delay (which depended on toxin concentration), a sharp increase of MEPSP rate from 10 to 50–60 per s, although application of 50 nM α-LT was completely ineffective (Fig. 5). Pretreatment of fly preparations with a large concentration (50 nM) of α-LT did not prevent the effect of subsequent application of 5 nM α-LIT. Pretreatment of frog preparations with excess of α-LIT did not prevent the typical effect of α-LT. These data and the results from binding experiments [34] provide evidence that there are prominent difference in the structure of toxin receptors on the presynaptic membrane of vertebrates and insects.

On the contrary, the analysis of the mode of the presynaptic effect of α-LIT

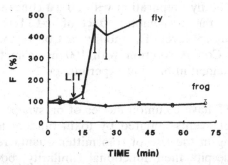

FIG. 5. Effects of α-latroinsectotoxin (LIT) on the frequency of miniature EPSPs in fly larvae and frog preparations.[34] Arrow, application into the bath of LIT (4 nM) on fly preparations and 50 nM on frog preparations. Means ±SE from eight experiments on fly and six on frog. Abscissa, time in min. Ordinate, frequency of miniature EPSPs given as a percentage of the control level.

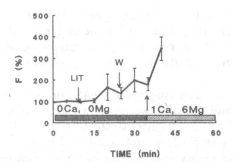

FIG. 6. Influence of the change of concentration of divalent ions in bathing medium on the effects of α-latroinsectotoxin.[34] LIT—application of 4 nM on fly preparations. The media used are shown by bars. W, removal of free LIT from the perfusing solution which was free of divalent ion and contained 1 mM EGTA. Means ±SE from 15 experiments.

in insects revealed a great deal in common with α-LT effect in vertebrates:

(1) the complete removal of divalent cations (M^{2+}) from the perfusion solution prevented the increase in MEPSPs induced by α-LIT, but the toxin retained its ability to bind selectively to insect presynaptic membranes. It was possible to evoke the usual α-LIT effect by addition of Ca^{2+} or Mg^{2+} after the preceding perfusion with M^{2+}-free saline and the complete removal of free toxin (Fig. 6). Experiments with $[^{125}I]$-α-LIT confirmed the independence of toxin binding from the Ca^{2+} concentration.

(2) α-LIT revealed its effect in the absence of Ca^{2+} when Mg^{2+} ions were present. However, the substitution of Ca^{2+} for Mg^{2+} resulted in a smaller effect of α-LIT which could be enhanced by addition of Ca^{2+};

(3) pretreatment of the fly preparation with 50 μM concanavalin A (Con A) completely prevented the presynaptic effect of α-LIT. MEPSP frequency remained stable at control level. This ability of Con A could be blocked by pretreatment of the Con A solution with 100 mM α-methylmannoside. A similar result was obtained in binding experiments.[34]

Thus α-LT and α-LIT have common modes of presynaptic action. For both molecules, the chain of events initiated by toxin binding to the presynaptic membrane and resulting in the rise of transmitter quanta release, is possibly the same. However, despite their functional similarity, both toxins interact with structurally different receptors. Such a difference may explain the selectivity of the presynaptic action of α-LT and α-LIT on vertebrates and insects, respectively.[34]

CONCLUSION

Spider venoms are rich in neurotoxins, which possess presynaptic (α-LT, α-LIT from *Latrodectus mactans tredecimguttatus*) and postsynaptic (JSTX, family of low-molecular-weight toxins isolated from Araneidae spider venoms), effects. The high potency and especially the unique specificity of these toxins make them a promising tool for the investigation of synaptic transmission. Toxins may be used both in physiological and neurochemical experiments for the elucidation of the key mechanisms of signal transduction.

REFERENCES

1. Mayer, M. L. & Westbrook, G. L., The physiology of excitatory amino acids in the vertebrate central nervous system. *Progr. Neurobiol.*, **28** (1987) 197–276.
2. Shinozaki, H., Pharmacology of the glutamate receptor *Progr. Neurobiol.*, **30** (1988) 399–435.
3. Monaghan, D. T., Bridges, R. J. & Cotman, C. W., The excitatory amino acid receptors: their classes, pharmacology and distinct properties in the function of the central nervous system. *Ann. Rev. Pharmacol. Toxicol.*, **29** (1989) 365–402.
4. Jackson, H. & Usherwood, P. N. R., Spider toxins as tools for dissecting elements of excitatory amino acid transmission. *Trends Neurosci.*, **11** (1988) 278–83.
5. Kawai, N., Niwa, A. & Abe, T., Spider venom contains specific receptor blocker of glutamergic synapses. *Brain Res.*, **247** (1982) 169–71.
6. Miwa, A., Kawai, N., Saito, M., Pan-Hou, H. S. & Yoshika, M., Effects of spider toxin (JSTX) on excitatory postsynaptic current at neuromuscular synapse of spiny lobster. *J. Neurosphysiol.*, **58** (1987) 319–26.
7. Bateman, A., Boden, P., Del, J. R., Quicke, D. L. J. & Usherwood, P. N. R., Postsynaptic block of glutamatergic synapse by low molecular weight fractions of spider venom. *Brain Res.*, **339** (1985) 237–44.
8. Budd, T., Clinton, P., Dell, A., Duce, I. R., Johnson, S. J., Quicke, D. L. J., Taylor, G. W., Usherwood , P. N. R. & Usoh, G., Isolation and characterization of glutamate receptor antagonists from venoms of orb-web spiders. *Brain Res.*, **448** (1988) 30–39.
9. Grishin, E. V., Volkova, T. M., Arseniev, A. S., Reshetova, O. S., Onoprienko, V. V., Magazanik, L. G., Antonov, S. M. & Fedorova, I. M., Structure-functional characterizations of argiopine—an ion channel blocker from the venom of spider *Argiope lobata. Bioorg. Khimia*, **12** (1986) 1121–4 (in Russian).
10. Magazanik, L. G., Antonov, S. M., Fedorova, I. M., Volkova, T. M. & Grishin, E. V., Effects of the spider *Argiope lobata* crude venom and its low molecular weight component, arglopine, on the post-synaptic membrane. *Biologicheskie Membr.*, **3** (1986) 1204–19. (In Russian).
11. Magazanik, L. G., Antonov, S. M., Fedorova, I. M., Volkova, T. M. & Grishin, E. V., Argiopine—a naturally occurring blocker of glutamate-sensitive synaptic channels. In *Receptors and Ion Channels*, ed. F. Hucho. & Yu. A. Ovchinnikov. de Gruyter, Berlin, 1987, pp. 305–12.
12. Antonov, S. M., Grishin, E. V., Magazanik, L. G., Shupliakov, O. V., Vesselkin, N. P. & Volkova, T. M., Argiopine blocks the glutamate responses and sensorimotor transmission in motoneurones of isolated frog spinal cord. *Neurosci. Lett.*, **83** (1987) 179–84.

13. Antonov, S. M., Dudel, J., Franke, C. & Hatt, H., Argiopine blocks glutamate-activated single-channel currents on crayfish muscle by two mechanisms. *J. Physiol. (Lond.),* **419** (1989) 569–87.

14. Grishin, E. V., Volkova, T. M. & Arseniev, A. S., Antagonists of glutamate receptors from the venoms of *Argiope lobata* spider. *Bioorg. Khimia,* **14** (1989) 832–8 (In Russian).

15. Kiskin, N. I., Krishtal, O. A., Tsyndrenko, A. V., Volkova, T. M. & Grishin, E. V., Argiopine, argiopinines and pseudoargiopinines as glutamate receptor blockers in hippocampal neurons. *Neurofiziologia,* **21** (1989) 748–56 (In Russian).

16. Antonov, S. M., Kalinina, N. I., Kurchavyj, G. G., Magazanik, L. G., Shupliakov, O. V. & Vesselkin, N. P., Identification of two types of excitatory monosynaptic inputs in frog spinal motoneurones. *Neurosci. Lett.,* **109** (1990) 82–7.

17. Antonov, S. M., Magazanik, L. G., Kalinina, N. I., Kurchavyj, G. G. & Vesselkin, N. P., Glutamate receptors in excitatory synaptic inputs to frog spinal motoneurones. In *Excitatory Amino Acids,* ed. B. S. Meldrum, F. Moroni & J. H. Woods. Raven Press, New York, 1991, pp. 303–6.

18. Adams, M. E., Herold, E. E. & Venema, V. J., Two classes of channel-specific toxins from funnel web spider venom. *J. Comp. Physiol. A,* **164** (1989) 333–42.

19. Rosenthal, L. & Meldolesi, J., Alfa-latrotoxin and related toxins. *Pharm. Ther.,* **42** (1989) 115–34.

20. Kerry, C. J., Ramsey, R. L., Sansom, M. S. P. & Usherwood, P. N. R., Single channel studies of non-competitive antagonism of a quisqualate-sensitive glutamate receptor by argiotoxin-636—a fraction isolated from orb-web spider venom. *Brain Res.,* **459** (1988) 312–27.

21. Bolshakov, V. Yu., Gapon, S. A. & Magazanik, L. G., Different types of glutamate receptors in isolated and identified neurones of the mollusc *Planobarius corneus. J. Physiol. (Lond.),* **439** (1991) 15–35.

22. Saito, M., Sahara, Y., Miwa, A., Shimazaki, K., Nakajima, T. & Kawai, N., Effects of spider toxin (JSTX) on hippocampal CA1 neurons in vitro. *Brain Res.,* **481** (1989) 16–24.

23. Walker, R. J., The action of kainic acid and quisqualic acid on the glutamate receptors of three identifiable neurones from the brain of the snail, *Helix aspersa. Comp. Biochem. Physiol.,* **55C** (1976) 61–8.

24. Ascher, P., Nowak, L. & Kehoe, J. S., Glutamate-activated channels in molluscan and vertebrate neurones. In *Ion Channels in Neural Membranes.* Alan R. Liss, 1986, pp. 283–95.

25. Kostyuk, P. G., Ionic background of activity in giant neurons of molluscs. In *Symposium on Neurobiology of Invertebrates,* ed. J. Salanki. Plenum Press, New York, 1968, pp. 145–67.

26. Headley, P. M. & Grillner, S., Excitatory amino acids and synaptic transmission: the evidence for a physiological function. *Trends Pharmacol.,* **11** (1990) 205–11.

27. Fletcher, E. J., Martin, D., Aram, J. A., Lodge, D. & Honore, T., Quinoxaline-diones selectively block quisqualate and kainate receptors and synaptic event in rat neocortex and hippocampus and frog spinal cord in vitro. *Brit. J. Pharmacol.,* **95** (1988) 585–97.

28. Frontali, N., Ceccarelli, B., Gorio, A., Mauro, A., Siekevitz, P., Tzeng, M. C. & Hurlbut, W. P., Purification from black widow spider venom of a protein factor causing the depletion of synaptic vesicles at neuromuscular junctions. *J. Cell Biol.,* **68** (1976) 462–79.

29. Cull-Candy, S. G., Neal, H. & Usherwood, P. N. R., Action of black widow spider venom on an aminergic synapse. *Nature,* **241** (1973) 353–4.

30. Kawai, N., Mauro, A. & Grundfest, H., Effect of black widow spider venom on the lobster neuromuscular junctions. *J. Gen. Physiol.,* **60** (1972) 650–64.

31. Meldolesi, J., Scheer, H., Madeddu, L. & Wanke, E., Mechanism of action of alfa-latrotoxin: the presynaptic stimulatory toxin of the black widow spider. *Trends Pharmacol.*, **7** (1986) 151–5.
32. Magazanik, L. G., Fedorova, I. M., Antonov, S. M., Kovalevskaya, G. I., Bulgakov, O. V., Pashkov, V. N. & Grishin, E. V., Black widow spider venom contains different components selectively affecting presynaptic membranes of vertebrates and insects. *Biologicheskie Membr,* **7** (1990) 660–61 (in Russian).
33. Kovalevskaya, G. I., Pashkov, V. N., Bulgakov, O. V., Fedorova, I. M., Magazanik, L. G. & Grishin, E. V., Identification and isolation of protein α-latroinsectotoxin from the venom of spider *Latrodectus mactans tredecimguttatus.* *Bioorg. Khimia,* **16** (1990) 1013–18 (in Russian).
34. Magazanik, L. G., Fedorova, I. M., Kovalevskaya, G. I., Pashkov, V. N., Bulgakov, O. V. & Grishin, E. V., Selective presynaptic insectotoxin (α-latroinsectotoxin) isolated from black widow spider venom. *Neuroscience,* **46** (1992) 181–8.

3

Probing Calcium Channels with Venom Toxins

MICHAEL E. ADAMS, VYTAUTAS P. BINDOKAS & VIRGINIA J. VENEMA

Department of Entomology, University of California, Riverside, California 92521, USA

INTRODUCTION

The crucial roles played by ion channels in cellular function make them obvious targets for pesticides and drugs. Ion channel-directed insecticides such as the synthetic pyrethroids are a good example. Antagonism of ion channels in therapeutic applications is exemplified by dihydropyridine calcium channel blockers used for cardiac treatment. Given these precedents, it seems reasonable that basic research devoted to ion channel mechanisms and associated pharmacology would promote discovery of new drug or pesticide chemicals. Nevertheless, it has been argued, particularly at this conference, that high through-put, random screening programs, rather than basic research, have historically contributed to new insecticide discovery. What then, can basic research offer in future searches for newer, more selective chemical agents for drug and pesticide development?

Two clear insights, which ultimately may facilitate new chemical discovery, have emerged from basic studies on ion channels over the past 10 years. One is the profound diversity of ion channels that exists in biological systems. Although the extent of this diversity has yet to be precisely characterized, it appears to be extreme for potassium and calcium channels and moderate for sodium channels. Understanding the functional significance of different channel subtypes will allow precise physiological targets to be identified for drug or insecticide design. The second insight emerging from basic studies is that ion channel subtypes are differentially sensitive to drugs and toxins. This differential ligand sensitivity is useful because it allows for selective pharmacological recognition of channel subtypes, thereby aiding both in their characterization as well as assignment of functional significance. Toxins have provided crucial information on the properties of the nicotinic receptor channel, the GABA-activated chloride channel, and voltage activated sodium and calcium channels.

Recently, dramatic applications have resulted from selective affinities of natural toxins to insect control. The insect selective scorpion toxin, AaIT,[1,2] which targets neuronal sodium channels, has been expressed in an 'engineered pathogen', resulting in enhanced efficacy.[3] A similar example using an acarine toxin appeared simultaneously.[4] These reports strongly suggest that selective ion channel-directed toxins hold promise for development of future generations of insect control agents. It is particularly noteworthy that the insect specific pathogens used in these models constitute an ideal delivery system for the introduction of peptides into pest insects.

This paper summarizes our investigations into the chemical basis of insect paralysis caused by spider venoms. These studies were undertaken with the expectation that the biochemical prey capture strategies used by insect predators might provide clues to vital physiological mechanisms not yet exploited as insecticide targets. Chemical analysis of venom from the funnel web spider *Agelenopsis aperta,* in combination with electrophysiological and biochemical assays, has generated new information that is organized in this paper under two themes, both of which involve calcium channels: 1) the biochemical basis for prey capture, and 2) the use of toxins as probes for ion channel characterization.

THREE CLASSES OF ION CHANNEL TOXINS IN *AGELENOPSIS APERTA* VENOM

The American funnel web spider, *Agelenopsis aperta,* paralyzes its prey by injecting a venom composed of acylpolyamine and peptide toxins. Fractionation of the venom by reversed-phase high performance liquid chromatography (HPLC) yielded the profile shown in Fig. 1. We first screened various venom fractions for effects on insect neuromuscular transmission using the large body wall muscles of the pre-pupal house fly, *Musca domestica.*[5,6] Intracellular microelectrode recording of the neurally evoked excitatory junctional potential (EJP) and the ionophoretic glutamate potential (IP) provided information about presynaptic versus postsynaptic actions of the venom fractions (see Refs. 7–9 for details). Fractions from the venom produced three distinct types of synaptic effects (Fig. 1, 2). Each effect is associated with the action of a unique class of toxin on a different ion channel target (Fig. 2). The α–agatoxins, a group of low molecular weight acylpolyamines, produce rapid and reversible block of both the neurally evoked excitatory junctional potential (EJP) and the ionophoretic glutamate potential (IP) (Fig. 1). These effects result from use-dependent block of transmitter-activated cation channels on the post-synaptic membrane.[7] The quick, transient paralysis produced in whole insects upon injection of the α–agatoxins is consistent with these observed neuromuscular actions. Chemical structures of the α–agatoxins[10,11] show that they resemble the acylpolyamine toxins from orb weaver spider venoms.[12–15]

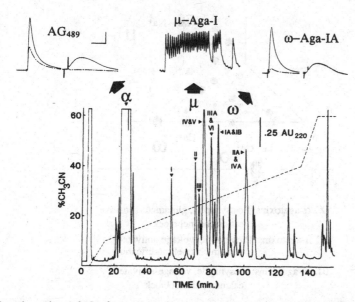

FIG. 1. Fractionation of *Agelenopsis aperta* venom by reversed-phase high performance liquid chromatography. Assay of fractions for effects on insect neurotransmission revealed three distinct types of action, each associated with a different class of venom toxin. AG$_{489}$ an α–agatoxin (0·8 μM), blocks both the excitatory junctional potential (EJP) and ionophoretic glutamate potential (IP). The peptide toxin μ–Aga–IV, one of six identified μ–agatoxins (0·1 μM), causes repetitive activity in a motor neuron upon superfusion at a concentration of 0·1 μM (upper middle). ω–Aga–IA, one of many ω–agatoxins, suppresses the EJP, but has no effect on the IP. HPLC conditions: Brownlee C8 column; acetonitrile/water in constant 0·1% trifluoroacetic acid, 1 ml/min flow rate. Calibration for electrophysiological recordings: 10 mV, 50 msec (upper left, upper right); 10 mV, 200 msec (upper middle).

A strikingly different action, that of repetitive motor unit activity, was registered in response to superfusion with fractions from the 'μ' region of the chromatogram (Fig. 1). This action is caused by 36–37 amino acid 'μ–agatoxins', a group of highly disulfide bridged peptides. Nanomolar concentrations of the μ–agatoxins produce both repetitive motor unit activity and increased spontaneous transmitter release in house fly motor units.[7,16] Both of these effects are consistent with perturbation of sodium channel function similar to that caused by excitatory scorpion toxins.[16,17] The effects of the μ–agatoxins persist for hours after washout of the toxin from the bath. This persistent action on motor nerves is correlated with irreversible paralysis and death in insects injected with the μ–agatoxins. Interestingly, the μ–agatoxins appear to be highly selective for insect sodium channels and have no known effects on vertebrate ion channels. In this respect, they are similar in action to the insect–specific excitatory scorpion toxins AaIT.[1]

A third qualitatively different type of neuromuscular action was caused by fractions from the 'ω' region of the chromatogram, which contains a heterogeneous group of polypeptide ω–agatoxins. These toxins produce

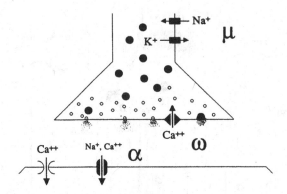

α: α-agatoxins: Postsynaptic, glutamate-sensitive
cation channel block

μ: μ-agatoxins: Presynaptic, voltage-activated
sodium channel modification

ω: ω-agatoxins: Presynaptic, voltage-activated
calcium channel block

Fig. 2. Schematic diagram of an insect neuromuscular junction, showing sites of action for three classes of venom toxins from the funnel web spider, *Agelenopsis aperta*. The α–agatoxins are postsynaptic antagonists of the transmitter-activate cation channel on the postsynaptic membrane. The μ–agatoxins are presynaptic activators, probably altering the kinetics of sodium channels in the terminal branches of motor axons. The ω–agatoxins are presynaptic antagonists, blocking voltage-dependent calcium channels in motor nerve terminals and thus preventing transmitter release. While the ω–agatoxins are potent neuronal calcium channel antagonists, they do not affect insect skeletal muscle calcium channels.

synaptic block, but the characteristics of this block are clearly different from that caused by the α–agatoxins. While the α–agatoxins block both EJPs and IPs[8,18,19] (Fig. 1). Since the ω–agatoxins block EJPs without affecting IPs[8,18,19] (Fig. 1). Since the ω–agatoxins do not alter sensitivity of the muscle to the neurotransmitter, we hypothesized that synaptic block was presynaptic. Ongoing studies of the ω–agatoxins from *A. aperta* venom (see below) have shown them to be a heterogeneous group with overlapping selectivities for calcium channels in insect, avian and mammalian neurons.

DIFFERENTIAL BLOCK OF INSECT CALCIUM CHANNELS BY ω–AGATOXIN SUBTYPES

Many ω–agatoxins producing presynaptic block of fly neuromuscular transmission have been isolated from *A. aperta* venom. These toxins also block calcium channels in insect neuronal cell bodies.[18,19] The chemical characteristics as well as differential block of calcium channels set them apart from the other toxins occurring in the venom.

Unlike the μ–agatoxins, whose amino acid sequences were found to be quite similar,[9] the ω–agatoxins were heterogeneous both in terms of chemical and pharmacological characteristics. Two groups of ω–agatoxins could be distinguished based on differences in molecular size and primary amino acid sequences.[8] Type I ω–agatoxins (ω–Aga–IA, ω–Aga–IB) ran on SDS–PAGE gels at between 7·1–7·5 kDa. Amino acid composition analyses showed them to contain 9 cysteines. Complete amino acid sequencing of ω–Aga–IA showed it to be an unblocked, 66 amino acid peptide. The N–terminal sequence of ω–Aga–IB showed high similarity to ω–Aga–IA. The 'Type II' toxin ω–Aga–IIA, a 9 kDa peptide, showed low similarity to the Type I toxins.

Type I and Type II ω–agatoxins differentially block insect neuromuscular transmission. We had previously observed that saturating concentrations of either Type I or Type II ω–agatoxins, when applied under low extracellular calcium concentrations ($[Ca]_o$), suppress EJPs up to 97%.[8,18] Under these conditions, a small remnant of the EJP persisted in all experiments, indicating that a small amount of transmitter release remained. Upon elevation of $[Ca]_o$ to 5 mM, the amplitude of this persistent EJP increased considerably.[19] In other words, the toxins produce only incomplete block under higher $[Ca]_o$. It was under these conditions that differences in Type I and Type II block were accentuated.

Under conditions of elevated $[Ca]_o$ (5 mM), saturating concentrations of Type I toxins maximally blocked up to 30–40%, while Type II toxins blocked up to 60–70%. Clearly, the two subtypes of ω–agatoxins produced different characteristic levels of inhibition and, in each instance, a substantial portion of the EJP was resistant to the effects of saturating toxin concentrations. When the two toxins were applied sequentially, we found that the effects of each toxin were largely additive, such that almost the entire EJP was blocked after exposure to both Type I and Type II toxins (Fig. 3).

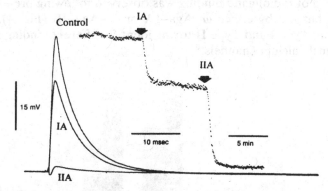

IG. 3. Differential action of two ω–agatoxins subtypes on the excitatory junctional potential recorded from house fly body wall muscle. Under conditions of 5 mM $[Ca]_o$, a saturating concentration of ω–Aga–IA (100 nM) causes a maximum of 25–30% reduction in the amplitude of the EJP. A substantial component of the EJP remains, which must result from significant residual presynaptic 'Type I insensitive' calcium conductance. Subsequent application of ω–Aga–IIA blocks most of this residual synaptic response. Thus, the combined application of both Type I and Type II ω–agatoxins blocks >95% of the EJP.

This differential block by Type I and Type II ω–agatoxin subtypes could be explained by one or more mechanisms. One scenario could involve differential block of distinct subtypes of Ca channels in nerve terminals (i.e., Type I and Type II sensitive calcium channels). Block of one channel subtype would leave the second unaffected, thus allowing a significant component of calcium entry and EJP to remain. Alternatively, conductance states of a uniform population of channels could be differentially affected by each toxin subtype. While we cannot presently distinguish between these alternative hypotheses, receptor binding data (see next section) provide clear evidence that Type I and Type II toxins have distinct binding sites on calcium channels.

ANTAGONISM OF AVIAN Ca CHANNELS BY ω–AGATOXINS AND ω–CONOTOXIN

Studies on the avian brain provided the first evidence indicating that a subset of ω–agatoxins bind to vertebrate calcium channels. In collaborative experiments with Dr. Baldomero Olivera of the University of Utah, the ω–agatoxins were tested for ability to inhibit high affinity binding of ^{125}I–ω–conotoxin GVIA (ω–CgTX) to chick brain synaptosomal calcium channels.[20] ω–CgTX, a toxin from the fish-eating cone snail, *Conus geographus,* was first shown to block presynaptic calcium channels at the amphibian neuromuscular junction,[21] and since then has become the diagnostic ligand for 'N-type' calcium channels described by Tsien and colleagues[22-24] in chick sensory neurons.

To test for ability of ω–agatoxins to block ^{125}I–ω–CgTX binding, chick synaptosomal membranes were pre-incubated with either ω–Aga–IA, ω–Aga–IB or ω–Aga–IIA. Upon subsequent addition of ^{125}I–ω–CgTX, a clear-cut inhibition of radioligand binding was observed following pre-exposure to ω–Aga–IIA, but not by either ω–Aga–IA or ω–Aga–IB (Fig. 4). These data indicated that Type I and Type II toxins identify different binding sites on chick synaptosomal calcium channels.[8]

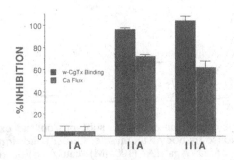

FIG. 4. Antagonism of chick brain synaptosomal calcium channels by subtypes of ω–agatoxins is correlated with inhibition of ^{125}I–ω–CgTX binding to synaptosomal membranes. ω–Aga–IIA and ω–Aga–IIIA block both ^{125}I–ω–CgTX binding and ^{45}Ca entry, while ω–Aga–IA is inactive in both assays. All toxins were applied at saturating concentrations to achieve maximal effect.

These results led us to systematically test fractions from *A. aperta* venom for ability to inhibit ^{125}I–CgTX binding to chick synaptosomal calcium channels, whereupon we found many fractions that mimic the inhibitory action of ω–Aga–IIA. Using inhibition of ^{125}I–CgTX binding as a bioassay, we isolated and sequenced ω–Aga–IIIA, which represents a third subtype of ω–agatoxins.[25] ω–Aga–IIIA shows considerable amino acid sequence similarity to ω–Aga–IIA, but is nevertheless smaller (76 versus about 92 amino acids) and also is about 5–10 times more abundant in the venom. ω–Aga–IIIA also differs from ω–Aga–IIA in being completely ineffective as an antagonist of presynaptic Ca channels at the house fly neuromuscular junction.[19] We have since isolated several more 'Type III' ω–agatoxins of similar size and abundance in the venom, and preliminary toxin gene cloning results suggest that the Type III family is quite large (unpublished data).

Differences in ability to inhibit ^{125}I–ω–CgTX binding proved to be highly predictive for functional antagonism of [45]Ca flux in chick synaptosomes (see below). Both ω–Aga–IIA and ω–Aga–IIIA are potent blockers of potassium-induced [45]Ca entry into chick synaptosomes (Figs 4, 5). In contrast, ω–Aga–IA, which was completely ineffective in blocking ^{125}I–ω–CgTX binding, also proved inactive as an inhibitor of potassium-induced [45]Ca entry into synaptosomes. These results provided further indication that subtypes of toxins may be selective for different calcium types.

The curves describing concentration dependence of calcium channel block in chick synaptosomes (Fig. 5) suggest that mechanistic differences underly block by ω–agatoxins and ω–conotoxin. ω–Aga–IIA and ω–Aga–IIIA are about 30-fold more potent than ω–CgTX in blocking potassium-stimulated Ca entry (IC$_{50}$ of about 1 nM, versus 38 nM for ω–CgTX). Unlike ω–CgTX, which blocks 100% of calcium entry at saturating concentrations, the ω–agatoxins block about 60–70% maximally. While the level of ω–CgTX block in chick synaptosomes is consistent with complete block of calcium channels in this

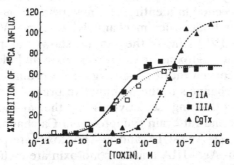

FIG. 5. Block of depolarization-induced [45]Ca entry into chick brain synaptosomes by ω–agatoxins and ω–CgTX. The marine cone snail toxin ω–CgTX blocks 100% of [45]Ca entry with an IC$_{50}$ of about 38 nM. ω–Aga–IIA and ω–Aga–IIIA are about 30–40 fold more potent (IC$_{50}$ = 1 nM), but maximally block only 60–70% of the flux response. Differences in potency and maximum level of block suggest that different mechanisms are involved in calcium channel antagonism by these peptides.

tissue (see also Refs 26, 27), the ω–agatoxins appear to produce incomplete block. Alternatively, the ω–agatoxins could produce complete block of only one of two channel subtypes in chick synaptosomes, leaving the other unaffected.

This proportion of block in chick synaptosomes caused by ω–Aga–IIA and ω–Aga–IIIA is remarkably similar to the proportion of the insect EJP blocked by ω–Aga–IIA (see Fig. 3). This suggests that ω–Aga–IIA may recognize a common binding site on insect and chick calcium channels and that the mechanism of block in the two tissues may be quite similar.

Finally, it should be pointed out that ω–Aga–IIIA blocks both N- and L-type calcium current in mammalian dorsal root ganglion neurons, but leaves T-type channels unaffected.[28] ω–Aga–IIIA thus appears to be a potent and selective blocker of high threshold calcium channels in vertebrate neurons.

NEW ω–AGATOXIN ANTAGONISTS OF MAMMALIAN Ca CHANNELS

Although ω–Aga–IIIA and ω–CgTX are effective antagonists of Ca channels in chick brain synaptosomes, we found both toxins to be ineffective at blocking rat brain synaptosomal calcium channels under the same conditions. Previous reports also have shown that mammalian synaptosomal calcium channels are resistant to block by ω–CgTX,[29-32] although Reynolds *et al.*[31] showed a significant component of ω–CgTX-sensitive calcium entry in rat brain synaptosomes.

Despite the ineffectiveness of ω–Aga–IIIA against rat synaptosomal calcium channels, we observed whole *A. aperta* venom to produce substantial block, indicating the presence of other components in the venom with different pharmacological selectivities. We therefore re-screened fractions from *A. aperta* venom for ability to block potassium-induced ^{45}Ca flux into rat brain synaptosomes, and succeeded in identifying a new peptide, ω–Aga–IVA, with novel selectivity for rat brain calcium channels. ω–Aga–IVA, which represents a family of 'Type IV' toxins in the venom, shows no apparent sequence similarity to the other ω–agatoxins or to ω–conotoxin and blocks rat brain Ca channels at nanomolar concentrations[34]. ω–Aga–IVA promises to be a valuable new probe for calcium channels in the mammalian brain.

The distinct but overlapping selectivities of the ω–agatoxins and ω–conotoxin against chick and rat brain synaptosomal Ca channels are illustrated in Fig. 6. ω–Aga–IIA, ω–Aga–IIIA and ω–conotoxin are potent blockers in chick brain. Whereas ω–Aga–IIIA and ω–conotoxin are ineffective against rat brain synaptosomal Ca channels, both ω–Aga–IVA and ω–Aga–IIA are potent blockers or rat brain synaptosomal Ca channels. ω–Aga–IVA is only weakly active in the chick preparation, while ω–Aga–IIA is active in both chick and rat brain, and blocks insect neurotransmission as well.

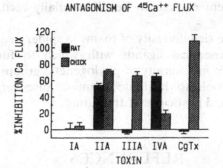

Fig. 6. Selectivities of ω–agatoxins and ω–CgTX in blocking potassium-stimulated
^{45}Ca entry into chick and rat synaptosomes. Shown are maximum levels of block by
saturating concentrations of each toxin. ω–Aga–IIIA and ω–CgTX are potent blockers
of ^{45}Ca entry into chick, but not rat synaptosomes. ω–Aga–IVA has a reversed
selectivity, blocking in rat, but only weakly in chick synaptosomes. ω–Aga–IIA is
active against both chick and rat synaptosomes, and ω–Aga–IA is inactive in both
systems.

SUMMARY AND CONCLUSIONS

Our analyses of peptide toxins in *A. aperta* venom have yielded an unexpec-
tedly rich diversity of Ca channel specific toxins. Four separate agatoxin
subtypes have been designated as Types I–IV. Competitive binding experiments suggest further that some of these toxins (Types II and III) interact with
ω–conotoxin binding sites, while others do not (Types I and IV). In terms of
functional Ca channel block, the overlapping selectivities of these toxins are
listed in Table 1.

These overlapping selectivities of the ω–agatoxins and ω–conotoxin suggest
that spiders have evolved toxins that interact with several distinct types of
binding sites on Ca channels. These selectivities suggest that Ca channels in
different tissues and phylogenetic groups are related, but nevertheless have

TABLE 1

Selectivities of ω–Agatoxins and ω–Conotoxin GVIA Against Insect and
Vertebrate Calcium Channels

Toxin	Insect Neuromuscular Transmission[8,18,19]	Synaptosomes		Rat DRG
		Chick[25]	Rat*	
ω–Aga–IA	+	−	−	+[33]
ω–Aga–IIA	+	+	+	?
ω–Aga–IIIA	−	+	−	+[28]
ω–Aga–IVA	−	−	+	?
ω–CgTX	−	+	−	+[28]

* Adams and Venema (unpublished).

surface features divergent enough to be differentially recognized by peptide toxins.

We anticipate that the rich diversity of toxins in spider venoms will provide a number of useful biochemical ligands with which to further classify and characterize Ca channels in a variety of phylogenetic groups. Hopefully, these ligands will serve as tools to facilitate chemical discovery relevant to the development of drugs and pesticides of the future.

REFERENCES

1. Zlotkin, E., Rochat, H., Kupeyan, C., Miranda, F. & Lissitzky, S., Purification and properties of the insect toxin from the venom of the scorpion *Androctonus australis* Hector. *Biochimie,* **53** (1971) 1073–& follow.
2. Zlotkin, E., The interaction of insect-selective neurotoxins from scorpion venoms with insect neuronal membranes. In *Neuropharmacology of Pesticide Action,* ed. M. G. Ford, G. G. Lunt, R. C. Reay & P. N. R. Usherwood. Ellis Horwood, London, 1986, pp. 352–383.
3. Stewart, L. M. D., Hirst, M., Ferber, M. L., Merryweather, A. T., Cayley, P. J. & Possee, R. D., Construction of an improved baculovirus insecticide containing an insect-specific toxin gene. *Nature,* **352** (1991) 85–8.
4. Tomalski, M. D. & Miller, L. K., Insect paralysis by baculovirus–mediated expression of a mite neurotoxin gene. *Nature,* **352** (1991) 82–5.
5. Irving, S. N. & Miller, T. A., Aspartate and glutamate as possible transmitters at the 'slow' and 'fast' neuromuscular junctions of the body wall muscles of *Musca* larvae. *J. Comp. Physiol.,* **135** (1980) 299–314.
6. Irving, S. N. & Miller, T. A., Ionic differences in 'fast' and 'slow' neuromuscular transmission in body wall muscles of *Musca domestica* larvae. *J. Comp. Physiol.,* **135A** (1980) 291–8.
7. Adams, M. E., Herold, E. E. & Venema, V. J., Two classes of channel–specific toxins from funnel web spider venom. *J. Comp. Physiol. [A],* **164** (1989a) 333–42.
8. Adams, M. E., Bindokas, V. P., Hasegawa, L. & Venema, V. J., ω–Agatoxins: Novel calcium channel antagonists of two subtypes from funnel web spider (*Agelenopsis aperta*) venom. *J. Biol. Chem.,* **265** (1990) 861–7.
9. Skinner, W. S., Adams, M. E., Quistad, G. B., Kataoka, H., Cesarin, B. J., Enderlin, F. E. & Schooley, D. A., Purification and characterization of two classes of neurotoxins from the funnel web spider, *Agelenopsis aperta. J. Biol. Chem.,* **264** (1989) 2150–5.
10. Quistad, G. B., Suwanrumpha, S., Jarema, M. A., Shapiro, M. J., Skinner, W. S., Jamieson, G. C., Lui, A. & Fu, E. W., Structures of paralytic acylpolyamines from the spider *Agelenopsis aperta. Biochem. Biophys. Res. Commun.,* **169** (1990) 51–6.
11. Jasys, V. J., Kelbaugh, P. R., Nason, D. M., Philips, D., Saccomano, N. A., Stroh, J. G. & Volkmann, R. A., Isolation, structure elucidation and synthesis of novel hydroxylamine-containing polyamines from the venom of the *Agelenopsis aperta* spider. *J. Am. Chem. Soc.,* **112** (1990) 6696–6704.
12. Grishin, E. V., Volkova, T. M., Arseniev, A. S., Reshetova, O. S., Onoprienko, V. V., Magazanic, L. G., Antonov, S. M. & Federova, I. M., Structure–functional characterization of argiopine—an ion channel blocker from the venom of spider, *Argiope lobata. Bioorg. Khim.,* **12** (1986) 110–12.

13. Aramaki, Y., Yasuhara, T., Higashijima, T., Yoshioka, M., Miwa, A., Kawai, N. & Nakajima, T., Chemical characterization of spider toxin, JSTX and NSTX. *Proc. Japan Acad.*, **62B** (1986) 359–62.

14. Adams, M. E., Carney, R. L., Enderlin, F. E., Fu, E. T., Jarema, M. A., Li, J. P., Miller, C. A., Schooley, D. A., Shapiro, M. J. & Venema, V. J., Structures and biological activities of three synaptic antagonists from orb weaver spider venom. *Biochem. Biophys. Res. Comm.*, **48** (1987) 678–83.

15. Toki, T., Yasuhara, T., Aramaki, Y., Kawai, N. & Nakajima, T., A new type of spider toxin, Nephilatoxin, in the venom of the Joro spider, *Nephila clavata*. *Biomed. Res.*, **9** (1988) 75–9.

16. Adams, M. E., Bindokas,, V. P. & Zlotkin, E., Synaptic toxins from arachnid venoms: probes for new insecticide targets. In *Insecticide Action: From Molecule to Organism*, ed. T. Narahashi & J. E. Chambers. Plenum Press, New York, 1989b pp. 189–203.

17. Walther, C., Zlotkin, E. & Rathmayer, W., Action of different toxins from the scorpion *Androctonus australis* on a locust nerve–muscle preparation. *J. Insect Physiol.*, **22** (1976) 1187–94.

18. Bindokas,, V. P. & Adams, M. E., ω–Aga–I: a presynaptic calcium channel antagonist from venom of the funnel web spider, *Agelenopsis aperta. J. Neurobiol.*, **20** (1989) 171–88.

19. Bindokas, V. P., Venema, V. J. & Adams, M. E., Differential antagonism of transmitter release by subtypes of ω–agatoxins. *J. Neurophysiol.*, **66** (1991) 590–601.

20. Cruz, L. J. & Olivera, B. M., Calcium channel antagonists. ω–conotoxin defines a new high affinity site, *J. Biol. Chem.*, **261** (1986) 6230–33.

21. Kerr, L. M. & Yoshikami, D., A venom peptide with a novel presynaptic blocking action. *Nature*, **308** (1984) 282–4.

22. Nowycky, M. C., Fox, A. P. & Tsien, R. W., Three types of neuronal calcium channel with different calcium agonist sensitivity. *Nature*, **316** (1985) 440.

23. Fox, A. P., Nowycky, M. C. & Tsien, R. W., Kinetic and pharmacological properties distinguishing three types of calcium currents in chick sensory neurones, *J. Physiol. (Lond.)*, **394** (1987) 149–72.

24. Fox, A. P., Nowycky, M. C. & Tsien, R. W., Single channel recordings of three types of calcium channels in chick sensory neurones, *J. Physiol. (Lond.)*, **394** (1987) 173–200.

25. Venema, V. J., Swiderek, K., Lee, T. D., Hathaway, G. M. & Adams, M. E., Antagonism of synaptosomal calcium channels by subtypes of ω–agatoxins. *J. Biol. Chem.*, **267** (1992) 2610–15.

26. Aosaki, T. & Kasai, H., Characterization of two kinds of high-voltage-activated Ca–channel currents in chick sensory neurons. Differential sensitivity to dihyropyridines and omega-conotoxin GVIA. *Pflugers Arch*, **414** (1989) 150–6.

27. Kasai, H., Aosaki, T. & Fukuda, J., Presynaptic Ca–antagonist omega–conotoxin irreversibly blocks N-type Ca-channels in chick sensory neurons. *Neurosci. Res.*, **4** (1987) 228–35.

28. Mintz, I. M., Venema, V. J., Adams, M. E. & Bean, B. P., Inhibition of N- and L-type Ca^{2+} channels by the spider venom toxin ω–Aga–IIIA. *Proc. Natl. Acad. Sci. U.S.A.*, **88** (1991) 6628–31.

29. Suszkiw, J. B., Murawsky, M. M. & Fortner, R. C., Heterogeneity of presynaptic calcium channels revealed by species differences in the sensitivity of synaptosomal ^{45}Ca entry ω–conotoxin. *Biochem. Biophys. Res. Comm.*, **145** (1987) 1283–6.

30. Suszkiw, J. B., Murawsky, M. M. & Shi, M., Further characterization of phasic calcium influx in rat cerebrocortical synaptosomes: Inferences regarding calcium channel type(s) in nerve endings. *J. Neurochem.*, **52** (1989) 1260–9.

31. Reynolds, I. J., Wagner, J. A., Snyder, S. H., Thayer, S. A., Olivera, B. M. & Miller, R. J., Brain-sensitive calcium channel subtype differentiated by ω-conotoxin fraction GVIA. *Proc. Natl. Acad. Sci.*, **83** (1986) 8804–7.
32. Lundy, P. M., Frew, R., Fuller, T. W. & Hamilton, M. G., Pharmacological evidence for an ω-conotoxin, dihydropyridine-insensitive neuronal Ca^{2+} channel. *Eur. J. Pharmacol.*, **206** (1991) 61–8.
33. Scott, R. H., Dolphin, A. C., Bindokas,, V. P. & Adams, M. E., Inhibition of neuronal Ca^{2+} channels by the funnel web spider toxin ω–Aga–IA. *Mol. Pharm.*, **38** (1990) 711–18.
34. Adams, M. E., Mintz, I. M., Venema, V. J., Swiderek, K. M., Lee, T. D. & Bean, B. P., Block of P-type calcium channels by the funnel web spider toxin, ω-Aga-IVA. *Nature* (1992) (in press).

4

Conus Peptides and Biotechnology

Baldomero M. Olivera,[a] Lourdes J. Cruz,[a,c] Richard A. Myers,[a]
David R. Hillyard,[b] Jean Rivier[d] & Jamie K. Scott[e]

[a]Department of Biology, [b]Department of Pathology, University of Utah, Salt
Lake City, Utah 84112, USA
[c]Marine Science Institute, University of the Philippines, Quezon City 1101,
Philippines
[d]Clayton Foundation Laboratories for Peptide Biology, Salk Institute, La Jolla,
California 92057, USA
[e]Division of Biological Sciences and Department of Medicine, University of
Missouri, Columbia, Missouri 65211, USA

INTRODUCTION

The peptide toxins present in the venom of the *Conus* marine snails
('conotoxins') have distinctive properties which make them of particular
interest for biotechnological applications at the present time. Although there is
a remarkable pharmacological diversity of conotoxins, each peptide toxin is
very specifically targeted.[1-3] In certain cases, conotoxins are able to discriminate between closely related subtypes of a receptor target.

Conus peptides are unusually small, typically 10–30 amino acids in length.
Most conotoxins have constrained conformations because of multiple disulfide
cross-links. Their small size makes direct chemical synthesis feasible, and a
determination of three-dimensional conformation relatively accessible using
multidimensional NMR. Operationally, these are defined chemicals which can
be synthesized, and studied by presently available chemical technologies.
Nevertheless, because these small peptides are the direct translation products
of genes, they can also be manipulated by the techniques of molecular
genetics.

In this article, we will include a brief description of the ecology and
evolution of the *Conus* marine snails, to provide a background biological
context. In addition, we will focus on the issue of the phylogenetic specificity
of these peptides, and explore the potential for identifying invertebrate-specific
toxins in these venoms. Finally, we discuss the prospects for using a molecular
genetic approach to generate conotoxin-like peptides *in vitro*.

ECOLOGY AND EVOLUTION OF *Conus*

The biology and evolutionary history of *Conus* is essential background information for understanding the unique properties of the *Conus* venom peptides. The cone snails are placed by most taxonomists in a single large genus comprising approximately 500 living species. The fossil record suggests that the genus has evolved rapidly in the Cenozoic Period,[4] probably in response to an evolutionary opportunity provided by the mass extinction (i.e. of dinosaurs on land, of ammonites in the ocean) at the end of the Mesozoic. With the reemergence of new marine communities, cone snails radiated to all biotically-rich, tropical marine environments. Unlike the cosmopolitan stem group, the Turrids, from which they almost certainly arose, the cone snails are presently restricted to tropical and semi-tropical waters. However, they are among the dominant predators in environments such as coral reef communities; as many as 27 *Conus* species have been found in a single Indo-Pacific reef.[5]

Conus generally use their venom as the primary tool in capturing prey. In rich tropical marine environments, rapid prey capture becomes a critically important factor in the success of a venomous predator. There are a large number of other animals that can potentially prey on gastropod snails: *Conus* snails may be particularly vulnerable while they are feeding. Thus, a major selective force in venom evolution is almost certainly the speed with which a *Conus* venom can effect prey immobilization. In this respect, the cone snails have achieved considerable success—certain fish-hunting *Conus* can literally immobilize their prey 1–2 s after injection of venom, and effect complete paralysis a few seconds later.

We have postulated elsewhere that some of the unique properties of the toxins in *Conus* venoms may well be a response to the relentless selection for rapid paralysis.[1,3] The unusually small size of the peptides may have evolved to facilitate efficient dissemination of the toxins through the body of the prey; a small peptide of 10–30 amino acids would cross permeability barriers (such as the blood vessels of fish, for example) much more quickly than a typical proteinaceous toxin of snakes or scorpions (c. 50–90 amino acids in length).

A more subtle consideration is the pharmacology of the peptide toxins. An effective toxin must, on the one hand, have reasonably high affinity for the intended targets but at the same time, exhibit negligible affinity for closely related receptor subtypes that may be encountered before the toxin gets to the physiologically relevant receptor. Since speed is a dominant consideration, if a toxin spent even a very brief residence time bound to a closely related (but physiologically irrelevant) receptor subtype, it would seriously detract from its efficacy as a biologically relevant component of the venom. Thus, the pursuit of ever-more-rapid prey immobilization could also be a selective force for toxins which exhibit strong receptor subtype discrimination. Certain conotoxins (i.e. the ω-conotoxins of fish-hunting cones) can show remarkable discrimination indices (i.e. $>10^8$-fold) between what are apparently closely related subtypes of the same receptor class (in this case, voltage-sensitive Ca²⁺ channels).[6,7]

Finally, it is worth noting that in terms of species diversity and prey taxa, the cone snails are an unusual evolutionary success story. Most *Conus* species are quite specialized in the prey attacked; however, as a whole, the cone snails envenomate at least five different phyla of prey. This is somewhat unusual; most venomous animals generally specialize on envenomating their own

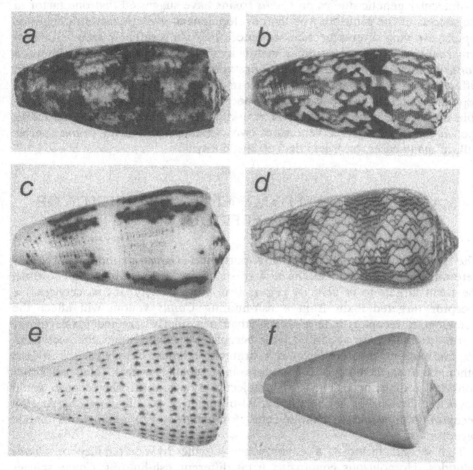

FIG. 1. *Conus* species specializing in different prey. Shown are the shells of six species of *Conus* which feed on four different phyla of prey. All specimens shown were collected in the Philippines. The top row are *Conus* species which feed exclusively on fish (Phylum Chordata): (a) *C. geographus* (7·9 cm); (b) *C. striatus* (9·9 cm); (c) *C. magus* (5·8 cm). Toxins from the venoms of these species are discussed in this article (see Table 1). The lower row includes *Conus* species which prey on invertebrates: (d) *C. textile* (7·5 cm) which feeds on other gastropods (Phylum Mollusca); (e) *C. leopardus* (10·3 cm) which feeds on enteropneust worms (Phylum Hemichordata); and (f) *C. Quercinus* (5·4 cm) which feeds on polychaete worms (Phylum Annelida). Some species of vermivorous *Conus* feed on worms belonging to the Phylum Echiurida.

phylum, i.e. snakes attack other vertebrates, scorpions other arthropods. There are 50–75 species of cone snails that do attack other molluscs, but an approximately equal number which feed exclusively on fish, and an even larger number which feed on three different phyla of marine worms (see Fig. 1 for some examples). In addition, all of these diverse specialist species have evolved in a relatively short evolutionary period; snakes, scorpions and spiders are all more ancient groups.

Molecular genetic studies on *Conus* toxins have suggested that one factor in the success of the genus may be the development of evolutionary strategies for rapidly evolving diverse peptide sequences.[3,8] As a result, not only is there a remarkable diversity of peptides within a single *Conus* venom, but between *Conus* venoms as well. When two *Conus* venoms are compared, peptides with different receptor target specificities are invariably found. However, even peptides which appear to have the same physiological function show remarkable divergence in sequence. Thus, presynaptic Ca^{2+} channel targeted toxins (i.e. ω-conotoxins) in the venoms of two different fish-hunting *Conus* species will, in many cases, have less than 50% identity.[9]

VARIATION IN PHYLOGENETIC SPECIFICITY OF *Conus* PEPTIDES

The phylogenetic specificity of any particular *Conus* peptide is probably determined by two factors. The first is selectivity, i.e. the receptor to which the toxin targets to *in vivo*. A priori, it is more likely that an acetylcholine receptor targeted toxin from a fish-hunting *Conus* venom will affect the homologous receptor in amphibians, than a similarly targeted toxin from a worm-hunting *Conus* venom. This is because we would expect the acetylcholine receptors of different lower vertebrates to be more closely related to each other than to homologous receptors in annelid worms. Thus, all other considerations being equal, a toxin selected to act on a fish receptor would have a much higher probability of also targeting to the homologous frog receptor than would a toxin selected for activity on a mollusc or annelid receptor.

The second factor is a consequence of the hypervariability of *Conus* peptides. Homologous conotoxins from different fish-hunting *Conus* species presumably target to very similar, if not identical, sets of receptor targets. Nevertheless, homologous toxins from different fish-hunting *Conus* species exhibit considerable sequence variation (see, for example, Table 1). Although all of the toxins are highly potent on fish, there can be considerable variation in their activity on homologous receptors in other vertebrates.

One of the better characterized examples is the major ω-conotoxins from *Conus geographus* and *Conus magus*, ω-conotoxins GVIA and MVIIA. These are both highly effective paralytic toxins on fish, with very similar potency.

TABLE 1
Some Homologous Toxins from Fish-Hunting Cone Snails

Conotoxin	Species	Sequence
Calcium channel inhibitors: ω–contoxins		
GVIA	C. geographus	C K S P G S S C S P T S Y N C C R–S C N P Y T K R C Y*
MVIIA	C. magus	C K G K G A K C S R L M Y D C C T G S C–R–S G K C*
Acetylcholine channel inhibitors: α-conotoxins		
GI	C. geographus	E C C N P A C G R H Y S C*
GII	C. geographus	E C C H P A C G K H F S C*
MI	C. magus	G R C C H P A C G K N Y S C*
SI	C. striatus	I C C N P A C G P K Y S C*
SIA	C. striatus	Y C C H P A C G K N F D C*

The standard one letter code is used except for hydroxyproline which is represented by *P*. The asterisks indicate that the peptides are amidated at the C-terminus. The sequences are taken from Refs 1 and 14.

However, if tested on frogs, GVIA is at least two orders of magnitude more effective than MVIIA.[10]

Another example is the major acetylcholine receptor targeted toxins in *Conus geographus* and *Conus striatus*, α-conotoxins GI and SI.[11] As would be expected, both toxins are highly potent in fish. However, on other vertebrate taxa, α-conotoxin GI shows a much broader phylogenetic range, being a potent paralytic toxin in all vertebrates tested, including mammals. In contrast, in terms of phylogenetic specificity, α-conotoxin SI is a much more 'narrowly focused' toxin. Thus, although the major paralytic toxins of fish-hunting cone snails all act on the relevant piscine receptors, these could be 'narrowly focused' or 'broadly focused' in terms of phylogenetic range. In most cases, the phylogenetic range of any particular toxin is presumably a result of chance—a paralytic toxin in a fish-hunting *Conus* venom is selected for its effects on the fish receptor target and whether or not it will also block a homologous mammalian target is probably irrelevant as a selective criterion.

However, such differences in phylogenetic range could account in part for the well documented observation that *Conus geographus* is the 'deadliest' *Conus* species. Both *Conus geographus* and *Conus striatus* are fairly common, large fish-hunting species distributed over a broad geographic range. The fact that a high fraction of stings from the former has caused human fatality, while no verifiable cases of fatal stings have been reported for *Conus striatus* could well be because the major paralytic toxins of *Conus geographus* are of the 'broadly focused' type, and act effectively on neuromuscular receptors in both fish *and* man, while the homologs in *Conus striatus* are ineffective on mammalian receptors.

Our studies of natural conotoxins from piscivorous *Conus* snail venoms demonstrate that several fish-hunting species have evolved a set of homologous toxins, all highly potent on fish receptors but with considerable variability in their activity on homologous receptors in other vertebrate systems.

INVERTEBRATE-SPECIFIC TOXINS

A potential application of *Conus* peptides is their use as pesticides. In principle, since they are the products of genes, *Conus* peptide encoding genes could be incorporated into plants; obviously stringent phylogenetic specificity would be required before such applications can be considered. Would *Conus* venoms provide a source of such appropriately targeted invertebrate-specific agents?

A number of general features of *Conus* venoms strongly suggest that the *Conus* system should be a rich source for toxins which do not act on vertebrates. Thus, only a small fraction (10–15%) of all *Conus* species prey on fish, the vast majority being invertebrate-specific. If partially purified components of a fish-hunting *Conus* venom are tested on mice, the majority cause some alteration in the behavior of mice.[3] However, if similar fractions of a vermivorous (worm-hunting) *Conus* venom are tested, most are inactive on mice (it is of interest that a small number do, in fact, exhibit biological activity when tested on mice; a subset of these may not only have a 'broadly focused' phylogenetic specificity, but might be universal ligands for a particular receptor, essentially phylogenetically non-discriminating).

Since most of our studies of *Conus* venoms so far have focused on the fish-hunting *Conus*, there is relatively little detailed biochemical information on particular invertebrate-specific toxins. One of the few that has been characterized is the King-Kong peptide,[12] a conotoxin from the beautiful snail-hunting species, *Conus textile* (the 'cloth-of-gold' cone). This is a typical conotoxin, 27 amino acids in length with three disulfide bonds. The peptide has no demonstrable activity on vertebrates, but it has potent biological effects on a number of invertebrates. In lobsters, it causes the animal to assume an exaggerated dominant posture (for this reason, Professor E. Kravitz dubbed the conotoxin the 'King-Kong peptide'). The peptide causes profound motor dysfunction in snails.

It is clear that a large number of peptides from the venoms of *Conus* species which prey on invertebrates will prove to be invertebrate-specific. However, how 'narrowly focused' such peptides are will probably vary a great deal, as has been found for the peptides from fish-hunting *Conus*.

SCREENING FOR 'DESIGNER CONOPEPTIDES'

Although the natural spectrum of peptides provided by the *Conus* venom system is highly diverse, specific applications will begin to require toxins with phylogenetic specificities which are unlikely to be found in the set of natural *Conus* peptides. For example, it may be desirable to have toxins that specifically kill only one order of insects but do not affect other arthropods, nor animals in other phyla. Such specificity is unlikely to be found in the set of *Conus* venoms because the toxins are targeted *in vivo* either to vertebrates, molluscs, or three phyla of worms, and not to insect receptors.

In principle, however, it should become possible to carry out 'phylogenetic focusing' *in vitro*. Once a peptide structure has been found which targets to a particular receptor type, it should be possible to use molecular genetics to select variants of a desired phylogenetic range. Thus, a lead structure might be a 'broadly focused' or even a phylogenetically non-discriminating conotoxin structure which acts on the target insect. A library could then be constructed, each containing a variant of the lead conotoxin sequence. Variants can be selected to bind the target insect receptor, but show no binding to homologous receptors in other taxa.

The conceptual and experimental basis for using an in-vitro approach for phylogenetic focusing of conotoxins is, in principle, already in place. This is the fUSE phage system[13] into which conotoxin-like modules have been cloned. The sequence of the conotoxin-like module can be varied by simply inserting oligonucleotides with random nucleotide sequences in selected positions. The variant conotoxin sequences can then be readily screened for binding to receptors, since these conotoxin modules ('conotopes') are exposed on the surface of the phage.

In principle, the screening procedure is simple: candidate phage which bind to the relevant receptor of the target insect, but not to homologous cloned mammalian receptors, and not to receptors of other invertebrates can be selected. Since ~10^9 different phage clones can be screened in a single assay, any variant that fits the required binding characteristics can be further enriched by successive rounds of affinity selection.

In order for the fUSE cloning system to be used, it must first be demonstrated that conotoxin-like modules can be successfully cloned into a coat protein of a fUSE phage, and that the phage still retains infectivity. This has recently been demonstrated using the fUSE-5 system and an α-conotoxin MI module. This experiment is detailed in Fig. 2.

The experiment analyzes the pIII coat protein of the fUSE phage which has an insert with a conotoxin module (the α-conotoxin MI sequence—see Table 1) as well as the target protease site for factor X protease. Treatment of the phage fusion protein with factor X causes a more rapid mobility, consistent with excision of the conotoxin module. Since this phage is infective, the result indicates that a conotoxin-like module has been inserted in the pIII coat protein of the fUSE phage, without grossly compromising the infectivity of the virus. These results indicate that the insertion of conotoxin modules into fUSE phage should be a promising technology for 'phylogenetic focusing'.

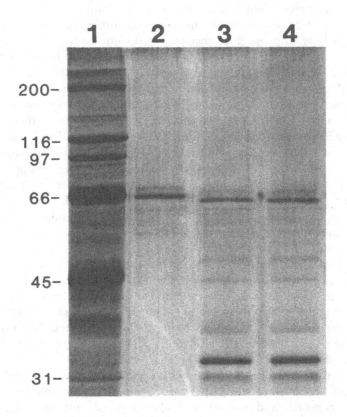

Fig. 2. Proteolytic cleavage of an expressed conotoxin module from a phage tail protein fusion. DNA encoding the following amino acid sequence (A DGRCCHP A CGKNYSCGGIEGR A G) was inserted into the cloning site of the fUSE 5 vector;[13] the insert would thus be expressed as an N-terminal fusion to the pIII coat protein of the phage. The insert contains both the α-conotoxin MI sequence (underlined), as well as the factor Xa protease recognition site, IEGR. After transfection of the DNA by electroporation, infective phage are cloned, identified by DNA sequencing in the insert region, and isolated and purified.[13] We refer to these as fUSE 5/α-conotoxin-MI phage.

One hundred μg (protein) of the fUSE 5/α-conotoxin MI phage were digested in 50 μl 20 mM Tris–HCl (pH 8·0), 100 mM NaCl, 2 mM CaCl$_2$, with or without 0·2% Triton X-100. Incubations were with 2 units factor Xa protease (New England Biolabs) for 6 h at room temperature. One volume 4% SDS sample buffer (with 10% β-mercaptoethanol) was added and then boiled for 5 min. Ten microliters were run on an 8% polyacrylamide gel (prepared as described by Laemmli[15]) and silver stained. Lane 1 shows MW markers; lane 2 is a control reaction mixture which was not digested; lanes 3 and 4 were digested without and with Triton, respectively. The 42 kDa pIII protein under consideration runs at an apparent M_r of c. 70 kDa on Laemmli gels. The other bands in lanes 3 and 4 are from the factor Xa protease preparation. The results are consistent with proteolytic cleavage of the N-terminal conotoxin module by the factor Xa protease from the pIII fusion protein.

DISCUSSION

In this article, we examined the potential of the *Conus* peptide system for biotechnology. We have focused in particular on the possibility of using *Conus* peptide-like agents as pesticides. There is substantial promise in directly screening *Conus* venom for agents with desired pharmacological characteristics. As discussed above, the venoms of *Conus* contain peptides with considerable pharmacological diversity. For each pharmacological target, because of the hypervariability of *Conus* peptide sequences, a large homologous sequence set is available. As a consequence of the variance in phylogenetic range, the probability of obtaining a large number of peptides that are invertebrate-specific in their biological activity seems very high. Since the nucleic acids encoding such agents can be introduced into plant genomes, this constitutes one potential line of attack.

However, because *Conus* peptides are relatively small and conformationally rigid, a second line of development is also promising: peptidomimetics based on the conformation of *Conus* peptides. Once a *Conus* peptide has been identified as targeting to a receptor of interest (say in an insect larva that is a major food crop pest), there would still be a major barrier to using the peptide directly. This is the case for all receptor targets which are not in external sensory organs, nor in the digestive system; most peptides are unlikely to penetrate the walls of an insect gut. However, the three-dimensional structure of *Conus* peptides can be determined in a relatively straightforward fashion. Since the peptides are quite small, they could serve as guide structures for designing peptidomimetics. These in turn can be modified appropriately so that they are effective when taken orally. A basic understanding of which structural features are critical for biological activity at the receptor level can be obtained from studies on the *Conus* peptide. Since homologs with variant sequences will generally be available from multiple *Conus* venoms, it should be possible to start with quite refined structural insight into the requirements for biological activity before starting the design of derivative peptidomimetics.

A final line of development which perhaps shows the greatest promise of all is to use a combination of our understanding of conotoxin biochemistry with a screening technology such as the fUSE phage system. This would permit the development of *Conus*-like peptides which are not found naturally in *Conus* venoms. Most importantly, this opens the door to obtaining peptides with specificities not selected for in nature. In principle, it should be possible to generate *Conus*-like agents which are even more finely tuned with respect to receptor subtype specificity and phylogenetic range. As the public demand increases for pesticides which are minimally disruptive from an ecological point of view, there will be a need to have such specifically targeted ligands. With the scenario for *Conus* peptides we have sketched above, it may be possible to apply a conotoxin-derived agent which will kill or immobilize a single pest species with no effect at all on any other species in the same biological community.

We believe that because of the unusual features of their biochemistry, *Conus* peptides can play a unique role in the future of biotechnology.

ACKNOWLEDGEMENTS

The work described was primarily supported by a grant from the National Institute of General Medical Sciences, GM22737. Additional support from the Office of Naval Research (to B.M.O.), Grant NS27219 from the National Institute of Neurological Disorders and Stroke (to L.J.C. and D.R.H.) and from the International Foundation for Science, Stockholm, Sweden (to L.J.C.) is also gratefully acknowledged. The work at the University of Missouri was supported by GM41478 (to G. P. Smith), and grant GM13772 as well as support from Protos Corporation and the Molecular Biology Program and the Department of Medicine, University of Missouri (to J.K.S.).

REFERENCES

1. Olivera, B. M., Gray, W. R., Zeikus, R., McIntosh, J. M., Varga, J., Rivier, J., Santos, V. de & Cruz, L. J., Peptide neurotoxins from fish-hunting cone snails. *Science,* **230** (1985) 1338–43.
2. Gray, W. R., Olivera, B. M. & Cruz, L. J., Peptide toxins from venomous *Conus* snails. *Annu. Rev. Biochem.,* **57** (1988) 665–700.
3. Olivera, B. M., Rivier, J., Clark, C., Ramilo, C. A., Corpuz, G. P., Abogadie, F. C., Mena, E. E., Woodward, S. R., Hillyard, D. R. & Cruz, L. J., Diversity of *Conus* neuropeptides. *Science,* **249** (1990) 257–63.
4. Kohn, A. J., Tempo and mode of evolution in Conidae. *Malacologia,* **32** (1990) 57–67.
5. Kohn, A. J. & Nybakken, J. W., Ecology of *Conus* on Eastern Indian Ocean fringing reefs: diversity of species and resource utilization. *Marine Biology,* **29** (1975) 211–34.
6. Cruz, L. J., Johnson, J. L., Imperial, J. L., Griffin, D., LeCheminant, G. L., Miljanich, G. P. & Olivera, B. M., ω-Conotoxins and voltage-sensitive Ca channels subtypes. *Curr. Top. Memb. Transp.,* **3** (1988) 417–29.
7. Olivera, B. M., Cruz, L. J., Santos, V. de, LeCheminant, G., Griffin, D., Zeikus, R., McIntosh, J. M., Galyean, R., Varga, J., Gray, W. R. & Rivier, J., Neuronal Ca channel antagonists. Discrimination between Ca channel subtypes using ω-conotoxin from *Conus magus* venom. *Biochemistry,* **26** (1987) 2086–90.
8. Woodward, S. R., Cruz, L. J., Olivera, B. M. & Hillyard, D. R., Constant and hypervariable regions in conotoxin peptides. *EMBO Journal,* **9** (1990) 1015–20.
9. Myers, R. A., McIntosh, J. M., Imperial, J., Williams, R. W., Oas, T., Haack, J. A., Hernandez, J. F., Rivier, J., Cruz, L. J. & Olivera, B. M., Peptides from *Conus* venoms which affect Ca^{++} entry into neurons. *J. Toxicol.—Toxin Reviews,* **9** (1990) 179–202.
10. Yoshikami, D., Bagabaldo, Z. & Olivera, B. M., The inhibitory effects of omega-conotoxins on Ca channels and synapses. *Ann. N.Y. Acad. Sci.,* **560** (1989) 230–48.
11. Zafaralla, G. C., Ramilo, C., Gray, W. R., Karlstrom, R., Olivera, B. M. & Cruz, L. J., Phylogenetic specificity of cholinergic ligands: α-Conotoxin SI. *Biochemistry,* **27** (1988) 7102–5.

12. Hillyard, D. R., Olivera, B. M., Woodward, S., Corpuz, G. P., Gray, W. R., Ramilo, C. A. & Cruz, L. J., A molluscivorous *Conus* toxin: conserved frameworks in conotoxins. *Biochemistry*, **28** (1989) 358–61.
13. Scott, J. K. & Smith, G. P., Searching for peptide ligands with an epitope library. *Science*, **249** (1990) 386–90.
14. Myers, R. A., Zafaralla, G. C., Gray, W. R., Abbot, J., Cruz, L. J. & Olivera, B. M., α-Conotoxins, small peptide probes of nicotinic acetylcholine receptors. *Biochemistry*, **30** (1991) 9370–7.
15. Laemmli, U. K., Cleavage of structural proteins during the assembly of the head of bacteriophage T4. *Nature*, **227** (1970) 680–5.

5

Natural and Synthetic Toxins at Insect Receptors and Ion Channels: the Search for Insecticide Leads and Target Sites

JACK A. BENSON*

R & D Plant Protection, Agricultural Division, Ciba-Geigy Ltd, CH-4002 Basel, Switzerland

INTRODUCTION

The target sites known to date within the insect nervous system for current commercial insecticides are more or less limited to the sodium channel, the components of the nicotinic cholinergic synapse, and, to a much lesser and rapidly diminishing extent, the GABA and octopamine receptors. The insecticide market is dominated by just three classes of compounds, the pyrethroids, the carbamates and the organophosphates. This means that the number of important groups of fast-acting insecticides, and their corresponding target sites, usually in the nervous system, is astonishingly small in comparison with the enormous number of synthetic compounds that have been assayed for useful insecticidal activity. This observation evokes a couple of obvious but nonetheless important questions. Is the small number of commercial insecticides and targets a true reflection of the possibilities or merely of our successes, methods and knowledge? Are there empirical or even *a priori* 'rules' concerning leads and neuronal targets that we could usefully apply to make the task of lead discovery easier or faster? I would like to discuss some tentative thoughts on these questions in relation to natural and synthetic toxins, and their effects on some insect neurotransmitter receptors and ion channels. Firstly, I support Casida's proposal that the number of potential neuronal target sites really is small.[1] Secondly, I would like to suggest that close attention to the mode of action of receptor- and channel-directed, insecticidal, natural toxins can help us identify targets hitherto unexploited. Thirdly, the natural toxins have been a major source of new leads and should continue to be so. *De novo* synthesis has yielded insecticidal compounds of considerable target site potency and great commercial value, but they are few in relation to the huge expenditure of resources needed to discover them.

* Address for correspondence: Dr. J. A. Benson, Pharmacology Institute, University of Zurich, Gloriastrasse 32, CH-8006 Zurich, Switzerland.

ELECTROPHYSIOLOGICAL RECEPTOR ASSAYS

By localised microapplication of neurotransmitter, Callec and Boistel,[2] and Kerkut *et al.*[3,4] were able to make the observation that the somata of insect neurones *in situ* respond to acetylcholine and GABA. This was quite surprising since the somata of these neurones are remote from the synapses. The latter are concentrated in a dendritic arborisation connected to the cell body by a neurite that is often very long. Usherwood and his colleagues[5-8] demonstrated that when physically isolated with only a very short neuritic stump left attached, the neuronal somata of the locust thoracic ganglia continue to respond in a characteristic manner to a wide range of transmitters. No satisfactory explanation has been proposed for this prolific expression of receptors at a site where they are apparently without a physiological rôle. Possibly the phenomenon merely reflects the manufacture of receptors in the soma, either in excess or before transport to the synaptic region. In any case, these isolated somata provide an excellent assay system for most, if not all, of the important insect neurotransmitter receptors. We found that they can be maintained under voltage clamp for several hours[9] and we have been able to define the pharmacology and voltage-dependence of responses mediated by nicotinic (ACh1) and muscarinic (ACh2) cholinergic receptors[10-14] as well as receptors for GABA,[9,12,14,15] serotonin[16] and octopamine.[17] They also respond to neuropeptides from the corpora cardiaca.[18] The isolated neuronal somata are extremely easy to prepare, and they compare favorably with *Xenopus* oocytes, the alternative example of a large, spherical and easily-clamped vehicle for receptor assays since, of course, they do not have to be injected with laboriously prepared mRNA and the combinations of receptor subunits expressed are guaranteed to be authentic. Glutamate responses seem rather variable and in locusts are probably best recorded in the neuropile,[19-21] a more difficult procedure. Consequently, for ease of recording when assaying compounds against glutamate receptors, we utilised the method of Irving and Miller[22] to record nerve-evoked EJPs intracellularly at the larval neuromuscular junction of the fly, *Musca domestica*.

NEURONAL TARGET SITES: PROVEN AND POTENTIAL

As new discoveries over the past decade dramatically increased the number of identified and physiologically–characterised insect peptide hormones, thereby doubling and trebling the total number of neurotransmitters and hormones known in insects, it was easy to conclude that the receptors, as well as the precursor processing pathways and re-uptake (or metabolic) breakdown pathways, for these peptide hormones represented new and perhaps highly selective insecticide targets. However, the lack of probes for peptide receptors has meant that a firm knowledge of the pathologies that might result from interference with these peptide-mediated systems is still lacking. Since it was speculated that many of the neuropeptides and peptide hormones are modulators with slow and often rather slight effects, the idea was that interference with their receptors would produce 'subtle effects' that would ultimately result in the death of the insect. Unfortunately, insects, like other animals, including humans, are remarkably resilient to metabolic malfunction. Subtle effects are not enough to prevent crop damage or even to control pest insect populations over a longer time span. Even growth regulators have had relatively little success in the market. The most successful insecticides seem to be those that cause convulsive activity and other neural traumas by acting at the small group of target sites mentioned at the beginning of the Introduction.

Looking at these known target receptors (the sodium channel, the components of the nicotinic cholinergic synapse, and the GABA and octopamine receptors), we can ask whether they have features in common that distinguish them from other elements of the nervous system, features that we might seek to identify elsewhere in our search for novel targets. They are discrete in structure and function like other receptors and channels, and they play important or even crucial rôles in the normal function of the nervous system, but so do several other channels, receptors and second messenger systems. Does this mean that all receptors and channels are likely targets? Can we narrow the choice? One clue is that some but not all receptors are the targets of toxins that are known or proposed to be natural pesticides. Consistent with this is the fact that two of the three most important classes of insecticide, the pyrethroids and carbamates, have natural prototypes (pyrethrins from *Chrysanthemum* (*Pyrethrum*) flowers and physostigmine (eserine) from the Calabar bean, *Physostigma venosum*), and at least two examples of the third class, the organophosphates, have been isolated from bacteria.[23] All three classes of insecticides thus act at targets utilised by natural toxins, but it appears that not all receptors and channels are the site of action of natural toxins.

Table 1 lists some potential neurotoxic insecticide target sites, based on the sites of action of past and present commercial synthetic insecticides and of natural toxins known or assumed to be produced as a defence against pests. To my knowledge, the neuroactive hormone receptors are the sites of action of few if any natural pesticides, and are consequently not included. For the same reason, the histamine and dopamine receptors are absent from the table, despite their undoubted importance in insects, as are the arachidonic acid and phosphotidylinositol second messenger pathways. The nicotinic, muscarinic and glutamatergic receptors are discussed in detail below. The GABA receptor ligand recognition site is the target of muscimol, a potent agonist with insecticidal properties. It was initially isolated from the poisonous 'fly fungus', *Amanita muscaria*, extracts of which were used to kill flies. The Cl⁻ ionophore of the GABA receptor is also a target, because it is the site of action not only of picrotoxin, a mixture (active component picrotoxinin) from the seed of *Anamirta cocculus* and various Menispermaceae, but also of several synthetic

TABLE 1
Potential Neuronal and Muscular Insecticide Target Sites Based on the Sites of Action of
Insecticidal Natural Products

Target	In vivo activator	Insecticidal natural product	Commercial insecticide
Neurotransmitter receptor ligand recognition sites			
Cholinergic			
Nicotinic	Acetylcholine	Nicotine	Nicotine
		Nereis toxin	Cartap
		Anabasine	Nitroguanidines
		Methyllycaconitine	Nitroenamines
		Coniine	
Muscarinic	Acetylcholine	Muscarine	None
		Arecoline	
Glutamatergic	Glutamate	Domoic acid	None
		Acromelic acid	
		δ–philanthotoxin	
Octopaminergic	Octopamine	None	Formamidines
GABAergic	γ–Aminobutyric acid	Muscimol	None
Ion channels			
Na channel	Depolarisation	Pyrethrins	Pyrethroids
		Veratridine	DDT
		N-alkylamides	
Cl channel			
GABA	γ–Aminobutyric acid	Picrotoxinin	Cyclodienes
passive	—	Avermectins	Avermectins
		Milbemycins	Milbemycins
Second messenger systems			
Cyclic AMP	e.g. Octopamine, 5HT	Methylxanthines	None
Transmitter re-uptake and breakdown systems			
Cholinesterase	Acetylcholine	Organophosphates	Organophosphates
		Physostigmine	Carbamates
Mitochondrial respiration			
Ox-phos	—	Rotenoids	Rotenone
			2,4-Dinitrophenol
			Diafenthiuron
Muscle			
Contraction	Depolarisation	Ryanodine	Ryania extract

insecticides, the best known being the cyclodienes. The octopamine receptor is
most likely the site of action of the formamidines, such as desmethylchlor-
dimeform, but not of any well-known natural insecticide. Yohimbine, an
indole alkaloid from *Corynanthe johimbe* and also *Rauwolfia* root, is octopaminergic but insecticidal activity has not been reported. Serotonin

receptors might also be candidates for addition to the list if the ergot alkaloids should prove to be insecticidal.

The Na channel is the target of numerous natural and synthetic insecticides and also a wide range of other plant- and animal-derived compounds. DDT and the pyrethroids show cross-resistance, but veratridine and the *N*-alkylamines[24] appear to act on a separate site in the channel and show negative cross-resistance with the pyrethroids.[25] The avermectins and milbemycins are of bacterial origin and consequently did not evolve as a defence against insects. Although these compounds have an effect on the Cl⁻ ion channels of the GABA receptor, the density of their binding sites in *Caenorhabditis* is so great that the far more numerous passive Cl⁻ channels found in most cell membranes are more likely to be the primary target.[26]

To date, cyclic AMP is the only second messenger system that seems likely to have become a natural insecticide target. The methylxanthines apparently have their insecticidal effect by inhibiting phosphodiesterase, resulting in the accumulation of cyclic AMP.[27] Similarly, it seems that only acetylcholinesterase, of the many transmitter breakdown and re-uptake systems present in insects, has been selected as a target for natural insecticides. Inhibition of mitochondrial respiration is the mode of action of discontinued products such as 2,4-dinitrophenol and Rotenone, the latter based on a compound extracted from derris root and other plant species. Insect toxic synthetic compounds with this mode of action are very common, but these products generally also show high mammalian and fish toxicity so that very few of them are candidates for the market. It is possible that a carbodiimide which is the active metabolite of diafenthiuron, a thiourea insecticide and acaricide, acts by mitochondrial inhibition.[28] Ouabain, obtained from the seeds of *Strophanthus* and other species, and an inhibitor of the neuronal Na pump ATPase, is apparently without insecticidal action. Finally, ryanodine, a neutral alkaloid isolated from the stem and root of *Ryania speciosa,* a member of the Flacourtiaceae native to Trinidad, used to be sold in the form of an insecticidal extract. Its site of action is the insect muscle where it affects contraction, perhaps via an inhibition of the sequestration of intracellular Ca.[29]

Table 1 is necessarily only a tentative abstract of current knowledge since the number of neurally-active natural products is very large and there are no doubt unpublished non-commercial synthetic insecticides with identified neuronal targets. Nevertheless, even taking into account products now considered to be too toxic for use in the field, the number of targets is small: five transmitter receptors, possibly three ion channels, a second messenger pathway, only one transmitter removal system, and one metabolic process. There is no obvious clustering of synthetic versus natural insecticides with regard to target site.

The main point of Table 1 is that there is only one target, the octopamine receptor, for which there is an important synthetic insecticide but no definitive natural insecticide, while, on the other hand, there are several sites of action for natural insecticides with no corresponding commercial product. The latter very likely represent unexploited pesticide targets. The most important ones are probably the muscarinic and glutamate receptors. Below, I shall consider the properties of some of these targets in detail in order to compare the results of *de novo* 'blue sky' synthesis with synthesis based on natural products. Some targets might be out of reach of commercial exploitation because of intrinsic toxicity problems arising from lack of divergence between insects and vertebrates at these sites. This problem is probably relatively unimportant. Many pyrethroids, to take the most obvious example, are highly potent at vertebrate sodium channels but display very low *in vitro* toxicity in vertebrates due to metabolic breakdown. Quantitative differences in receptor and ion channel pharmacology probably do also provide a basis for selectivity. The insensitivity of squid giant axon Na channels to DDT, for example, argues for optimism in this regard. The distribution of the commercial products listed in Table 1 also illustrates Casida's observation that there are overworked and underworked targets.[1]

Nicotinic acetylcholine receptors

The nicotinic synapse is the most important target in the insect nervous system for modern commercial insecticides, and consequently it is also surely an overworked target. One component of the synapse, the acetylcholine receptor ligand-binding site, serves well to illustrate the main points of this paper. This is the site of action, for example, of nicotine itself, used for many years as a greenhouse fumigant, and Cartap, the active form of which has a structure identical to nereistoxin, a component of the venom of the marine polychaete, *Lumbriconereis*. The nicotinic receptor expressed by locust neuronal somata (the ACh1 receptor[10]) seems to be pharmacologically similar although not identical to the post-synaptic receptor.

The order of potency of a selection of nicotinic antagonists at the ACh1 receptor is: 1-(pyridin-3-yl-methyl)-2-nitromethylene-imidazoline (PMNI) $\gg$ α-bungarotoxin $\geq$ lobeline $\geq$ mecamylamine $>$ trimethaphan camsylate $>$ chlorisondamine $\geq$ d-tubocurarine $\geq$ hexamethonium $\geq$ gallamine $\geq$ tetra-ethylammonium.[30] Of these, PMNI, nicotine and anabasine have well-known topical or oral insecticidal effects. Nicotine and anabasine are both alkaloids from *Nicotiana* (cultivated tobacco), lobeline is from *Lobelia inflata* (wild tobacco) and d-tubocurarine is from the South American *Chondodendron* species. A non-specific antagonist is strychnine, a well-known toxin from the nuts of various *Strychnos* species, which is as active as α-bungarotoxin at the ACh1 receptor[30] but which is also, of course, highly toxic to vertebrates where it binds to both glycine and acetylcholine receptors. Two other natural products with potent activity at other receptors also have antagonistic effects at the ACh1 receptor. Bicuculline, an alkaloid from *Dicentra cucullaria*, various Fumariaceae and several *Corydalis* species, is the diagnostic antagonist at vertebrate GABA$_A$ receptors. It appears to be inactive at most insect 'GABA$_A$'-type receptors and also at the unusual GABA receptor in the heart of the primitive arthropod, *Limulus*.[31] Bicuculline is, however, a moderately effective blocker of the ACh1 response ($EC_{50} = 3 \cdot 1 \times 10^{-5}$ M).[14] Picrotoxin is a potent blocker of $GABA_A$ receptor-activated Cl^- channels both in vertebrates and insects,[9] and it has a weaker effect ($EC_{50} = 2 \cdot 9 \times 10^{-5}$ M) at the ACh1 receptor. Two other nicotinic ligands with contact insecticidal actions, not tested in this series, are methyllycaconitine,[32] from *Delphinium* seeds, and coniine,[33] from hemlock, *Conium maculatum*, and the pitcher plant, *Sarracenia flava*. Dihydro-β-erythroidine, an alkaloid from *Erythrina* seeds, and cytisine from the seeds of *Laburnum* and other legumes, are active at insect nicotinic receptors but insecticidal activity has not been reported to my knowledge. The mode of action of all the very potent ACh1-active compounds, with the possible exception of mecamylamine, appears to be competitive agonism or antagonism. The weaker and more non-specific compounds such as bicuculline and picrotoxin seem to have their effect by blocking the receptor-activated ion channel. This suggests a much greater similarity in the channel region than in the ligand recognition site among the ligand-activated receptor-channel complexes, as might be expected.

The nicotinic receptor has thus proven to be a very popular target with both Nature and commercial insecticide chemists. However, in addition it is possible to conclude that Nature has not necessarily always beaten us to it as far as potent and specific ligands with topical insecticidal activity are concerned. From the observations summarised above, it appears that the compound most active *in vivo* that also acts externally is PMNI (EC_{50} c. $0 \cdot 1$ nM), a synthetic antagonist without known natural analogues that acts as an agonist at higher concentrations.[34,35] The potency, mode of action and specificity of PMNI are illustrated in Fig. 1. Bath application of 1 nM PMNI rapidly reduced to about 25% of the control level the response evoked by 100 ms pulses of acetylcholine in a voltage-clamped, isolated, locust neuronal soma. The response to muscarine was unaffected and there was no agonistic shift in the holding current. The threshold for agonistic activity by this compound is about 3×10^{-8} M.[34] The only alkaloid very potent at the ACh1 receptor is lobeline (EC_{50} c. 40 nM), which, although rather selective for the insect ACh1 receptor, is without topical activity. At its target, a 'blue sky' synthetic compound can thus exceed in activity the most potent known plant derivative.

Muscarinic acetylcholine receptors

The physiological role of muscarinic receptors in the insect nervous system is only now being investigated in any detail,[36,37] despite many years of successful use of binding assays to characterise insect neuronal muscarinic binding sites. This class of receptors provides an example of a target site for natural toxins, especially alkaloids, some of which are natural insecticides, that, so far as we know, has yet to be exploited as a target for a commercial insecticide.

The model muscarinic receptor used in our studies is the locust thoracic neurone somal ACh2 receptor.[10] Its pharmacology has been well

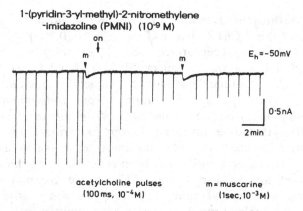

FIG. 1. Nicotinic cholinergic antagonistic effect of the synthetic insecticide, PMNI (1-(pyridinyl-3-yl-methyl)-2-nitromethylene-imidazol). Chart record showing the membrane current recorded from an isolated locust neuronal soma voltage-clamped at a holding potential (E_h) of -50 mV. Pulses of acetylcholine (100 ms; 10^{-4} M) and muscarine (1 s; 10^{-3} M) were applied to the soma via a pressure micropipette at regular intervals, resulting in transient inward currents (downward deflections). PMNI (10^{-9} M) was bath-applied (beginning at the arrow) resulting in a decrease in the amplitude of the responses to acetylcholine but with no change in the amplitude of the muscarinic response or in the holding current. At higher concentrations, PMNI exhibits nicotinic cholinergic agonism.

characterised.[30] The order of potency of a selection of muscarinic antagonists at the ACh2 receptor is: QNB $\geq$ scopolamine $>$ atropine $>$ 4-DAMP $\geq$ benactyzine $\geq$ HHSiD $\geq$ pirenzepine. QNX, AF-DX 116, gallamine triethiodide and methoctramine are almost or completely inactive. The active agonists include muscarine, oxotremorine, arecoline and pilocarpine. McN-A-343 is inactive.[30] Interestingly, many of these compounds, including those without effect at the muscarinic ACh2 receptor, are active as antagonists at the nicotinic ACh1 receptor, with EC_{50} values of around 10^{-5}–10^{-6} M.[30] Of these compounds, muscarine itself, pilocarpine, scopolamine and arecoline are alkaloids. Muscarine, like muscimol, comes from *Amanita muscaria*. This presence of multiple toxins in plants and fungi is common, and probably points towards synergism in their action on pests, an aspect to be taken into account during lead optimisation. Arecoline, a component of betel nuts, was occasionally used as a vermifuge in veterinary medicine, but like pilocarpine, from the leaves of the South American plant *Pilocarpus*, and scopolamine, a tropane alkaloid from various Solanaceae, it has had no agricultural applications. The most potent antagonist was QNB (quinuclidinyl benzilate) which is not a natural product. In contrast to the mixed agonist/antagonist PMNI, it is without insecticidal properties. As is usually the case at excitatory receptors, agonists seem to make the best insecticides. Excitation, consequent massive release of transmitter, and convulsive and repetitive neural activity seem to be more reliable indicators of incipient mortality than quiescence in the case of neural insecticides.

The muscarinic and nicotinic receptors have both been attractive to Nature s pesticide targets, but only the nicotinic receptor has been exploited to date y industry. The muscarinic receptor is an under-worked target for which there re leads among insecticidal natural products as well as among the large range f novel structures currently being synthesised in the search for leads for the harmaceutical market.

;lutamate receptors

'he third receptor type that I would like to discuss is somewhat problematic. 'o my knowledge, no commercial pesticide acts at the glutamate receptor.)espite the existence of some natural glutamatergic compounds with insec-cidal properties, especially in arthropod venoms, there is no decisive evidence 1at the glutamate receptor is the target site of compounds that evolved to rotect plants from insects. In principle, based on criteria such as those iscussed above and elsewhere,[34] the glutamate receptor certainly seems to be potential target for insecticides. For example, there is no doubt that it plays n extremely important part in the moment-to-moment life of the insect, 1ediating most, if not all, neuromuscular synaptic transmission, and some entral transmission as well.[19-21] However, in comparison with the potency of cetylcholine and some of the cholinergic agonists and antagonists at the icotinic and muscarinic receptors, glutamate itself is rather weakly active 'hen applied *in vitro* to the *Musca* neuromuscular junction preparation. The 1me is true for quisqualic acid, from the seeds of *Quisqualis indica,* and kainic :id, from the marine red alga, *Digenea,* both of which show anthelminthic 1ther than topical or oral insecticidal activity. Similarly, synthetic δ– hilanthotoxin (PTX-4.3.3), one of the toxic polyamine fractions from the :nom of the bee wolf, *Philanthus triangulus,*[38,39] has an EC_{50} value of c.) μM on the *Musca* preparation. Can industrial chemists hope to do any better : this receptor? In the case of PTX-4.3.3, analogues showing enhanced *vitro* activity have been synthesised, but they are without topical or oral .secticidal activity. Like PTX-4.3.3 and probably also because they are 1annel blockers, the synthetic analogues are rather non-selective, acting at)th glutamate and nicotinic receptor ion channels.[40] Although PTX-4.3.3 and ΓX-3.4.3 have similar potencies at the *Musca* glutamate and ACh1 receptors, •me of the analogues are more potent at the locust ACh1 receptor (e.g. Fig. .[41] Synthetic analogues of PTX-4.3.3 show *in vitro* potency enhanced by up • 8-fold on the *Musca* preparation,[42] by up to 33-fold on locust retractor 1guis neuromuscular transmission,[43] and 75-fold on the ACh1 receptor.[41] ainic acid and related compounds may provide better leads. For example,)moic acid, a kainate-like anthelminthic from the red marine alga, *Chondria 'mata,* blocks glutamatergic synapses and has been reported to show contact :tivity with a potency comparable to that of DDT.[44] It would be of great terest to be certain that the glutamate receptor is the site of insecticidal :tion of this compound. As with several kainoids, the troublesome synthesis ` domoic acid makes it both expensive to obtain in adequate quantities for

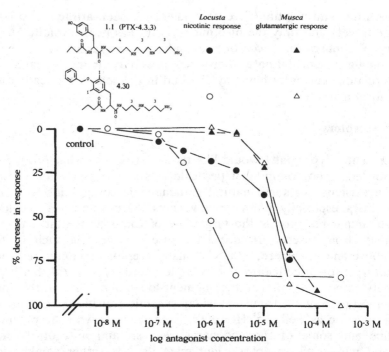

FIG. 2. Dose response curves for the blocking action of PTX-4.3.3 (δ-philanthotoxin-4.3.3) (filled symbols) and an iodinated analogue (empty symbols) at the nicotinic (ACh1) receptor of *Locusta* thoracic somata (circles) and the *Musca* glutamatergic neuromuscular junction (triangles). Compound code numbers and data from Ref. 41.

electrophysiology and unattractive to industry as a lead. Despite the rather high doses required for many compounds to show effects at the glutamate receptor, there does seem to be sufficient evidence of insecticidal activity from apparently glutamatergic natural products to include this receptor on the list of commercially useful and hitherto unexploited targets.

TOXICITY AND RESISTANCE

One or two other aspects of the search for new leads and targets in relation to natural toxins are worth touching on here. The first is mammalian toxicity. This is a serious problem. Many of the natural pesticides mentioned above are very toxic to vertebrates and, indeed, may have evolved for that very reason. There are, however, also natural insecticides without mammalian, fish or bird toxicity. Whether both groups will be considered equally good sources of leads will depend on perceptions of how successful reduction of mammalian and fish toxicity is as an aspect of product development. On the other hand, hytotoxicity is likely to be less of a problem with leads that are plant erivatives. An attractive feature of natural toxins as leads is that for every uch toxin, a natural breakdown pathway has presumably co-evolved, since ccumulation of natural toxins in the environment seems to be the exception ather than the rule and is always short-term rather than persistent. This is in ontrast to some of the synthetic insecticides where persistence is perceived as problem, DDT being the most obvious example. The other side of the coin , of course, that some degree of persistence is essential in a successful esticide and many natural products, especially peptides, are very labile when hey are no longer in the intracellular medium.

With regard to resistance, Casida drew a very important conclusion from his roposal that pesticide target sites are finite in number.[1] If insecticide target ites are few, they become a critical and irreplaceable resource subject to verutilisation. This results in the development of resistance and cross-esistance among different important insecticides acting at the same target.)ne way to conserve the target sites is to redirect our attention to the eceptors and channels in the insect nervous system that have been targeted by Jature but which are so far underused or totally unexploited commercially.

CONCLUSIONS

he number of neural target sites for insecticides, known and potential, is mall, but it is not as limited as the modes of action of the commercial nsecticides might lead us to believe. There are still several unexploited targets 1 the insect nervous system that have been identified by mode of action tudies on natural, insecticidal compounds that probably evolved as plant-rotection mechanisms. It is possible that the number of identified targets can e expanded to a limited extent by further natural product mode of action esearch.

'Blue sky' synthesis has contributed some remarkable substances to the small ange of truly successful commercial insecticides, but this number is very small 1 comparison with the number of such synthetic compounds screened for ctivity. The much smaller number of natural products screened for activity as been equally or more productive of successful leads.

Two conclusions can be drawn from these observations. The first is that ffort and resources on a scale similar to those applied to *de novo* synthesis ould be usefully applied to the search for natural insecticides, primarily as a ource of leads *per se* but also to identify and characterise neuronal target sites. econdly, it is practical for chemical synthesis to have a much greater 'directed' omponent, drawing heavily on the structures of natural compounds known to e topically or orally insecticidal, or known to be extremely potent at the nportant insecticide target sites. This should cover the whole spectrum from harmacophore modelling to synthesis of analogue series based on natural roduct leads. As in nature, synergism by mixtures is likely to be important.

By working from compounds that have evolved as plant protection mechanisms, the problems of penetrability and persistence inherent in pure receptor and pharmacophore modelling approaches are incorporated into the lead optimisation process from the start. It is hard to believe that research directed at targets and leads conveniently identified and provided for us by Nature can be less successful than random synthesis and mass screening.

REFERENCES

1. Casida, J. E., Pesticide mode of action: evidence for and implications of a finite number of biochemical targets. In *Pesticides and Alternatives*, ed. J. E. Casida, Elsevier Science Publishers, Amsterdam, 1990, pp. 11–22.
2. Callec, J.-J. & Boistel, J., Les effets de l'acétylcholine aux niveaux synaptique et somatique dans le cas du dernier ganglion abdominal de la Blatte *Periplaneta americana*. *C. R. Séances Soc. Biol. Paris*, **161** (1967) 442–6.
3. Kerkut, G. A., Pitman, R. M. & Walker, R. J., Sensitivity of neurones of the insect central nervous system to iontophoretically applied acetylcholine or GABA. *Nature, Lond.*, **222** (1969) 1075–6.
4. Kerkut, G. A., Pitman, R. M. & Walker, R. J., Iontophoretic application of acetylcholine and GABA onto insect central neurones. *Comp. Biochem. Physiol.*, **31** (1969) 611–33.
5. Usherwood, P. N. R., Giles, D. & Suter, C., Studies of the pharmacology of insect neurones *in vitro*. In *Insect Neurobiology and Pesticide Action* (*Neurotox '79*). Society of Chemical Industry, London, 1980, pp. 115–28.
6. Suter, C. & Usherwood, P. N. R., Action of acetylcholine and antagonists on somata isolated from locust central neurones. *Comp. Biochem. Physiol.*, **80C** (1985) 221–9.
7. Giles, D. & Usherwood, P. N. R., The effects of putative amino acid transmitters on somata isolated from neurones of the locust central nervous system. *Comp. Biochem. Physiol.*, **80C** (1985) 231–6.
8. Suter, C., The action of octopamine and other biogenic amines on locust central neurones. *Comp. Biochem. Physiol.*, **84C** (1986) 181–7.
9. Lees, G., Beadle, D. J., Neumann, R. & Benson, J. A., Responses to GABA by isolated insect neuronal somata: pharmacology and modulation by a benzodiazepine and a barbiturate. *Brain Res.*, **401** (1987) 267–78.
10. Benson, J. A. & Neumann, R., Nicotine and muscarine evoke different responses in isolated, neuronal somata from locust thoracic ganglia. *Soc. Neurosci. Abs.*, **13** (1987) 938.
11. Benson, J. A., Pharmacology of a locust thoracic ganglion somal nicotinic acetylcholine receptor. In *Nicotinic Acetylcholine Receptors in the Nervous System*, *NATO ASI Series H, Vol. 25*, ed. F. Clementi, C. Gotti & E. Sher. Springer, Berlin, Heidelberg, 1988, pp. 227–40.
12. Benson, J. A., Transmitter receptors on insect neuronal somata: GABAergic and cholinergic pharmacology. In *The Molecular Basis of Drug and Pesticide Action—Neurotox '88*, ed. G. G. Lunt. Elsevier Biomedical, Amsterdam, 1988, pp. 193–206.
13. Benson, J. A., M1-like muscarinic receptors mediate cholinergic activation of an inward current in isolated neuronal somata from locust thoracic ganglia. *Soc. Neurosci. Abs.*, **15** (1989) 365.
14. Benson, J. A., Bicuculline blocks the response to acetylcholine and nicotine but not to muscarine or GABA in isolated insect neuronal somata. *Brain Res.*, **458** (1988) 65–71.

15. Neumann, R., Lees, G., Beadle, D. J. & Benson, J. A. (1987) Responses to GABA and other neurotransmitters in insect central neuronal somata *in vitro*. In *Sites of Action for Neurotoxic Pesticides,* ed. R. M. Hollingworth & M. B. Green. American Chemical Society, Washington, DC, pp. 25–43.
16. Bermudez, I., Beadle, D. J. & Benson, J. A., Multiple serotonin-activated currents in isolated, neuronal somata from locust thoracic ganglia. *Soc. Neurosci. Abs.,* **16** (1990) 857.
17. Kaufmann, L. & Benson, J. A., Pharmacology and voltage-dependence of the response to octopamine in isolated, voltage-clamped *Locusta* neurones. *Soc. Neurosci. Abs.,* **17** (1991) 277.
18. Bermudez, I., Hietter, H., Trifilieff, E., Beadle, D. J. & Luu, B., Effects of novel peptides from the *corpora cardiaca* of *Locusta* on insect central neurones. *Pesticide Sci.,* **32** (1991) 523–4.
19. Sombati, S. & Hoyle, G., Glutamatergic central nervous transmission in locusts. *J. Neurobiol.,* **15** (1984) 507–16.
20. Dubas, F., Inhibitory effect of L-glutamate on the neuropile arborizations of flight motoneurones in locusts. *J. exp. Biol.,* **148** (1990) 501–8.
21. Dubas, F., Actions of putative amino acid neurotransmitters on the neuropile arborizations of locust flight motoneurones. *J. exp. Biol.,* **155** (1991) 337–56.
22. Irving, S. N. & Miller, T., Ionic differences in 'fast' and 'slow' neuromuscular transmission in body wall muscles of *Musca domestica* larvae. *J. Comp. Physiol.,* **135** (1980) 291–8.
23. Neumann, R. & Peter, H. H., Insecticidal organophosphates: Nature made them first. *Experientia,* **43** (1987) 1235–7.
24. Ottea, J. A., Payne, G. T. & Soderlund, D. M., Action of insecticidal *N*-alkylamines at Site 2 of the voltage-sensitive sodium channel *J. Agric. Food Chem.,* **38** (1990) 1724–8.
25. Elliott, M., Farnham, A. W., Janes, N. F., Johnson, D. M., Pulman, D. A. & Sawicki, R. M., Insecticidal amides with selective potency against resistant (*Super-kdr*) strain of houseflies (*Musca domectica* L.) *Agric. Biol. Chem.,* **50** (1986) 1347–9.
26. Martin, R. J. & Pennington, A. J., A patch-clamp study of effects of dihydroavermectin on *Ascaris* muscle. *Br. J. Pharmacol.,* **98** (1989) 747–56.
27. Nathanson, J. A., Caffeine and related methylxanthines: possible naturally occurring pesticides. *Science,* **226** (1984) 184–7.
28. Ruder, F. J., Guyer, W., Benson, J. A. & Kayser, H., The thiourea insecticide/acaricide diafenthiuron has a novel mode of action—inhibition of mitochondrial respiration by its carbodiimide product. *Pestic. Biochem. Physiol.,* **41** (1991) 207–19.
29. Jenden, D. J. & Fairhurst, A. S., The pharmacology of ryanodine. *Pharmacol Rev,* **21** (1969) 1–25.
30. Benson, J. A., Electrophysiological pharmacology of the nicotinic and muscarinic acetylcholine responses of isolated neuronal somata from locust thoracic ganglia. *J. of Exp. Biol.* (in press).
31. Benson, J. A., A novel GABA receptor in the heart of a primitive arthropod, *Limulus polyphemus. J. exp. Biol.,* **147** (1989) 421–38.
32. Jennings, K. R., Brown, D. G. & Wright, D. P., Methyllycaconitine, a naturally occurring insecticide with high affinity for the insect cholinergic receptor. *Experientia,* **42** (1986) 611–3.
33. Mody, N. V., Henson, R., Hedin, P. A., Kokpol, U. & Miles, D. H., Isolation of the insect paralyzing agent coniine from *Sarracenia flava. Experientia,* **32** (1976) 829–30.
34. Benson, J. A., Insect nicotinic acetylcholine receptors as targets for insecticides. In *Progress and Prospects in Insect Control,* ed. N. R. McFarlane. BCPC Monograph No. 43, British Crop Protection Council, Farnham, UK, 1989, pp. 59–70.

35. Benson, J. A., Nitromethylene heterocycle insecticides are antagonists at insect neuronal nicotinic receptors. In *Seventh International Congress of Pesticide Chemistry*, Vol. 1, ed. H. Frehse, E., Kesseler-Schmitz & S. Conway. International Union of Pure and Applied Chemistry, Hamburg, 1990, p. 348.
36. Trimmer, B. A. & Weeks, J. C., Effects of nicotinic and muscarinic agents on an identified motoneurone and its direct afferent inputs in larval *Manduca sexta*. *J. exp. Biol.*, **144** (1989) 303–37.
37. Hue, B., Lapied, B. & Malecot, C. O., Do presynaptic muscarinic receptors regulate acetylcholine release in the central nervous system of the cockroach *Periplaneta americana*? *J. exp. Biol.*, **142** (1989) 447–51.
38. Piek, T., Fokkens, R. H., Karst, H., Kruk, C., Lind, A., van Marle, J., Nakajima, T., Nibbering, N. M. M., Shinozaki, H., Spanjer, W. & Tong, Y. C., Polyamine like toxins—a new class of pesticides? In *Neurotox '88: Molecular Basis of Drug and Pesticide Action*, ed. G. G. Lunt. Elsevier Science Publishers, Amsterdam, 1988, pp. 61–76.
39. Eldefrawi, A. T., Eldefrawi, M. E., Konno, K., Mansour, N. A., Nakanishi, K., Oltz, E. & Usherwood, P. N. R., Identification and synthesis of a potent glutamate receptor antagonist in wasp venom. *Proc. Natl. Acad. Sci. USA*, **85** (1988) 4910–3.
40. Piek, T., Hue, B., Pelhate, M., David, J. A., Spanjer, W. & Veldsema-Currie, R. D., Effects of the venom of *Philanthus triangulum* F. (Hym. Sphecidae) and β- and δ-philanthotoxin on axonal excitability and synaptic transmission in the cockroach CNS. *Arch. Insect Biochem. Physiol.*, **1** (1984) 297–306.
41. Benson, J. A., Kaufmann, L., Hue, B., Pelhate, M. Schürmann, F., Gsell, L. & Piek, T., The physiological action of analogues of philanthotoxin–433 at insect nicotinic receptors (in preparation).
42. Benson, J. A., Schürmann, F., Kaufmann, L., Gsell, L. & Piek, T., Inhibition of dipteran larval neuromuscular synaptic transmission by analogues of philanthotoxin–433: a structure-activity study. *Comp. Biochem. Physiol.* (in press).
43. Bruce, M., Bukownik, R., Eldefrawi, A. T., Eldefrawi, M. E., Goodnow, R., Kallimopoulos, T., Konno, K., Nakanishi, K., Niwa, M. & Usherwood, P. N. R., Structure-activity relationships of analogues of the wasp toxin philanthotoxin: non-competitive antagonists of quisqualate receptors. *Toxicon*, **28** (1990) 1333–46.
44. Maeda, M., Kodama, T., Tanaka, T., Ohfune, Y., Nomoto, K., Nishimura, K. & Fujita, T., Insecticidal and neuromuscular activities of domoic acid and its related compounds. *J. Pesticide Sci.*, **9** (1984) 27–32.

6

Avermectins: Idiosyncratic Toxicity in a Subpopulation of Collie Dogs

J. M. Schaeffer,* S. P. Rohrer, D. Cully, & J. Arena

Department of Biochemical Parasitology, Merck Sharp & Dohme Research Laboratories, PO Box 2000, Rahway, New Jersey 07065, USA

INTRODUCTION

Avermectins are a family of macrocyclic lactones isolated from *Streptomyces avermitilis*[1,2] which have potent anthelmintic[3] and insecticidal activity.[4,5] The name 'avermectin' reflects their efficacy against worms or 'vermes' as well as their activity against ectoparasites. The avermectins are composed of a 16-membered ring and are distinguished from other macrocyclic antibiotics by their characteristic spiroketal, hexahydrobenzofuran unit and the disaccharide component at the 13-position (Fig. 1). The naturally occurring avermectins are separated into four major components (A_{1a}, A_{2a}, B_{1a} and B_{2a}), of which the B series are generally more biologically active. Abamectin is the non-proprietary name for a marketed miticide and insecticide composed of avermectin B_{1a} (>80%) and avermectin B_{1b} (<20%). Ivermectin is a semi-synthetic analog, (>80% 22,23-dihydroavermectin B_{1a} and <20% 22,23-dihydroavermectin B_{1b}) widely used as an endectocide for farm animals. Other avermectin analogs have been reported to have good efficacy against various insect species.[6] Milbemycins are natural products closely related to the avermectins, differing only in their lack of a C-13 disaccharide substituent.

The mode of action of avermectins in target species is not completely understood,[7] however, specific, high affinity ivermectin binding sites have been identified and characterized in the free-living nematode, *Caenorhabditis elegans*.[8–10] There is a clear correlation between the binding affinities and *in vivo* efficacy of a series of avermectin analogs, indicating that the binding site is physiologically important.[8,10] Specific avermectin binding sites have also been identified in mammalian brain tissue,[8,11,12] however the affinity is approximately 100-fold lower in brain tissue which may account for the

---

* To whom correspondence should be addressed at: Department of Biochemical Parasitology, RY80T-132, PO Box 2000, Rahway, New Jersey 07065.

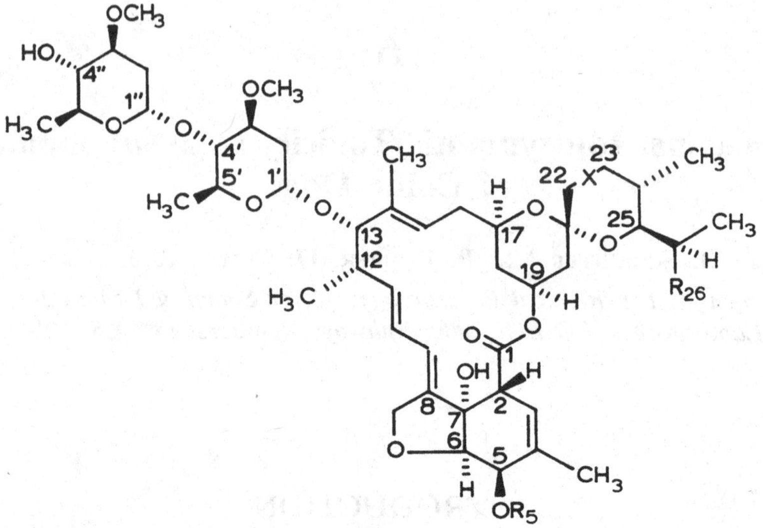

FIG. 1. Structures of the naturally occurring avermectins. Components A: $R_5 = CH_3$; components B: $R_5 = H$; components a: $R_{26} = C_2H_5$; components b: $R_{26} = CH_3$; components 1: $X = \!-\!CH\!=\!CH\!-\!$; components 2: $X = \!-\!CH_2CH\!-\!OH$.

relatively low mammalian toxicity of this family of compounds. Electrophysiological studies support a role for the avermectins in modulation of GABA-gated chloride channels in mammalian neuronal tissues.[13]

Ivermectin is also widely used in veterinarian medicine and provides an effective prophylactic treatment for *Dirofilaria immitis*, the causative agent of canine heartworm.[14] Ivermectin has proven to be a safe compound with few reports of adverse reactions. However, idiosyncratic reactions have been reported to occur from the extralabel use of ivermectin in a limited population of collie dogs.[15,16] Although the mechanism of ivermectin toxicosis in the sensitive collie is not clear, the presence of increased levels of ivermectin in brain tissue of dogs displaying symptoms of ivermectin toxicosis, suggests involvement of the central nervous system.[17–19]

In the sensitive collie, mild signs of ivermectin toxicity are elicited at doses as low as 75 μg/kg in an oral dosage formulation (this is 10-fold greater than the recommended therapeutic dosage). As the dosage increases, the severity of the toxicity increases. Comparable signs of ivermectin toxicity are elicited in non-sensitive dogs at ivermectin dosages of 4000–12 000 μg/kg. The physiological difference between sensitive and non-sensitive dogs is not known, however several possible mechanisms may be postulated, including differences in metabolism, binding to serum proteins, ability of the drug to cross the blood–brain barrier, specific ivermectin receptors and/or modulation of the GABA-gated chloride channel.

In this report we describe experiments designed to characterize the interaction between ivermectin and specific ivermectin binding sites in addition to the interaction between ivermectin and GABA-gated ion channels in brain tissues isolated from beagles and sensitive collies.

RESULTS

Specific ivermectin binding sites in beagle and sensitive collie brain

Synaptosomal membranes prepared from either sensitive collie or beagle cerebral cortex were incubated for 60 min at 22°C in the presence of increasing concentrations of [³H] ivermectin with (nonspecific) or without (total binding) a 200-fold molar excess of unlabeled ivermectin. Specific [³H] ivermectin binding to the synaptosomal membranes was saturable with increasing concentrations of [³H] ivermectin (Fig. 2). Scatchard analysis of these data yielded a straight line consistent with the existence of a single class of binding sites. The dissociation constant, K_d, and capacity, B_{max}, were not statistically different (54–64 nM and 2·1–3·1 pmol/mg, respectively). Table 1 presents the parameters of ivermectin binding to four regions of both beagle and sensitive collie brain. There is no significant difference in the apparent dissociation

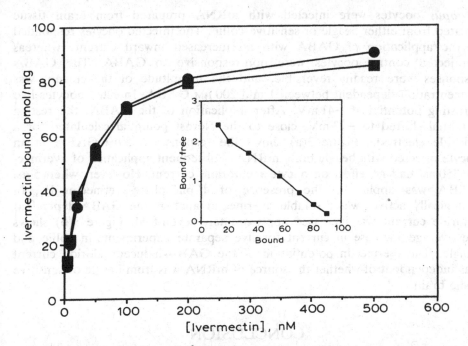

FIG. 2. Equilibrium binding of [³H] to dog brain cerebral cortex synaptosomes. Increasing concentrations of [³H] ivermectin were incubated with the synaptosomes and specific binding was determined[20] for both beagle (●) and sensitive collie (■). A Scatchard analysis (inset) of the saturation data is shown.

TABLE 1
Ivermectin Binding Sites in Beagle and Sensitive Collie Brain

Brain region	Beagle		Sensitive collie	
	K_d	B_{max}	K_d	B_{max}
Cerebral cortex	64	3·1	54	2·1
Cerebellum	41	1·6	48	1·9
Hypothalamus	60	1·4	44	1·6
Caudate nucleus	69	1·5	28	1·0

Specific [^{3}H] ivermectin binding to synaptosomes prepared from various brain regions of beagles and sensitive collies was determined as described by Olsen and Snowman.[20]

constant of the regions examined in beagle and sensitive collie (28–69 nM). Similarly, the binding capacity remained constant in all samples examined (1·0–3·0).

Avermectin potentiates the GABA response in oocytes injected with *m*RNA isolated from beagle and sensitive collie brain

Xenopus oocytes were injected with mRNA prepared from brain tissue isolated from either beagle or sensitive collie. The injected oocytes responded to the application of GABA with an increased inward current, whereas uninjected control oocytes were non-responsive to GABA. The GABA responses were readily reversible, and the magnitude of the current was concentration dependent between 1 and 200 μM GABA. Injected oocytes had a resting potential of −41 mV. After application of the GABA, the resting potential shifted to −23 mV, close to the Nernst potential calculated for a chloride electrode. Figure 3(a) shows the response to 5 μM GABA in an oocyte injected with beagle brain mRNA. Subsequent application of avermectin 250 nM had no effect on oocyte membrane current. However, when 5 μM GABA was applied in the presence of 4″-phosphate-avermectin B$_{1a}$ (a biologically active, water soluble avermectin analog) the GABA-dependent chloride current was increased approximately seven-fold. Figure 3(b) shows the average increase in current for five separate experiments in collie and beagle. The ivermectin potentiation of the GABA-induced chloride current was independent of whether the source of mRNA was from beagle or sensitive collie brain.

CONCLUSION

The results presented in this paper suggest that the ivermectin toxicosis observed in some collies is not due to altered ivermectin binding sites nor

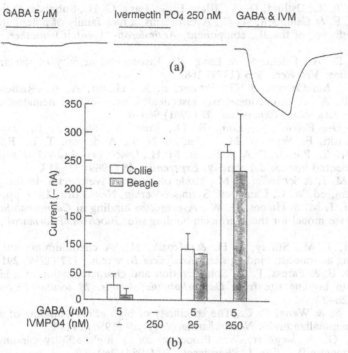

FIG. 3. Avermectin potentiates the GABA response in oocytes injected with mRNA isolated from beagle and sensitive collie. (a) Membrane currents from oocytes injected with beagle mRNA in the presence of 5 μM GABA; 250 nM 4″-PO₄-avermectin and both GABA (5 μM and 4″-PO₄-avermectin (250 nM). (b) Membrane currents were measured in beagle and sensitive collie mRNA-injected oocytes and the response to various concentrations of GABA and/or 4″-PO₄-avermectin were recorded.

altered GABA binding sites. It is possible that alterations in the blood–brain barrier or in an unknown ivermectin transport system is responsible for this condition.

REFERENCES

1. Burg, R. W., Miller, B. M., Baker, E. E., Birnbaum, J., Currie, J. A., Hartman, R., Kong, V.-L., Monaghan, R. L., Olson, G., Putter, I., Tunac, J. P., Wallick, H., Stapley, E. O., Miwa, R. & Omura, S., Avermectins, a new family of potent anthelmintic agents: producing organism and fermentation. *Antimicrob. Agents Chemother.*, **15** (1979) 361–7.
2. Miller, T. W., Chaiet, L., Cole, D. J., Flor, J. E., Goegelman, R. T., Gullo, V. P., Joshua, H., Kempf, A. J., Krellwitz, W. R., Monaghan, R. L., Ormaond,ˌR. E., Wilson, K. E., Albers-Schonberg, G. & Putter, I. Avermectins, a new family of potent anthelmintic agents: isolation and chromatographic properties. *Antimicrob.*

3. Egerton, J. R., Ostlind, D. A., Blair, L. S., Eary, C. H., Suhayda, D., Cifelli, S., Riek, R. F. & Campbell, W. C., Avermectins, a new family of potent anthelmintic agents: efficacy of the B_{1a} component. *Antimicrob. Agents Chemother.*, **15** (1979) 372–8.

4. Ostlind, D. A., Cifelli, S. & Lang, R., Insecticidal activity of the antiparasitic avermectins. *Vet. Rec.*, **105** (1979) 168.

5. Putter, I., MacConnell, J. G., Preiser, F. A., Haidri, A. A., Ristich, S. S. & Dybas, R. A., Avermectins: novel insecticides, ascarids and nematoicides from a soil micro-organism. *Experentia*, **37** (1981) 963–4.

6. Mrozik, H., Eskola, P., Linn, B. O., Lusi, A., Shih, T. L., Tischler, M., Waksmunski, F., Wyvratt, M. J., Hilton, N. J., Anderson, T. E., Babu, J. R., Dybas, R. A., Preiser, F. A. & Fisher, M. H., Discovery of novel avermectins with unprecedented insecticidal activity. *Experentia*, **45** (1989) 31–317.

7. Turner, M. J. & Schaeffer, J. M., Mode of action of ivermectin. In *Ivermectin and Abamectin*, ed. W. C. Campbell. Springer-Verlag, New York, 1989, pp. 73–88.

8. Schaeffer, J. M. & Haines, H. W., Avermectin binding in *Caenorhabditis elegans*: A two-state model for the avermecin binding site. *Biochem. Pharmacol.*, **38** (1989) 2329–38.

9. Schaeffer, J. M., Stiffey, J. H. & Mrozik, H., A chemiluminescent assay for measuring avermectin binding sites. *Analytical Biochem.*, **177** (1989) 291–5.

10. Cully, D. F. & Paress, P. S., Solubilization and characterization of a high affinity ivermectin binding site from *Caenorhabditis elegans*. *Molecular Pharmacol.*, **40** (1991) 326–32.

11. Pong, S. S. & Wang, C. C., The specificity of high affinity binding of avermectin B_{1a} to mammalian brain. *Neuropharmacology*, **19** (1980) 311–17.

12. Drexler, G. & Sieghart, W., Properties of a high affinity binding site for [^{3}H]avermectin B_{1a}. *Eur. J. Pharmacol.*, **99** (1984) 269–77.

13. Sigel, E. & Baur, R., Effect of avermectin B_{1a} on chick neuronal γ-aminobutyrate receptor channels expressed in *Xenopus oocytes*. *Molecular Pharmacol.*, **32** (1978) 749–52.

14. Campbell, W. C., Ivermectin: An update. *Parasitology Today*, **1** (1985) 10–16.

15. Easby, S. M., Ivermectin in the dog. *Vet. Rec.*, **115** (1984) 45.

16. Houston, D. M., Parent, J. & Matushek, K., Ivermectin toxicosis in a dog. *J. Am. Vet. Med. Assoc.*, **191** (1987) 78–80.

17. Pulliam, J. D., Seward, R. L., Henry, R. T., Steinberg, S. A., Investigating ivermectin toxicity in collies. *Vet. Med.*, **80** (1985) 36–40.

18. Paul, A. J., Tranquilli, W. J., Seward, R. L., Todd, K. S. & DiPietro, J. A., Clinical observations in collies given ivermectin orally. *Am. J. Vet. Res.*, **48** (1987) 684–5.

19. Tranquilli, W. J., Paul, A. J. & Seward, R. L., Ivermectin plasma concentrations in collies susceptible to ivermectin induced toxicosis. *Am. J. Vet. Res.*, **50** (1989) 769–70.

20. Olsen, R. W. & Snowman, A. M., Avermectin B_{1a} modulation of gamma-aminobutyric acid/benzodiazepine receptor binding in mammalian brain. *J. Neurochem.*, **44** (1985) 1074–82.

SECTION 2

Molecular and Cellular Physiology of Insects and Nematodes

7

The GABA Activated Cl⁻ Channel in Insects as Target for Insecticide Action: a Physiological Study

HARALD C. VON KEYSERLINGK

Schering AG, Agrochemical Research, P.O. Box 65 03 11, D-1000 Berlin 65, FRG

&

R. JOHN WILLIS

Schering Agrochemicals Ltd, Saffron Walden, Essex CB10 1XL, UK

INTRODUCTION

For the past 200 years the majority of insect control technologies have been and still are based on molecules which interfere with the voltage activated Na⁺ channel, the cholinergic synapse and the GABAergic synapse. The idea that one major inhibitory system in insects, the GABA activated Cl⁻ channel is an important target of insecticide action is based on the observation that cyclohexane and cyclodiene insecticides antagonise GABA activated Cl⁻ currents[1,2] and that avermectins agonise inhibitory Cl⁻ currents in most insect preparations.[3–6] This GABA activated Cl⁻ channel in insect nerve and muscle cell membranes has since become a major focus of attention in both academic and industrial research institutions.[7,8] GABA receptors associated with Cl⁻ ion channels have been found in insect nerve and muscle cells (Table 1).

Neuronal GABA receptors have been localised in the postsynaptic, dendritic membranes in the neuropile, but also in the membranes of nerve cell somata where their function is not really clear. These nerve cell bodies do not receive inhibitory, GABA emitting nerve terminals. In most cases these somata are not actively involved in electrical signalling at all but seem to be responsible for housekeeping support for the electrically active dendritic, axonal and terminal ramifications of the neuron. Some but not all skeletal muscle fibres in insects are endowed with GABA activated Cl⁻ channels and some of these have been well characterised.[9–13] Less information is available on GABA receptors in visceral and autonomous nerve–muscle systems such as pulsating muscles regulating the flow of body fluids, ventilatory muscles regulating respiration, reproductive organs and the stomatogastric system. GABA

79

TABLE 1
Established and Putative Localisations of
GABA Activated Cl⁻ Channels in Insects

Neuron	
Dendritic, postsynaptic	
Soma	function ??
Terminal, presynaptic	??
Skeletal muscle	
Postsynaptic	
Extrasynaptic	
Other muscles	??

receptors with a unique pharmacological profile have recently been well characterised in the limulus heart.[14]

The main physiological role of GABAergic innervation in central nervous and in peripheral neuromuscular systems seems to be the organisation of simultaneous and consecutive excitatory events into coordinated output. Coordinated behaviour is possible only through a delicately tuned balance between excitatory and inhibitory synaptic events resulting in patterns of action potentials which encode meaningful messages for the receiving effector systems.[15,16]

Molecular biologists have recently identified a bewildering variety of different GABA$_A$ receptor subunits in mammalian nervous systems based on varying degrees of sequence homologies, differing regional localisations and differing pharmacological profiles.[17-25] Which ones of these 12 or 14 different proteins associated with GABA activated Cl⁻ channels in vertebrates match GABA receptor subunits from insect sources is currently a major focus of research activities (Fig. 1).

GABA RECEPTOR PHARMACOLOGY

This GABA activated Cl⁻ ionophore complex is characterised by at least five major separate binding domains for agonists, antagonists, modulators and channel blockers [Fig. 1(c) and Table 2].

The success of every structure–activity–optimisation endeavour depends entirely on the reliability and relevance of the activity data the chosen assay can feed back. Which assays give the clearest information on the pharmacology of GABA receptors in insects?

Experimental methods

Nerve cell bodies dissociated from locust thoracic ganglia under continuous, rapid saline perfusion have proved to be a very stable and useful preparation [Fig. 2(b)]. GABA activated Cl⁻ channels are found on every soma, as has

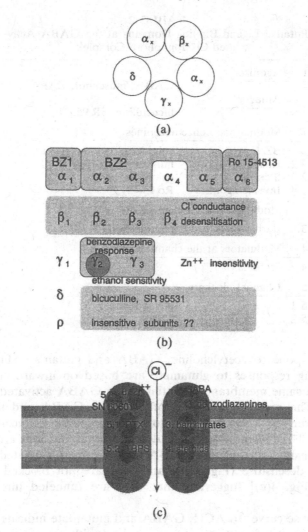

Fɪɢ. 1. The GABA activated Cl⁻ ionophore complex. (a) The hypothetical pentameric structure of the channel. (b) Putative associations between protein structure and pharmacological profile or function. (c) The major ligand binding domains.

previously been described.[26,27] These nerve cell bodies are not from identified cells. Pharmacological differences found may be due to different receptor densities or receptor types in these different cells rather than to different properties of the investigated ligands. The experimental set-up allows recording from these cells for many hours and thus characterisation of voltage activated and ligand activated currents before each pharmacological experiment. In this way it is possible to do comparative pharmacological studies on a fairly homogeneous population of characterised cell types.

TABLE 2
Putative Ligand Binding Domains at the GABA Activated Cl⁻ Ionophore Complex

1.	Agonists	
		GABA, muscimol, ZAPA
	Antagonists	
		bicuculline, SR 95531
2.	Modulators: benzodiazepines	
2.1.	BZ 1	zolpidem, oxoquazepam
2.2.	Unspecific	flunitrazepam
2.3.	'Peripheral'	chlorodiazepam
2.4.	Inverse agonist	Ro 15-4513
3.	Modulators at the channel site	
3.	Barbiturates	pentobarbital
4.	Modulators at the channel site	
4.	Steroids	
5.	Channel blockers	
5.1.		picrotoxinin
5.2.		TBPS

All cells respond to acetylcholine, GABA and glutamate [Fig. 3(a)]. The hyperpolarising responses to glutamate are based on inward currents which reverse at the same membrane potential as the GABA activated current does [Fig. 4(a)]. The reversal potentials of both the GABA and the glutamate activated current are very sensitive to changes in the extracellular and the intracellular Cl⁻ concentration, thus suggesting that Cl⁻ ions are carrying the glutamate activated current too. GABA and glutamate activated Cl⁻ currents do not cross desensitise (Fig. 4(b)] and they are not blocked by the same compounds [Fig. 4(c)] suggesting that they are funneled through separate proteins.

Dose response curves to ACH, GABA and glutamate indicate that the cells differ in their sensitivities to these agonists and with respect to the maximum peak current that can be elicited by each individual agonist. Typical maximum currents induced by GABA pulses are between -0.5 and $-1.5\,nA$ at $-80\,mV$ holding potential. Fifty per cent maximum current can be induced by GABA pulse concentrations between 10 and 5000 μM. In most cells the inward current elicited by pressure pulses of GABA, $10^{-4}\,M$ inactivates within seconds after the end of the GABA pulse.

The dose response curves obtained by either increasing GABA concentrations [Fig. 5(a)] or by increasing the pulse duration at a fixed GABA concentration ($5 \times 10^{-5}\,M$) [Fig. 5(b)] are very steep. The GABA activated Cl⁻ currents are most sensitive to pharmacological modification in the range between 20% (potentiators) and 80% (attenuators) of their maxima.

Using this preparation as functional assay to study the pharmacology of the insect GABA activated Cl⁻ channel we have found the following results.

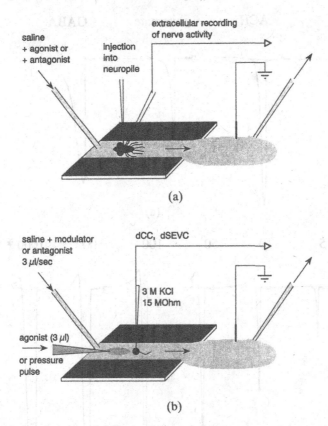

FIG. 2. The experimental set-up. (a) Recording of the spontaneous firing activity of the isolated synganglion of *Calliphora erythrocephala* wandering stage larvae. Neuropile injections of drugs are possible. (b) Arrangement for single electrode voltage clamp (SEVC) experiments on isolated nerve cell bodies dissociated from locust metathoracic ganglia. The middle groove for the saline flow between the two parafilm sheaths covering either side of the microscopic cover slip is so narrow that 1 μl saline covers the nerve cell and bridges both edges.

The agonist site

Several GABA agonists can be identified and there is little variation in the relative agonist sensitivity among the different cells (Fig. 6). Muscimol, ZAPA and its seleno-analogue are the most potent agonists but isoguvacin THIP and thiomuscimol are active as well. Although several of these agonists are potent compared to the endogeneous ligand GABA, 10^{-5} M is a relatively high concentration in general terms of drug activity.

As has been described before,[8,27] the competetive and diagnostic mammalian GABA$_A$ antagonists bicuculline and SR 95 531 show no effect up to 10^{-4} M concentrations.[26] One interesting aspect has emerged recently, when it was shown conclusively that in the granular layer of the mammalian cerebellum,

ACH   glutamate   GABA

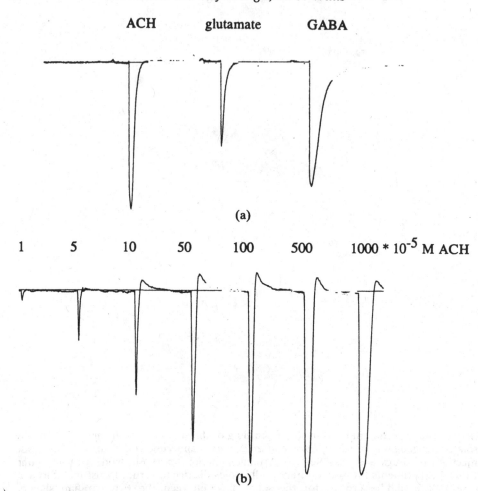

(a)

1   5   10   50   100   500   1000 * 10^{-5} M ACH

(b)

FIG. 3. Inward currents recorded from isolated locust nerve cell bodies evoked by hand applied 3-μl pulses of transmitter solutions. Calibration: 1 nA, 10 s. (a) Cell clamped at a holding potential (hp) of -80 mV. ACH, glutamate and GABA are dissolved in saline at 10^{-4} M. (b) Concentration response curve to ACH. Hp -100 mV (note biphasic response which occurs always at holding potentials more negative than -90 mV).

a region with a high density of GABA, muscimol and TBPS binding sites, no bicuculline and SR 95 531 binding can be detected, suggesting that in this part of the mammalian brain there is a GABA$_A$ receptor population whose pharmacological profile seems to match that of the insect GABA receptor.[22]

The modulatory site

Four putative benzodiazepine binding domains have been identified in vertebrate tissues, three in different CNS regions and one in the periphery.

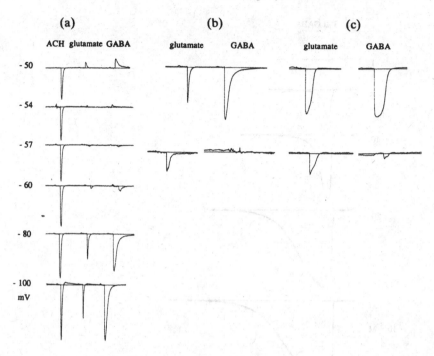

Fɪɢ. 4. Comparison of currents evoked by 3-μl pulses of glutamate and GABA at
10^{-4} ᴍ. Calibration: 1 nA, 10 s. (a) Glutamate and GABA evoked currents reverse at
precisely the same membrane potential (this cell −55 mV). Calibration: 0·5 nA at hp
between −50 and −60 mV. (b) No cross-desensitisation between GABA and glutamate
after 10 min superfusion with GABA $5 * 10^{-5}$ ᴍ. Calibration: 0·5 nA, 5 s. (c) Superfusion with picrotoxinin, 10^{-5} ᴍ, for 10 min suppresses the GABA activated current to a
much higher degree than the glutamate activated current. Calibration: upper trace,
1 nA, 5 s; lower trace, 0·5 nA, 5 s.

Ligand binding studies have shown that the most potent benzodiazepine in
insects was chlorodiazepam, Ro 05-4864, the selective ligand for the peripheral
benzodiazepine receptor in vertebrates which is not associated with a GABA
activated Cl⁻ channel complex.[8,28] Again, it has been shown recently that
Ro 05-4864 binding sites also occur in the piscine cerebellum.[29]

Physiological data from locust somata on potentiation or attenuation of
GABA activated Cl⁻ currents by ligands at the benzodiazepine site turned out
to be very variable between individual cells. Among the non-selective
benzodiazepine ligands diazepam is inactive but flunitrazepam consistently
potentiates the GABA response with the factor 1.4. The selective BZ 1 ligands
zolpidem, Cl 218872 and oxoquazepam gave capricious results, potentiating
GABA responses strongly in some cells but being inactive or even attenuating
them in others. The most potent but most capricious ligand in this respect was
zolpidem [Fig. 7(b)]. Benzodiazepine type 1 binding sites are found in
particularly high densities in the mammalian cerebellum.[29,30] The 'selective

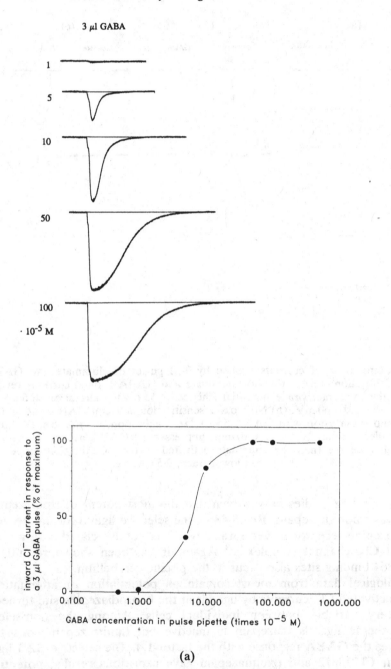

FIG. 5. Quantification of the response to GABA. SEVC, hp −80 mV. (a) Concentration response curve. Hand application of 3-μl GABA pulses. Calibration: 0·5 nA, 5 s. (b) Pulse duration response curve. Stimulator driven pressure pulses of GABA at 10^{-4} M. Calibration: 0·5 nA, 2 s.

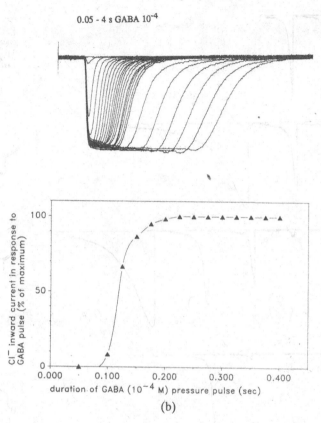

0.05 - 4 s GABA 10⁻⁴

(b)

Fig. 5—*contd.*

peripheral' BZ ligand, chlorodiazepam, potentiates the GABA responses to a significantly higher degree than flunitrazepam does although here, too, individual neurones differed substantially [Fig. 7(d)]. The inverse agonist, DMCM, attenuates the GABA responses in locust somata [Fig. 7(e)].

Thus, the benzodiazepine modulatory sites on the GABA receptor Cl⁻ ionophore complex which not long ago were believed to occur in vertebrates only as late specialisation in evolution seem to play a more important role in the regulation of insect activity and behaviour than has hitherto been recognised. These binding sites clearly warrant a more detailed examination. Subpopulations of benzodiazepine receptors with differing pharmacological profiles possibly exist within insect central nervous systems in a manner analogous to the vertebrate brain.

The channel modulation sites

It has been known for some time that pentobarbital significantly potentiates the GABA activated Cl⁻ current at concentrations above 5×10^{-6} M (Fig.

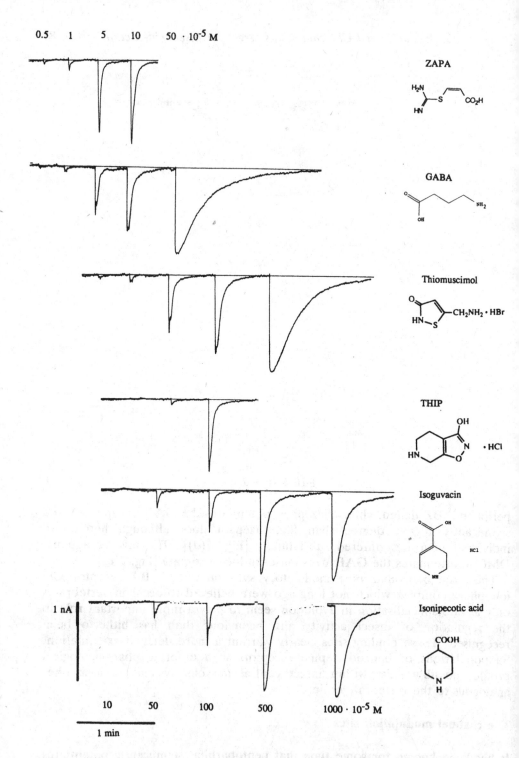

FIG. 6. Concentration response curves to various GABA agonists. SEVC, hp −80 mV. Calibration: 1 nA, 1 min.

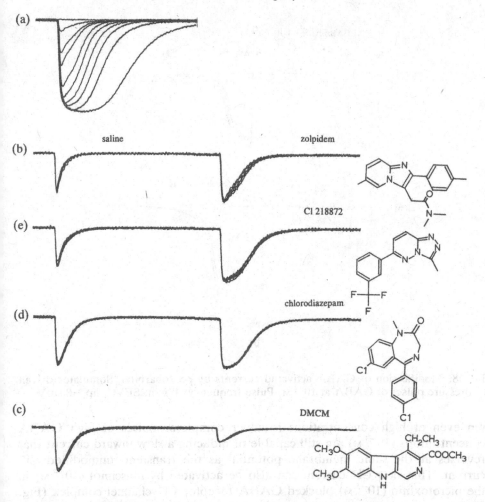

FIG. 7. Modification of GABA activated currents by ligands of benzodiazepine binding sites. Stimulator driven pressure pulses of GABA at 10^{-4} M. Pulse frequency: 1/min. SEVC, hp −80 mV. Calibration: 0·5 nA, 5 s. (a) Superimposed traces of inward currents evoked by GABA pressure pulses of increasing duration (10, 100, 150, 200, 300, 400, 500, 750, 1000, 2000 ms). (b)–(e) Five superimposed traces. Left hand side: saline. Right-hand side: 10–15 min after perfusion with the modulator, 10^{-5} M.

8).[8,26] The pharmacology of this binding site, however, has not yet been studied in any detail in insects. This lack of information is even more apparent with regard to the steroid binding site.

The channel blocker sites

Among the non-competitive Cl⁻ channel blockers picrotoxinin and the TBPS analogues belong to the most potent compounds producing ≈50% block at about 5×10^{-7} M. However, the non-competitive block induced by picrotoxi-

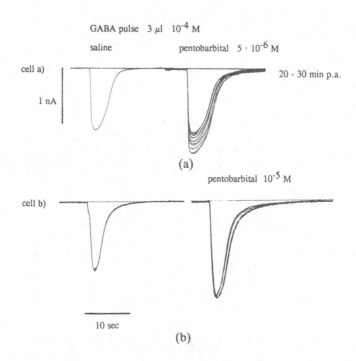

FIG. 8. Modification of GABA activated currents by pentobarbital. Stimulator driven
   pressure pulses of GABA at 10^{-4} M. Pulse frequency: 1/min. SEVC, hp -80 mV.

nin even at high concentrations is never complete. Pulses of high GABA
concentrations (10^{-3} M) are still capable of inducing a slow inward current that
reverses at the same membrane potential as the transient, unmodified Cl^-
current. This slow Cl^- current can also be activated by muscimol (10^{-4} M) in
the picrotoxinin (10^{-5} M) blocked GABA receptor Cl^- channel complex (Fig.
9). Thus some interaction between the agonist site and the PTX site seems to
occur at high agonist concentrations.

   In summary: there are several agonists, several modulators and several
non-competitive antagonists which are recognised by the GABA activated Cl^-
channel in locust somata. These molecules alter the functional behaviour of
this protein complex in concentration ranges which are not far from the action
of cyclohexane-, cyclodiene- and avermectin-insecticides. So apparently, on
the cell membrane we are unable to distinguish the action of an insecticide
from the action of one of these pharmacological tools.

   Could then perhaps one of these ligands serve as lead structure for
insecticide synthesis?

INSECTICIDE ACTION

The principal property of insecticides is the fact that these molecules invading
the organism from outside through the integument take over control from the

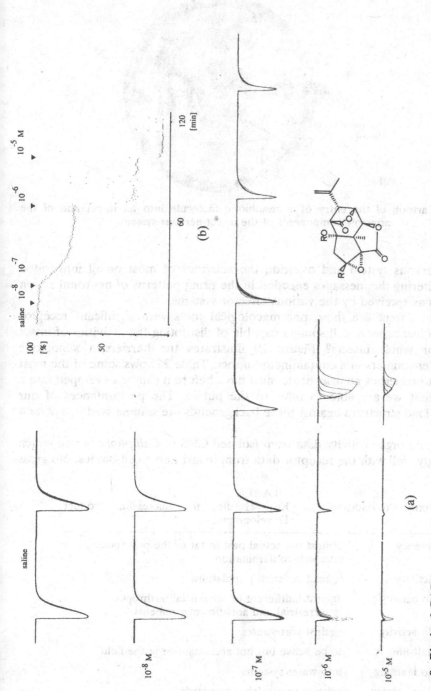

FIG. 9. Block of GABA activated currents by picrotoxinin. Stimulator driven pressure pulses of GABA at 10⁻⁴ M. Pulse frequency: 1/min. SEVC, hp −80 mV. Calibration: 0·5 nA, 5 s. (a) At least five traces superimposed. At 10⁻⁵ M and 10⁻⁶ M PTX, after reaching the plateau, the duration of the GABA pulse has been increased 2, 5, 10 and 20 times. Increasing GABA doses evoke increasing inward currents. This is not expected to happen with purely non competitive channel blockers. (b) The peak current amplitude of each GABA activated current as recorded during the entire experiment.

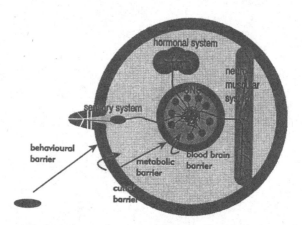

FIG. 10. Cartoon of the entry of a xenobiotic molecule into an insect and of the principle components of the insect nervous system.

central nervous system and override the activities of most or all innervated cells by altering the messages encoded in the firing patterns of neuronal action potentials as received by the various effector systems.

To what extent are these pharmacological tools with significant receptor activity in insect nerve cell somata capable of disrupting the activities of intact ganglia or whole insects? Figure 10 illustrates the barriers a xenobiotic molecule encounters in a contaminated insect. Table 3 shows some of the most important properties an insecticide must have before it can be developed into a product that we are able to offer to the public. The performances of our potential lead structures against these backgrounds are summarised in Tables 4 and 5.

The whole organ activity data from isolated CNS of *Calliphora* larvae match surprisingly well with the receptor data from locust nerve cell bodies. No gross

TABLE 3
Some Preconditions or Key Hurdles for Successful Product Development

Potency	against the actual pest instar of the pest species after self-contamination
Activity	against resistant populations
No activity	against indifferent or 'beneficial' arthropods in terrestrial and aquatic environments
No activity	against vertebrates
Half-life	to be active but not accumulative in the field
No leaching	into water systems
Cost efficacy	relative to available standards

TABLE 4
GABA Receptor Pharmacology, Activity in Cell– and Organ Preparations

GABA receptor pharmacology	Locust soma	Fly CNS	Locust muscle
		EC_{20} ($\times 10^{-5}$ M)	
1.1. GABA agonists			
Muscimol	1	—	1	10
Z-3-(amidinothio)propenoic acid	1	>> 50	2
GABA	2	>> 10 000	10
Z-3-(amidinoseleno)propenoic acid	5	— 500	10
E-3-(amidinothio)propenoic acid	10	>> 100	100
Thiomuscimol	20	— 10	1 000
Isoguvacine HCl	20	— 10	100
THIP	40	— 10	100
Isonipecotic acid	50	>> 100	200
Z-3-(2-benzimidazolylthio)propenoic acid	>> 100	>> 50	5
1.2. GABA antagonists			
Bicuculline methiiodide	>> 10	>> 10	>> 10
SR 95531	>> 10	>> 10	>> 10

	Potentiation at 10^{-5} M	EC_{20} ($\times 10^{-5}$ M)	
2.1. Benzodiazepines, BZ1			
Zolpidem	3·8	+ + 1	1
Cl 218872	2·2	+ + 100	>> 1
Oxoquazepam	1·6	+ + 1	>> 1
2.2. Benzodiazepines, mixed			
Flunitrazepam	1·4	rhythm 10	>> 10
2.3. Benzodiazepines, 'peripheral'			
Ro 05-4864 clorodiazepam	3·6	>> 50	>> 1
2.4. Benzodiazepines, 'atypical, 6'			
Ro 15-4513	1·3	>> 50	>> 1
2. 'Inverse agonists'			
DMCM	0·6	>> 50	>> 1
3. Channel site, barbiturates			
Pentobarbital	2·1	— 80	50

		EC_{20}($\times 10^{-5}$ M)	
4. Channel blockers			
Picrotoxinin	0·01	+ + 0·1	50
TBPS	0·05	+ + 0·1	>> 10
lindane	0·05	+ + 0·1	0·1

TABLE 5
GABA Receptor Pharmacology, *in vivo* Activity after Injection and Topical Application

GABA receptor pharmacology			*Musca domestica* adult female		*Heliothis virescens* larvae III	
			Injection	Topical	Injection	Topical
			ED_{50} (μg/insect)			
1.1. GABA agonists						
Muscimol	S^a	K	0.5	>> 10	>> 10	>> 10
Z-3-(amidinothio)propenoic acid		>>	0.5	>> 10	>> 1	
GABA		>>	100	>> 10		
Z-3-(amidinoseleno)propenoic acid		>>	1	>> 10	>> 10	
E-3-(amidinothio)propenoic acid		>>	2	>> 10		
Thiomuscimol	S^a	K	5	>> 10		
Isoguvacine HCl	S^a	S	10	>> 10		
THIP	S^a	K	5	>> 10	>> 10	
Isonipecotic acid		>>	10	>> 10		
Z-3-(2-benzimidazolylthio)propenoic acid		>>	1	>> 10		
1.2. GABA antagonists						
Bicuculline methiiodide	S^a	S	10	>> 10	>> 10	>> 10
SR 95531		>>	10			
2.1. Benzodiazepines, BZ1						
Zolpidem		>>	2	>> 10		
Cl 218872		>>	2.8			
Oxoquazepam		>>	3.7			
2.2. Benzodiazepines, mixed						
Flunitrazepam	S^a	(K)	10	>> 10		
2.3. Benzodiazepines, 'peripheral'						
Ro 05-4864 clorodiazepam	S^a	S	1.6	>> 10		
2.4. Benzodiazepines, 'atypical, 6'						
Ro 15-4513		>>	0.5			
2. 'Inverse agonists'						
DMCM		>>	0.5			
3. Channel site, barbiturates						
Pentobarbital	S^a	K	10	>> 10	>> 5	>> 50
4. Channel blockers						
Picrotoxinin	C^b	K	1	>> 10	>> 1	>> 100
TBPS	C^b	T	0.22	>> 2	>> 1	>> 2
lindane	C^b	T	0.1	T 0.1	T 1	T 5

[a] S = sedative.
[b] C = convulsive.

pharmacological differences can be detected between the GABA receptors in the locust neuronal soma membranes and the functional dendritic GABA receptors in isolated ganglia of the fly larvae. Only ZAPA and its analogues do not cross the blood–brain barrier. They are active, however, when injected directly into the neuropile.

receptors in the CNS and the peripheral GABA receptors in muscle membranes are the poor sensitivity of the peripheral GABA responses to Cl⁻ channel blockers such as picrotoxinin and TBPS, yet their relatively high sensitivity to agonists such as muscimol and ZAPA.

The housefly injection data (Table 5) indicate that most of the ligands which show significant potency at the receptor level induce symptoms or knock-down conditions.

Pharmacological modifications of GABA activated Cl⁻ current including potentiation, and attenuatioin or blockade, disrupt the voluntary neuromuscular system of insects to such an extent that complete paralysis may be induced for many hours or even days. This, however, is not necessarily associated with the death of the affected animal. These deeply intoxicated insects recover after hours or days, most of them without obvious residual lesions. In other words: pharmacological modification of GABA receptor functioning is not lethal on its own. Drastic potentiation or blockade of GABA activated Cl⁻ current for hours does not kill the affected isolated brain cell or the isolated ganglion. Secondary processes elsewhere in the organism must be initiated and sustained before the entire animal irreversibly collapses. These secondary intoxication processes are not very well understood. They cannot reliably be monitored in the voluntary neuromuscular system. Some functions of the autonomous nervous system regulating pulsating and ventilatory movements, hemocoel pressure and visceral muscles such as movements of the intestine, excretory and reproductive organs may provide more accurate information about the actual status of intoxication and the distance or proximity to death.

At first sight our data suggest that agonists like muscimol, THIP and ZAPA may be viewed as potential candidates as lead structures for insecticide synthesis. Among the non-competitive antagonists TBPS clearly stands out. Some benzodiazepine- and β carboline-type modulators can also be disruptive to the activities of whole ganglia and intact insects. Comparing the activity of these molecules in houseflies and lepidopterous larvae we find that none of these GABA agonists, antagonists and modulators shows any effect against one of the most important insect pests, *Heliothis virescens* larvae, although there is evidence that GABA activated Cl⁻ channels do exist in the central nervous system of this species. This observation is puzzling and casts doubt on the routine of using locusts, cockroaches and flies as models to extrapolate activity data across the evolutionary tree.

When a lead structure is selected, clear aims for optimisation have to be defined. In the case of a muscimol-type agonist receptor potency, irreversible, lethal action and cuticle penetration as well as activity against lepidopterous larvae have to be increased substantially before competitive insecticidal activity can be expected. None of the compounds discussed so far show insecticide activity after self-contamination or topical application in the most important insect pest species. This is a serious limitation since simultaneous optimisation of both potency at the receptor protein and diffusion through the arthropod integument has proved to be a difficult task. The toxicokinetic behaviour of integument has proved to be a difficult task. The toxicokinetic behaviour of these ligands of the GABA activated Cl⁻ channel does not permit the contamination of this vulnerable protein with a sufficiently high concentration for a sufficiently long time to induce irreversible lesions.

Since 1980 about 3600 patents claiming insecticide activity have been filed. We have not been able to identify structures which have been derived purely from receptor pharmacology and have been optimised to potent insecticides. Professor Casida's trioxabicyclooctane-type molecules developed from TBPS probably are the only outstanding exception.[31] This does not necessarily imply that it is impossible to achieve good cuticle penetration and to retain or improve receptor potency at the same time. It only points to the fact that these are separate optimisation routes following different structure activity rules. In general, however, we have experienced that insecticides have been discovered the other way round.

THE MODE OF ACTION OF A NOVEL AZOLE INSECTIDE

A few members of a herbicide series had shown some activity in an insecticide screen. Analogue synthesis around these lead structures succeeded in shifting

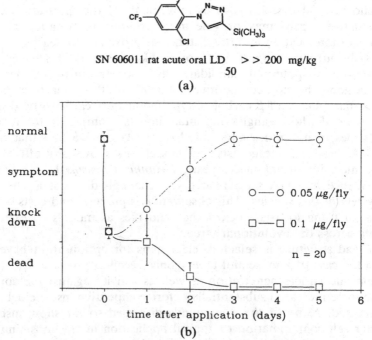

Fig. 11. Structure, mammalian toxicity and insecticide activity after topical application against female houseflies of SN 606011.

the activity away from herbicidal to insecticidal activity only. What is the mode of action of this molecule? Topically applied to houseflies it rapidly induces hyperexcitation and knock-down. This knock-down condition lasts for a day or two until, depending on the applied dose, either death or recovery occurs (Fig. 11). Recordings of the electrical activity of the flight muscle cells for several days during the entire intoxication process clearly show that nerve cell activity is significantly increased soon after topical application of the compound. The patterns of these muscle potentials show that during these many hours of continuous spiking activity the intervals between successive action potentials are reduced [Fig. 12(a), (b), (c)].

The isolated CNS preparation from *Calliphora* larvae responds to this compound with a gradual fusion of the cyclic waves of spontaneous firing activity resulting in uninterrupted firing for many hours at concentrations as low as 5×10^{-8} M (Fig. 13). In the isolated locust nerve cell body this

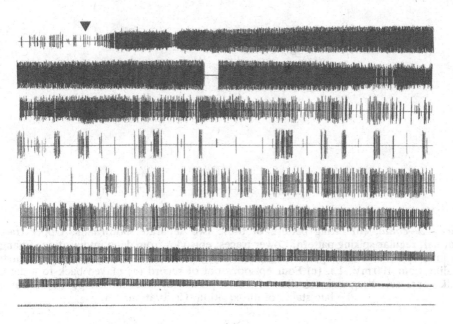

(a)

Fig. 12. Activity of SN 606011 against female houseflies. Simultaneous extracellular recording from two dorsolongitudinal flight muscles (4L display positive, 4R display negative). (a) Continuous record for 2 days. Arrow: topical application of 0·07 μg SN 606011. Note 1: rapid onset of neurotoxic symptoms indicating rapid penetration through the cuticle. 2. Convulsive hyperexcitation resulting in complete disruption of sensory and locomotor functions. 3. Almost perfect recovery and normal stimulus response behaviour. 4. Progressive decline of the amplitude of the muscle potentials associated with irreversible lesions resulting in the death of the insect. Calibration: 100 mV, 1 h.

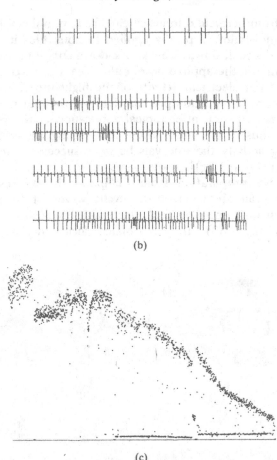

(b)

(c)

Fig. 12—*contd.* (b) Firing pattern of these two separate motorunits. Upper trace: normal, regular spiking pattern. Lower traces: episodes from deep intoxication. Change in inter spike interval: higher firing frequency and intermittent high frequency bursts. Calibration: 100 mV, 1 s. (c) Four episodes out of record (a) played back to a digital spike analyser. Progressive decline of the amplitude of these muscle potentials during the late stages of intoxication. Calibration: 1 h.

compound antagonised the GABA activated Cl^- current at 10^{-8} M concentrations (Fig. 14). However, detailed competition studies have to be done before the precise binding domain on this GABA receptor–Cl^- ionophore complex can be identified with certainty.

How specific is the activity of this molecule and which other proteins are affected? Table 6 summarises the effects of SN 606 011 on the different cellular functions looked at so far. These data show clearly that up to concentrations of 5×10^{-6} M no other system is affected. At 10^{-5} M, however, we see some effects in the neuromuscular periphery. The GABA responses of the locust extensor tibia are attenuated but not blocked (Fig. 15).

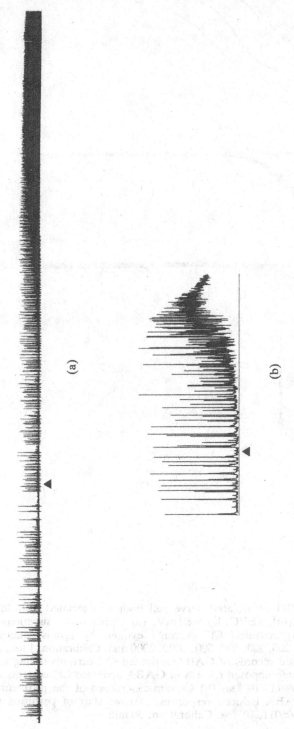

(a)

(b)

Fig. 13. Activity of SN 606011 in the isolated CNS preparation of *Calliphora* larvae [Fig. 2(a)]. (a) Time–frequency histogram. Arrow: start of perfusion with SN 606011, 10^{-7} M. Calibration: 500 spikes/s, 5 min. (b) Integral of total nerve activity (area under the curve). Arrow: start of perfusion with SN 606011, $5*10^{-8}$ M. Calibration: 5 min.

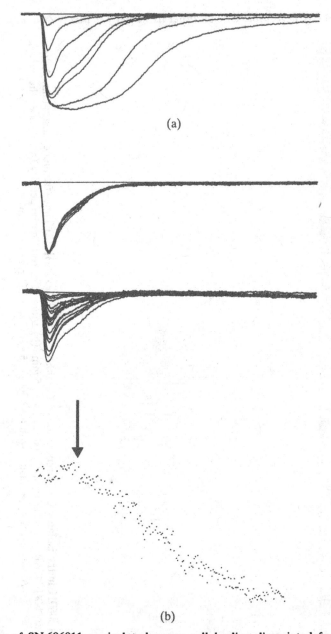

(a)

(b)

Fig. 14. Activity of SN 606011 on isolated nerve cell bodies dissociated from locust metathoracic ganglia [Fig. 2(b)]. SEVC, hp −80 mV. (a) Upper trace: superimposed records of GABA (10^{-4} M) activated Cl^- currents evoked by agonist pulses of increasing duration (10, 100, 200, 300, 400, 500, 1000, 2000 ms). Calibration: 1 nA, 10 s. Middle trace: 10 superimposed records of GABA activated Cl^- currents during saline perfusion. Lower trace: 20 superimposed records of GABA activated Cl^- currents after saline perfusion with SN 606011, 10^{-8} M. (b) Continuous record of the peak current amplitudes of successive GABA induced responses. Arrow: start of perfusion with SN 606011, 10^{-8} M. Calibration: 30 min.

TABLE 6
Actions of SN 606011 in Insect Neuromuscular
Systems

Neuron (dissociated locust nerve cell body) Responses to		10^{-x} M
GABA	block	8
ACH	> >	5
Glutamate	> >	5
Action potential	> >	5
Membrane resistance	> >	5
Resting potential	> >	5
Muscle (locust extensor tibiae muscle) Responses to		
GABA	block	5
Glutamate	> >	5
Octopamin	> >	5
Proctolin	> >	5
5-HT	> >	5
ACH	> >	5
Membrane resistance	> >	5
Resting potential	> >	5
Heartbeat (cockroach)		
Frequency	> >	5

CONCLUSIONS

Circumstantial, pharmacological evidence suggests that the insect GABA receptors described so far may be linked closest to those GABA receptor types which are found in high density in the vertebrate cerebellum (bicucullin and SR 59 531 insensitive; Ro 05-4868 sensitive, zolpidem sensitive, Ro 15-4513 sensitive). It seems likely that GABA receptors of different subunit composition exist, both within the insect central nervous system and the periphery.

The apparent insensitivity of lepidopterous larvae to the GABA pharmacology suggests differing toxikokinetics and requires more detailed investigation.

The novel azole insecticide, SN 606 011, belongs to the most potent antagonists of GABA activated Cl⁻ current in insects.

The ubiquitous presence of glutamate activated Cl⁻ channels in insect nerve cell body membranes which are not found in the mammalian CNS raises questions about their function, their pharmacology and their vulnerability in toxicological terms.

Finally, we would like to share a few ideas and speculations.

Blockade or uninterrupted firing of nerve action potentials for many hours can be survived without serious, long-lasting lesions. Which processes in the cascade of events during intoxication are lethal? Could it be that in insects the peripheral, voluntary neuromuscular system is a suitable target for paralysing posture and locomotion only, but that it is much less suited as target for the

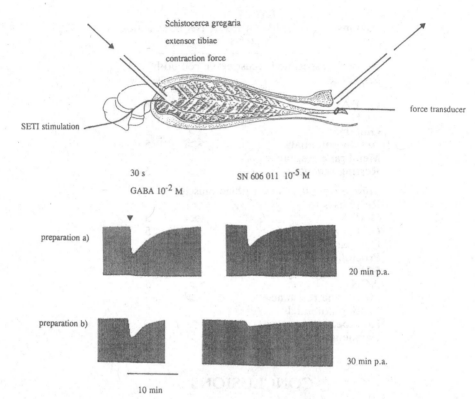

FIG. 15. Activity of SN 606011 (10^{-5} M) on the GABA evoked attenuation of the electrically induced contraction force in the locust extensor tibiae muscle. SETI (N3b) stimulation: 1·5 V, 0·8 ms, 50 Hz, 200 ms train duration, one train every 20 s.

disruption of life itself? And could it be that the neurosecretory system is a more relevant target in toxicological terms? Would it be possible to interfere with these homeostatic functions directly, leaving the components of membrane excitability undisturbed?

The selection of a lead structure with a high potential for success may not be trivial at all. Pharmacological potency alone certainly is not sufficient. Not only good neuropharmacology is required but enlightened insect toxicology as well. Could it be that one of the most important contributions of more detailed observations than screening data can provide lies in the precision with which error messages can be formulated? What is the key hurdle for the failure of this particular compound or idea? Every structure optimisation procedure is a trial and error endeavour which is exclusively guided by the relevance of the activity data that we are able to feed back. Trial and error is good as long as one is able to correct the error fast. The tip of the electrode is one superb and indispensable guide to identify precisely the hurdle one may have bumped against. This process of a first attempt, identifying and memorising the error

ind retrying a corrected version again is the operational principle which makes
even simple nervous systems so tremendously powerful. Neurobiologists call
such procedures orientation and learning. We should try to learn more from
he operational functioning of nervous systems.

ACKNOWLEDGEMENTS

We would like to thank in particular J. D. Turner for his interest, suggestions
ind help with regard to the benzodiazepine pharmacology. We thank for
echnical assistance D. Funke, B. Herz and M. Pöhlmann and I. Willecke for
>atient help with the manuscript.

REFERENCES

1. Matsumura, F. & Ghiasuddin, S. M., Evidence for similarities between cyclodiene
   type insecticides and picrotoxinin in their action mechanisms. *J. Environ. Sci.
   Health*, **18** (1983) 1–14.
2. Wafford, K. A., Sattelle, D. B., Gant, D. B., Eldefrawi, A. T. & Eldefrawi, M.
   E., Noncompetetive inhibition of GABA receptors in insect and vertebrate CNS by
   endrin and lindane. *Pestic. Biochem. Physiol.*, **33** (1989) 213–19.
3. Tanaka, K. & Matsumura, F., Action of avermectin B_{1a} on the leg muscles and the
   nervous system of the American cockroach. *Pestic. Biochem. Physiol.*, **24** (1985)
   124–35.
4. Abalis, I. M., Eldefrawi, A. T. & Eldefrawi, M. E., Actions of avermectin B_{1a} on
   the γ-aminobutyric acid$_A$ receptor and chloride channels in rat brain. *J. Biochem.
   Toxicol.*, **1** (1986) 69–82.
5. Bokisch, A. J. & Walker, R. J., The action of avermectin (MK 936) on identified
   central neurones from *Helix* and its interaction with acetylcholine and gamma-
   aminobutyric acid (GABA) responses. *Comp. Biochem. Physiol. (C)*, **84** (1986)
   119–25.
6. Hart, R. J., Lacey, G. & Wray, D. W., The postsynaptic actions of avermectin at
   locust thoracic ganglia. J. *Physiol. (London)*, **376** (1986) 43.
7. Lummis, S. C. R., GABA receptors in insects. *Comp. Biochem. Physiol. (C),
   Comp. Pharmacol. Toxicol.*, **95C** (1990) 1–8.
8. Sattelle, D. B., GABA receptors of insects. *Adv. Insect Physiol.*, **22** (1990) 1–113.
9. Cull Candy, S. G., Miniature and evoked inhibitory junctional currents and gamma
   aminobutyric-acid-activated current noise in locust *Schistocerca gregaria* muscle
   fibres. *J. Physiol. (London)*, **374** (1986) 179–200.
0. Scott, R. H. & Duce, I. R., Inhibition and gamma aminobutyric acid-induced
   conductance on locuse *Schistocerca gregaria* extensor tibiae muscle fibres. *J. Insect
   Physiol.*, **33** (1987) 183–90.
1. Bermudez, I. & Beadle, D. J., Electrophysiological and ultrastructural properties
   of cockroach (*Periplaneta americana*) myosacs. *J. Insect Physiol.*, **34** (1988) 881–9.
2. Murphy, V. F. & Wann, K. T., The action of GABA receptor agonists and
   antagonists on muscle membrane conductance in *Schistocerca gregaria*. *Br. J.
   Pharmacol.*, **95** (1988) 713–22.
3. Fraser, S. P., Djamgoz, M. B. A., Usherwood, P. N. R., O'Brien, J., Darlison, M.
   G. & Barnard, E. A., Amino acid receptors from insect muscle: electrophysiological characterisation in *Xenopus* oocytes following expression by injection of
   messenger RNA. *Mol. Brain. Res.*, **8** (1990) 331–42.

14. Benson, J. A., A novel GABA receptor in the heart of a primitive arthropod, *Limulus polyphemus. J. exp. Biol.,* **147** (1989) 421–38.
15. Watson, A. H. D. & Burrows, M., Immunocytochemical and pharmacological evidence for GABAergic spiking local interneurons in the locust. *J. Neurosci.,* **7** (1987) 1741–51.
16. Egelhaaf, M., Borst, A. & Pilz, B., The role of GABA in detecting visual motion. *Brain Research,* **509** (1990) 156–60.
17. Shivers, B. D., Killisch, I., Sprengel, R., Sontheimer, H., Köhler, M., Schofield, P.R. & Seeburg, P. H., Two novel GABA$_A$ receptor subunits exist in distinct neuronal subpopulations. *Neuron,* **3** (1989) 327–37.
18. Sieghart, W., Multiplicity of GABA$_A$-benzodiazepine receptors. *TiPS,* **10** (1989) 407–11.
19. Betz, H., Ligand-gated ion channels in the brain; the amino acid receptor superfamily. *Neuron,* **5** (1990) 383–92.
20. Blankenfeld, G. von, Ymer, S., Pritchett, D. B., Sontheimer, H., Ewert, M., Seeburg, P. H. & Kettenman, H., Differential benzodiazepine pharmacology of mammalian recombinant GABA$_A$ receptors. *Neurosci. Lett.,* **115** (1990) 269–273.
21. Draguhn, A., Verdoorn, T. A., Ewert, M., Seeburg, P. H. & Sakman, B., Functional and molecular distinction between recombinant rat GABA$_A$ receptor subtypes by Zn^{2+}. *Neuron,* **5** (1990) 781–8.
22. Olsen, R. W., McCabe, R. T. & Wamsley, J. K., GABA$_A$ receptor subtypes: autoradiographic comparison of GABA, benzodiazepine and convulsant binding sites in the rat central nervous system. *J. Chem. Neuroanat.,* **3** (1990) 59–76.
23. Verdoorn, T. A., Draguhn, A., Ymer, S., Seeburg, P. H. & Sakman, B., Functional properties of recombinant rat GABA$_A$ receptors depend upon subunit composition. *Neuron,* **4** (1990) 919–28.
24. Sigel, E., Baur, R., Trube, G., Möhler, H. & Malherbe, P., The effect of subunit composition of rat brain GABA$_A$ receptors on channel function. *Neuron,* **5** (1990) 703–11.
25. Lüddens, H. & Wisden, W., Function and pharmacology of multiple GABA$_A$ receptor subunits. *TiPS,* **12** (1991) 49–51.
26. Lees, G., Beadle, D. J., Neumann, R. & Benson, J. A., Responses to GABA by isolated insect neuronal somata: pharmacology and modulation by a benzodiazepine and a barbiturate. *Brain Res.,* **401** (1987) 267–78.
27. Benson, J. A., Transmitter receptors on insect neuronal somata: GABAergic and cholinergic pharmacology. In *Neurotox '88: Molecular Basis of Drug & Pesticide Action,* ed. G. G. Lunt. Elsevier Science Publishers BV (Biomedical Division), Amsterdam, 1988, pp. 193–206.
28. Lunt, G. G., Brown, M. C. S., Riley, K. & Rutherford, D. M., The biochemical characterisation of insect GABA receptors. In *Neurotox '88: Molecular Basis of Drug & Pesticide Action,* ed. G. G. Lunt. Elsevier Science Publishers BV (Biomedical Division), Amsterdam, 1988, pp. 185–92.
29. Eshleman, A. J. & Murray, T. F., Dependence on γ-aminobutyric acid of pyrethroid and 4'-chlorodiazepam modulation of the binding of t-[^{35}S]butylbicyclophosphorothionate in piscine brain. *Neuropharmacol.,* **29** (1990) 641–8.
30. Niddam, R., Dubois, A., Scatton, B., Arbilla, S. & Langer, S. Z., Autoradiographic localisation of [^{3}H]zolpidem binding sites in the rat CNS: comparison with the distribution of [^{3}H]flunitrazepam binding sites. *J. Neurochem.,* **49** (1987) 890–99.
31. Palmer, C. J., Smith, I. H., Moss, M. D. V. & Casida, J. E., 1-[4-[(trimethylsilyl)ethynyl]phenyl]-2,6,7,-trioxabicyclo[2.2.2]octanes: a novel type of selective proinsecticide. *J. Agric. Food Chem.,* **38** (1990) 1091–3.

8

Pharmacology of the *Ascaris* Nervous System

R. J. Walker, L. M. Colquhoun, H. R. Parri, R. G. Williams & L. Holden-Dye

Department of Physiology and Pharmacology, Bassett Crescent East, University of Southampton, Southampton SO9 3TU, UK

INTRODUCTION

Nematodes are members of a phylum which has made an important impact on medicine and agrochemistry. Out of a total of some 30 000 species described, about half are parasitic. Their occurrence world-wide is phenomenal: a quarter or more of the world's population are afflicted with nematode infections, they cause considerable losses in livestock, and the small plant parasitic nematodes are responsible for widespread crop damage.

In spite of the medical and economic importance of nematodes, until very recently, relatively little work had been undertaken on the basic physiology and pharmacology of their nervous systems. The bulk of the earlier evidence has come from biochemical, histochemical and immunocytochemical localization of putative transmitters and has been reviewed by Willett.[1] Basically acetylcholine (ACh), 4-aminobutyric acid (GABA) and certain amines, possibly including dopamine and 5-HT, may act as transmitters or modulators in this phylum. Much of our knowledge of nematode nervous systems relies on two 'model' systems; the free-living *Caenorhabditis elegans*, which can be readily cultured and has been used extensively in developmental studies, and the large parasitic nematode *Ascaris,* which has been used for functional studies on the peripheral and central nervous system. There is need for caution in this approach as there are known to be pharmacological differences between species of nematode. Nevertheless, nearly all the functional studies on nematodes to date have been performed on *Ascaris.* In the present paper we wish to highlight progress that has been made in the understanding of the functional and pharmacological properties of the peripheral and central nervous system of *Ascaris* over the last few years.

PERIPHERAL NERVOUS SYSTEM

Certain anthelmintics are known to have actions at peripheral neurotransmitter receptors in *Ascaris.* Levamisole, pyrantel and morantel have actions at the cholinoceptor and piperazine has actions at the GABA receptor. These actions may underly their anthelmintic potential. As well as being of interest from a chemotherapeutic viewpoint, the receptors are important phylogenetically, e.g. the nematodes are amongst the lowest of the phyla to use ACh as a transmitter and their properties are discussed in relation to similar receptors in insects and mammals.

The cholinergic system

In vertebrates there are at least three pharmacologically distinct types of nicotinic receptor differentially located in the nervous system at the neuromuscular junction, at ganglia and in the brain. Recent results from the cloning of nicotinic receptor subunits have suggested that even this estimate may be conservative, with up to seven different structures being reported for one of the nicotinic receptor subunits.[2] In this context we have elaborated upon the earlier observations that the excitatory receptor at the *Ascaris* neuromuscular junction is nicotinic in nature,[3-5] by using intracellular electrophysiological recording techniques to study the action of selective agonists and antagonists on the muscle bag cells. In particular, the actions of toxins have been studied as these are known to be potent and selective tools for the identification of nicotinic receptors in vertebrate preparations.[6]

In *Ascaris* muscle cells voltage clamped at resting membrane potential ACh causes an inward current.[7] Replacing extracellular Na^+ with glucosamine abolishes the depolarization (Colquhoun, L. M., Holden-Dye, L. & Walker, R. J., unpublished observations). The reversal potential has been estimated to be around -10 mV.[8] At the single channel level ACh (1–10 μM) activates cation selective channels.[9] The *Ascaris* ACh receptor thus resembles a nicotinic receptor-gated cation channel.

Nicotine itself mimicked the action of ACh on *Ascaris* muscle, though it was less potent than might be expected if it were acting at a typical nicotinic receptor (Table 1). Ganglionic nicotinic agonists, e.g. DMPP and HPPT (metahydroxyphenyl-propyltrimethylammonium), were by far the most potent agonists. Another ganglionic agonist, tetramethylammonium (TMA^+), was considerably less potent than ACh. Muscarinic agonists were very weak on this preparation.

A definitive range of ACh antagonists was tested on *Ascaris* (Table 2). What is most readily apparent from this table is that the *Ascaris* nicotinic receptor is not potently blocked by any single class of nicotinic receptor antagonist. Thus, mecamylamine, a ganglionic antagonist, and benzoquinonium, a neuromuscular junction blocker, are the two most potent antagonists with IC_{50}s in the high nanomolar range. The mechanism of action of mecamylamine would be of interest as it has been shown to block at the vertebrate neuromuscular junction in a voltage-dependent manner.[10] The classical ganglion blocker, hexamethonium, had very weak antagonist activity. Decamethonium is a depolarizing blocking agent at the vertebrate neuromuscular junction, and although not

TABLE 1

The Relative Potency of Agonists for the *Ascaris* ACh Receptor Compared to ACh

Agonist	*Relative potency*	
	Receptor site of action	*Depolarization*
HPPT	Nicotinic, ganglia	6·5
DMPP	Nicotinic, ganglia	3
ACh		1
Carbachol	Muscarinic and nicotinic	0·4
Nicotine	Nicotinic	0·2
Anatoxin	Nicotinic, neuronal	0·15
TMA$^+$	Nicotinic, ganglia	0·04
Suberyldicholine	Nicotinic	0·05
Muscarone	Muscarinic > nicotinic	0·01
Furtrethonium	Muscarinic	0·007
Cytisine	Nicotinic	0·002
Arecoline	Muscarinic	0·001
NMTHT	Insect neuronal ACh	No effect
McN-A-343	Muscarinic, ganglia	Weak depolarization
Bethanechol	Muscarinic	No effect
Methacholine	Muscarinic	No effect
Pilocarpine	Muscarinic	Weak hyperpolarization
Muscarine	Muscarinic	Weak hyperpolarization
Oxotremorine	Muscarinic	Without effect

The relative potency for each cell was determined by taking the ratio of the concentration of ACh to the concentration of agonist that produced the same depolarization, from parallel portions of the dose–response curve (therefore a relative potency >1 indicates a compound more potent than ACh). The compounds are described in terms of their most potent actions on cholinoceptors in vertebrates unless stated otherwise. The results are the mean of three to six determinations. HPPT, metahydroxy-phenylpropyltrimethylammonium; DMPP, 1,1-dimethyl-4-phenylpiperazinium; TMA, tetramethylammonium; NMTHT, 2(nitromethylene)tetrahydro 1,3,-thiazine.

very potent, decamethonium did depolarize and block the ACh response in *Ascaris* muscle in a dose-dependent manner. Two other neuromuscular junction blockers, tubocurare and pancuronium both blocked at low micromolar concentrations. Another vertebrate nicotinic receptor antagonist, dihydro-ß-erythroidine was devoid of activity at the *Ascaris* ACh receptor. A relatively potent antagonist, quinacrine, belongs to a group of non-competitive inhibitors (NCI) that have been shown to block the nicotinic-gated cation channel in vertebrates, including *Torpedo*.[11]

Overall the ACh receptor on *Ascaris* muscle cannot be readily classified in terms of the mammalian classification, although the potent action of the ganglionic agonists lends support to the idea that it may be similar to a vertebrate neuronal rather than muscle nicotinic receptor. This contention is further supported by the differential effects of two toxins that have been shown

TABLE 2
Antagonists at the Ascaris ACh Receptor

Drug	Description	IC_{50} (μM)
Benzoquinonium	Neuromuscular blocker	0·46
Mecamylamine	Ganglion > neuromuscular blocker	0·41
Histrionicotoxin	Blocks nicotinic channel	0·43
N-methyl-lycaconitine	Insect and neuromuscular blocker	0·65 ± 0·12
α-Bungarotoxin	Very potent neuromuscular blocker	2·2
Quinacrine	Blocks nicotinic channel	2·05[a]
Tubocurare	Neuromuscular > ganglion blocker	5·26
Pancuronium	Neuromuscular ≫ ganglion blocker	4·49
Trimethaphan	Ganglion blocker	17
Chlorisondamine	Ganglion blocker	65
Atropine	Muscarinic > ganglion blocker	52
Decamethonium	Neuromuscular blocker	96
Hexamethonium	Ganglion blocker at channel site	329
Dihydro-β-erythroidine	Neuronal nicotinic blocker	>100

The IC_{50} is the concentration of antagonist that was required to reduce the depolarization to a submaximal concentration of ACh by 50%. Values are the mean for three to four experiments. [a]The IC_{50} for this compound was determined from the block of the current elicited by ACh in voltage-clamped cells. The antagonists are described in terms of their most potent actions on vertebrate cholinoceptors, unless stated otherwise.

to discriminate between the muscle and neuronal subtypes of nicotinic receptor. α-Bungarotoxin is a highly selective and potent antagonist of the vertebrate muscle nicotinic recptor[12] but only blocks the Ascaris response by about 50% at 1 μM. On the other hand, neosurugatoxin is a potent and selective antagonist of the neuronal subtypes of nicotinic receptor[6] and this agent blocks the Ascaris response with an IC_{50} of 0·17 μM (Fig. 1).

The Ascaris peripheral nicotinic receptor differs in some important aspects from those studied in detail so far in insects, e.g. the fast coxal depressor motoneurone (D_f) of the cockroach, Periplaneta americana (Fig. 2). Thus, although these receptors share an inability to discriminate well between curare and atropine, and in common with many other invertebrate nicotinic receptors, they are not blocked by hexamethonium, the Ascaris receptor is potently activated by DMPP whereas the cockroach receptor is not.[13] Another agonist which has been shown to be active on insect central nicotinic receptors, but inactive on Ascaris muscle, is 2-(nitromethylene)tetrahydro 1,3,-thiazine (NMTHT[14]). Furthermore, dihydro-β-erythroidine is a potent antagonist of the somatic ACh receptor on the cockroach D_f neurone[13] and yet it was without effect on the Ascaris receptor. The receptor on the D_f neurone, unlike the Ascaris receptor, is α-bungarotoxin sensitive and in this respect resembles the vertebrate muscle receptor. Another insect nicotinic neuronal receptor, the receptor on the dorsal-unpaired median (DUM) cells, is less sensitive to α-bungarotoxin and it would be interesting to be able to make a comparison

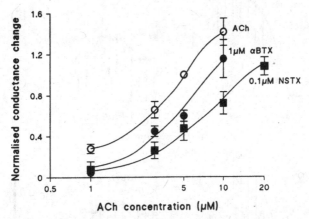

FIG. 1. The potency of two selective toxins on the *Ascaris* ACh receptor. The toxins were applied for 3 min prior to, and concurrent with, application of ACh. The responses are normalized with respect to the control response to 5 μM ACh. Control response to ACh (O; $N = 8 \pm$ SEM). Response to ACh in the presence of the neuromuscular junction nicotinic receptor antagonist α-bungarotoxin (α-BTX ●: 1 μM; $N = 3 \pm$ SEM). The antagonism by α-bungarotoxin was not increased after a 1-h exposure of the preparation to the toxin and was not reversed after a 1-h wash. Response to ACh in the presence of the selective neuronal nicotinic receptor antagonist neosurugatoxin (NSTX ■; 0·1 μM; $N = 5 \pm$ SEM). The antagonism was reversed following a 30-min wash period.

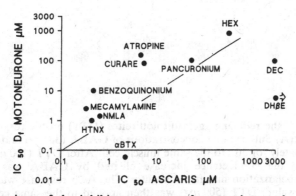

FIG. 2. A comparison of the inhibitory potency of antagonists at the *Ascaris* muscle cholinoceptor and the cholinoceptor on the D_f motoneurone of the cockroach *Periplaneta americana*. The IC_{50} is the concentration of antagonist required to reduce the response to either, bath applied ACh (5–10 μM) for *Ascaris*, or iontophoretically applied ACh for *Periplaneta*, by 50%.[13] The line indicates the position of points for a direct, positive correlation. α-BTX, α-bungarotoxin; HTNX, histrionicotoxin; NMLA, *n*-methyllycaconitine; HEX, hexamethonium; DEC, decamethonium; DHβE, dihydro-β-erthyoidine.

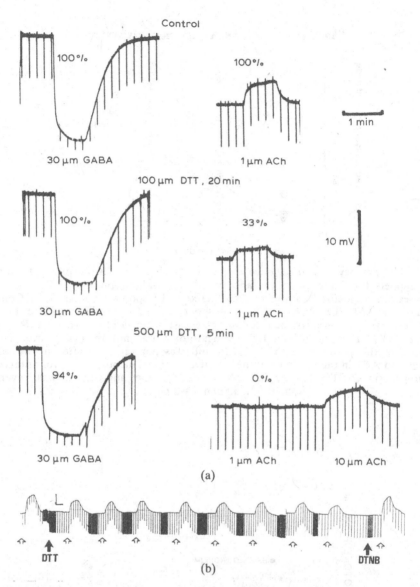

FIG. 3. The effect of the reducing agent dithiothreitol (DTT) on the response to ACh on *Ascaris* muscle. (A) Submaximal concentrations of GABA (30 μM) and ACh (1 μM) were applied alternately for 30 s to the same muscle cell. After DTT (100 μM) had been applied for 20 min the conductance increase elicited by GABA was unchanged, although the hyperpolarization had faded slightly. The ACh response was reduced to 33% of its control value. DTT (500 μM) was then applied to the same cell for 5 min. This resulted in a small decrease in the response to GABA whilst the response to ACh (1 μM) was negligible. Increasing the concentration of ACh to 10 μM elicited a small response. (B) An example of consecutive recordings from the same cell indicating the time course of the block of the ACh response by DTT (100 μM added at the first solid arrow). Applications of ACh (10 μM) are indicated by open arrows. Application of the oxidizing agent dithiobis-(2-nitrobenzoate) (1 mM; DTNB indicated by the second solid arrow) for 5 min reversed the block and restored the response to ACh. A similar effect was seen in four other preparations. The vertical scale bar is 10 mV; the horizontal scale bar is 30 s.

between the pharmacology of these receptors and *Ascaris* muscle nicotinic receptors.

Structural information on the *Ascaris* nicotinic receptor remains to be obtained. We have observed in *Ascaris* that the ACh, but not the GABA, response is sensitive to the action of the disulphide bond reducing agent, dithiothreitol, and that this effect is reversed completely by the oxidising agent dithiobis-(2-nitrobenzoate) (DTNB) (Fig. 3.). All nicotinic receptor α-subunits cloned and sequenced to date, including an insect α-subunit,[15] contain two definitive cysteine residues, which have been localized at the active site of the receptor.[16] Further work would be required to determine if this is the site of action of DTT on the *Ascaris* receptor. Nonetheless, this observation, together with the pharmacological evidence, suggests that there may well be a strong structural similarity between the *Ascaris* receptor and its vertebrate counterparts. The implication is that the nicotinic receptor has been remarkably conserved through the course of evolution.

The GABA system

GABA ($>1\,\mu M$) elicited a hyperpolarization accompanied by a conductance increase. Analysis of the dose–response relationship supports the hypothesis that two molecules of GABA are required for receptor activation, indicative of the positive cooperativity that seems to be a constant feature of ligand-gated receptors. The reversal potential for the response is near to the equilibrium potential for chloride ions indicating that the receptor gates a chloride channel.[17,18] The selectivity of this channel indicates that it is like most other

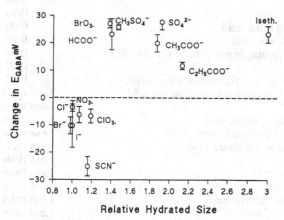

FIG. 4. The changes in the reversal potential for the GABA response (E_{GABA}) when the ion indicated is substituted for chloride. Negative shifts relative to chloride indicate anions more permeant than chloride through the GABA–chloride channel, while positive shifts indicate anions that are less permeant. The theoretical shift for a totally impermeant anion is +26 mV. Points are the mean of four to five determinations ±SEM. Reproduced from Parri *et al.*[19] with permission of *Experimental Physiology*.

GABA-operated chloride channels found in vertebrates with a cut-off diameter of between 0·29 and 0·33 nm[19] (Fig. 4). The GABA-A mimetics, muscimol, isoguvacine and imidazole acetic acid, also hyperpolarized the muscle though less potently than GABA. The GABA-B selective compound, baclofen, was inactive whilst another, new, GABA-B agonist 3-aminopropylphosphonic acid (CGA 147 823;[20,21]), was a very weak GABA-mimetic. From a range of agonists tested only two were found to be more potent than GABA itself, these were (S)-(+)-dihydromuscimol which was more than 5 times more potent and ZAPA ((Z)-3-[(aminoiminomethyl)-thio]-2-propenoic acid hydrochloride) which was 1·5 times more potent (Table 3). An analogue of ZAPA,

TABLE 3
Agonists at the *Ascaris* GABA Receptor

Drug	Relative potency
(S)-(+)-dihydromuscimol	7·53
ZAPA	1·87
GABA	1
(R)-(−)-Dihydromuscimol	0·85
trans-Aminocrotonic acid	0·55
Muscimol	0·42
(R)-(−)-3-OH-GABA	0·25
Isoguvacine	0·20
Imidazole acetic acid	0·20
cis-β-Carboxyvinylisoselenourea (CVI)	0·19
(S)-(+)-3-OH-GABA	0·13
δ-Aminovaleric acid	0·16
THIP	0·06
Guanidoacetic acid	0·16
β-Guanidopropionic acid	0·09
cis-Aminocrotonic acid	0·03
3-Aminopropylphosphonic acid (CGA 147 823)	0·03
Kojic amine	0·02
(S)-(−)-4-methyl-*trans*-aminocrotonic acid	0·01
(R)-(+)-4-methyl-*trans*-aminocrotonic acid	Less than 0·01
Thiomuscimol	Less than 0·01
Piperidine-4-suplphonic acid	Less than 0·01
Dihydropiperidine-4-sulphonic acid	Less than 0·01
3-Aminopropanesulphonic acid	Less than 0·01
4-Piol	Less than 0·01
5'Ethylmuscimol	Less than 0·01
Propylmuscimol	Less than 0·01
3-Thioimidazol-Acrylic acid	Less than 0·01
β-Alanine	Less than 0·01
Glycine	Less than 0·01
Taurine	Less than 0·01

The relative potency was determined from the ratio of the concentration of GABA to the concentration of agonist that produced equivalent responses. The results are the mean from 3 or more determinations.

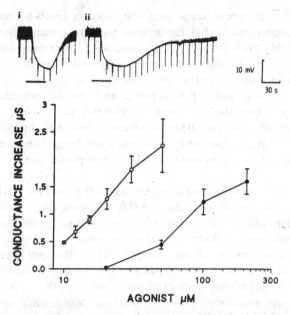

FIG. 5. *Cis-β*-carboxyvinylisoselenourea (CVI) is a moderately potent GABA-mimetic on *Ascaris* muscle cells. The conductance increase was elicited by bath applied agonist. $N = 3 \pm SEM$. The inset shows a sample response to GABA (20 µM) and CVI (200 µM). GABA ○; CVI ●. The bar indicates the duration of application of the drug. The current pulses causing the downward deflections were 20 nA.

cis-β-carboxyvinylisoselenourea (CVI), was also a quite potent GABA-mimetic (Fig. 5).

In spite of the inactivity of the sulphonic acid derivatives, the relative potencies of agonists on *Ascaris* muscle and their affinities for the GABA-A receptor binding site are correlated in a significant positive fashion.[18] Furthermore, as the receptor is a GABA-gated chloride channel and exhibits positive cooperativity it has many of the features of the mammalian GABA-A receptor. Here the similarity ends, however, as it is insensitive to bicuculline the competitive antagonist at the mammalian GABA-A receptor.[22] This is perhaps not too surprising as in other invertebrates, e.g. insects, bicuculline has little or no action.[23–25] The unusual feature of the *Ascaris* receptor is that all the known non-competitive antagonists of mammalian GABA-A receptors are either inactive or very weak antagonists receptor including picrotoxin (100 µM), TBPS (10 µM; *t*-butylbicyclophosphorothionate) and dieldrin (100 µM). This is in marked contrast even to the situation, reported so far, in insects where the chloride mediated GABA responses on neurones and muscle are sensitive to picrotoxin.[26–30] Other GABA-A antagonists such as pitrazepin, RU5135 and the pyridazinyl-GABA derivative SR95531 (2-[carboxy-3'-propyl]-3-amino-6-parametboxy-phenyl-pyridazinium-Br) and SR95103 also failed significantly to reduce *Ascaris* GABA responses except at high concentrations.[18,31,32] It would seem that the *Ascaris* GABA receptor recognizes none of the compounds that have been used as potent pharmacological tools for the identification of GABA-A mediated responses in mammalian systems. The modulatory sites frequently associated with the GABA-A receptor in the mammalian central nervous system are also not present in *Ascaris* as the benzodiazepine, flurazepam, and the barbiturate, pentobarbitone, do not modulate the response to GABA in *Ascaris*. However, a comprehensive list of possible benzodiazepine receptor ligands, including peripheral benzodiazepine type compounds, has not been tested in *Ascaris*. By comparison, some insect neuronal GABA receptors are modulated by flurazepam.[28]

Avermectins are the most potent antagonists of the *Ascaris* GABA receptor found to date. Ivermectin blocks the GABA response in a non-competitive manner.[28] The lipophilicity of avermectin compounds mean that the determination of an accurate affinity for the receptor is complicated by the time-dependent accumulation of the compounds in the cell membrane. We have estimated that ivermectin blocks the GABA response with an affinity in the low micromolar range after a short application (less than 1 h). Whether or not this underlies the anthelmintic action of these compounds, or contributes to it, remains the subject of debate.[33]

A COMPARISON OF THE *ASCARIS* GABA AND ACh RECEPTORS

Recently it has been shown that there is significant sequence homology between members of the ligand-gated receptor channel family.[34] It has been suggested that this is reflected in the ability of antagonists to cross-react with different receptors of the same receptor family.[25] Both the GABA and ACh

TABLE 4

A Comparison of the Ability of Antagonists to Block the *Ascaris* GABA and ACh Receptors on the Somatic Muscle Cells. The IC_{50} is the Concentration of Antagonist Required to Reduce the Response to 50% of its Control Value. Values are the Mean ± SEM for (*N*) Determinations

Drug	$IC_{50}(\mu M)$		
	GABA	*ACh*	
Bicuculline	>1 000	105 ± 65	(4)
Picrotoxin	428 ± 58 (5)	152 ± 24	(4)
Strychnine	326 ± 51 (4)	1·2 ± 0·29	(4)
Tetraphenylphosphonium	>100	1·9 ± 0·3	(3)
Curare	>100	3·1 ± 0·4	(3)
MK-801	Not tested	>100	

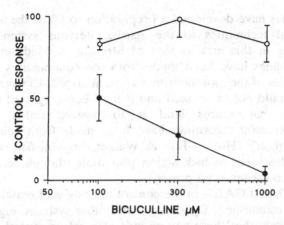

FIG. 6. Bicuculline, a GABA$_A$ receptor antagonist, is virtually ineffective at blocking the GABA response in *Ascaris,* however it is a moderately potent antagonist of the ACh response. Control responses were to (●) 10 μM ACh and (○) 30 μM GABA in the absence of bicuculline, $N = 3$–4, vertical bars give SEM.

receptor on *Ascaris* muscle cells appear to be ligand-gated receptor channels. As these receptors are located on the same cells we took the opportunity to compare the action of some classical ligand-gated receptor antagonists at these two sites (Table 4; Fig. 6). Remarkably, although the GABA receptor is not blocked by bicuculline even at high concentrations, bicuculline is a moderate antagonist of the ACh receptor. This is similar to the observation made on dissociated locust neurones by Benson.[25] More recently, this phenomenon has been looked for on mammalian intermediate pituitary lobe cells.[35] Picrotoxin is also a better antagonist of the ACh receptor than the GABA receptor in *Ascaris,* though the difference is not as marked as for bicuculline. Strychnine, known as a selective glycine antagonist, was also a good antagonist of the *Ascaris* ACh receptor. This compound is known to block at a number of ACh receptor sites including nicotinic receptors at vertebrate ganglia and on *Aplysia* neurones.[36] Tetraphenylphosphonium (TPP) is a non-competitive inhibitor that blocks both nicotinic receptors and GABA-stimulated chloride flux in mammalian brain.[37] Although this compound was a potent antagonist of the ACh response in *Ascaris* it was without effect on the GABA response at the concentrations tested. The non-competitive NMDA (N-methyl-D-aspartate) receptor antagonist, MK-801, has been reported to be an antagonist at vertebrate nicotinic receptors[38] but it did not appear to have an action at the *Ascaris* nicotinic receptor.

CENTRAL NERVOUS SYSTEM

Very little is known of the electrophysiological properties of *Ascaris* central neurones. However, significant progress has been made over the last few years and two laboratories have developed a preparation to allow the application of electrophysiological techniques to the *Ascaris* nervous system. The main laboratory working in this area is that of Stretton in Madison, Wisconsin. Intracellular recordings have been made from the commissures enabling the bioelectric properties of the motoneurones to be analysed. Classical all or none action potentials could not be evoked and it has been suggested that the high resistance of the commisures lend it to passive rather than active transmission.[39] Successful recordings have been made from neurones in the retrovesicular ganglion[40] (Holden-Dye & Walker, unpublished observations). These neurones also seem to lack action potentials, though certain types of oscillatory activity do seem to be present.

The role for ACh and GABA in the control of body wall muscle activity has been more or less established. By analogy with other systems, e.g. insects and gastropods, it is likely that these compounds also act as central transmitters, although the pharmacological profiles of these central receptors may differ from those on the body wall muscles. Two GABA monoclonal antibodies, as well as staining the inhibitory motoneurones and their commissures, stain four neurones in the nerve ring and three or four pairs of neurones in the ventral and lateral ganglia.[41] The GABAergic synapses in the central nervous system remain to be identified and no doubt will be of considerable interest as they represent a proposed site of action of avermectins.[42] Cholinergic neurones in the rostral ganglia have yet to be identified by immunocytochemistry however, ACh (1–30 μM) depolarizes neurones in the retrovesicular ganglion and it is most likely that ACh is a central transmitter.

Other classical transmitters are present in nematodes but their functional significance remains to be determined. Monoamines are present in nematode tissues;[1,43,44] dopamine-containing neurones have been identified in the nervous system of *Caenorhabditis*. There is also some evidence for the presence of 5-HT in nematodes though certain studies have failed to identify it.[1] The possible role of excitatory amino acid transmitters, such as glutamate and aspartate, in nematodes has yet to be examined. Information on this would be of interest in the light of the anthelmintic action of the excitatory amino acid receptor agonist, kainic acid.

Recently there have been several studies reporting the presence of peptides in nematode nervous tissue using immunocytochemical techniques, reviewed by Stretton.[45] Davenport et al.[46] found evidence for a FMRFamide-like peptide in *Ascaris* while Leach et al.[47] obtained immunoreactivity aganst ACTH and FMRFamide in *Goodeyus*. Extending this approach Cowden et al.[48] have sequenced a heptapeptide from *Ascaris* with the following amino acids; Lys-Asn-Glu-Phe-Ile-Arg-Phe-NH$_2$ (KNEFIRFamide or AF1). When applied at nanomolar concentrations this reversibly abolishes slow membrane potential oscillations of inhibitory motoneurones.[48] In addition to the heptapeptide several other, possibly related peptides are present in *Ascaris,* e.g. AF2 (KHEYLRF-NH$_2$) has been identified by Stretton's group. There is strong FMRF-NH$_2$-like immunoreactivity in *Ascaris* central nervous system

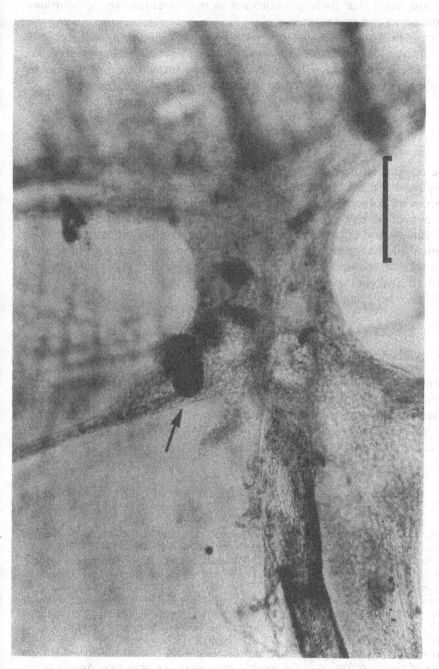

FIG. 7. Positive immunoreactivity to calcitonin-gene related peptide (CGRP) antisera in neurones of the ventral ganglion of *Ascaris*. The scale bar is 100 μm.

and it is most likely that these peptides act as neurotransmitters or neuromodulators on nerve and muscle cells.

The *Ascaris* nervous system is apparently rich in neuropeptide content with several neuropeptides being localized in the rostral ganglia (Holden-Dye, L., Williams, R. G., Spacyzski, R. & Walker, R. J., unpublished observations; Fig. 7) and some being colocalized to the same neurone.[49]

SUMMARY

Given the resources, there is no reason why the next 5 years or so should not see a rapid development in our knowledge of the functional and pharmacological properties of *Ascaris* nervous system. The opportunity now exists to apply standard neurobiological techniques to this preparation. This will be of immense value—not only to enhance our understanding of currently used anthelmintics and to further the development of new ones, but also in a much wider context to contribute to our understanding of the phylogeny of neurotransmitter receptor systems.

ACKNOWLEDGEMENTS

We are grateful to the SERC Invertebrate Neuroscience Initiative and the Jersey Government for financial support. Neosurugatoxin was a generous gift from Professor T. Kosuge. *Cis-β*-carboxyvinylisoselenourea and 3-thioimidazol-acrylic acid were kindly made for us by Professor R. Cookson. CGA 147 823 was a gift from Dr J. A. Benson. We also thank Professor G. J. Dockray for the gift of CGRP antibody (L271).

REFERENCES

1. Willett, J. D., Control mechanisms in nematodes. In *Nematodes as Biological Models*, ed. Zukerman. Academic Press, New York, 1980, pp. 197–225.
2. Deneris, E. S., Connolly J., Rogers, S. W. & Duvoisin, R., Pharmacological and functional diversity of neuronal nicotinic acetylcholine receptors. *Trends in Pharmacological Sci.*, 12 (1991) 34–40.
3. Baldwin, E. & Moyle, V., A contribution to the physiology and pharmacology of *Ascaris lumbricoides* from the pig. *Br. J. Pharmac.*, 4 (1949) 145–52.
4. Natoff, I. L., The pharmacology of the cholinoceptor in muscle preparations of *Ascaris lumbricoides*. *Br. J. Pharmac.*, 37 (1969) 251–7.
5. Rozhkova, E. K., Malyutina, T. A. & Shishov, B. A., Pharmacological characteristics of cholinoreception in the somatic muscles of the nematode *Ascaris suum*. *Gen. Pharmac.*, 11 (1980) 141–6.
6. Luetje, C. W., Wada, K., Rogers, S., Abramson, S. N., Tsuji, K., Heinemann, S. & Patrick, J., Neurotoxins distinguish between different neuronal nicotinic acetylcholine receptor subunit combinations. *J. Neurochem.*, 55 (1990) 632–40.

7. Colquhoun, L. M., Holden-Dye, L. & Walker, R. J., The pharmacology of cholinoceptors on the somatic muscle cells of the parasitic nematode *Ascaris suum*. *J. exp. Biol.*, **158** (1991) 509–30.
8. Harrow, I. D. & Gration, K. A. F., Mode of action of the anthelmintics morantel, pyrantel and levamisole on muscle cell membrane of the nematode *Ascaris suum*. *Pesticide Sci.*, **16** (1985) 662–72.
9. Pennington, A. J. & Martin, R. J., A patch-clamp study of acetylcholine-activated ion channels in *Ascaris suum* muscle. *J. exp. Biol.*, **154** (1990) 201–21.
10. Varanda, W. A., Arcava, Y., Sherby, S. M., Vanmeter, W. G., Eldefrawi, M. E. & Albuquerque, E. X., The acetylcholine receptor of the neuromuscular junction recognizes mecamylamine as a noncompetitive antagonist. *Molec. Pharmac.*, **28** (1985) 128–37.
11. Spivak, C. E. & Albuquerque, E. X., Trimethylphenylphosphonium blocks the nicotinic acetylcholine receptor noncompetitively. *Molec. Pharmac.*, **27** (1985) 246–55.
12. Chiappinelli, V. A., Actions of snake venom toxins on neuronal nicotinic receptors and other neuronal receptors. *Pharmac. Therap.*, **31** (1985) 1–32.
13. David, J. A. & Sattelle, D. B., Actions of cholinergic pharmacological agents on the cell body membrane of the fast coxal depressor motoneurone of the cockroach (*Periplaneta americana*). *J. exp. Biol.*, **108** (1984) 119–36.
14. Buckingham, S. B., Harrison, J. B., Leech, C. A. & Sattelle, D. B., Nicotinic acetylcholine receptor activation in dissociated neurones from *Periplaneta americana* and *Musca domestica*. *J. Physiol.* (*Lond.*), **418** (1989) 193.
15. Marshall, J., Buckingham, S. D., Shingai, R., Lunt, G. G., Goosey, M. W., Darlison, M. G., Sattelle, D. B. & Barnard, E. A., Sequence and functional expression of a single α subunit of an insect nicotinic acetylcholine receptor. *EMBO J.*, **9** (1990) 4391–8.
16. Kao, P., Dwork, A. & Kaldany, R., Identification of the α subunit half cysteine specifically labelled by an affinity reagent for the acetylcholine receptor binding site. *J. Biol. Chem.*, **259** (1984) 11662–5.
17. Martin, R. J., The effect of 4-aminobutyric acid on the input conductance and membrane potential of *Ascaris* muscle. *Br. J. Pharmac.*, **71** (1980) 99–106.
18. Holden-Dye, L., Krogsgaard-Larsen, P., Nielsen, L. & Walker, R. J., GABA receptors on the somatic muscle cells of the parasitic nematode *Ascaris suum*: Stereoselectivity indicates similarity to a GABA-A type agonist recognition site. *Br. J. Pharmac.*, **98** (1989) 841–51.
19. Parri, H. R., Holden-Dye, L. & Walker, R. J., Studies on the ionic selectivity of the GABA operated chloride channel on the somatic muscle bag cells of the parasitic nematode *Ascaris suum*. *Exp. Physiol.*, **76** (1991) 597–606.
20. Holden-Dye, L., Hewitt, G. M., Wann, K. T., Krogsgaard-Larsen, P. & Walker, R. J., Studies involving avermectin and the 4-aminobutyric acid (GABA) receptor of *Ascaris suum* muscle. *Pestic. Sci.*, **24** (1988) 231–45.
21. Benson, J. A., A novel GABA receptor in the heart of a primitive arthropod. *J. exp. Biol.*, **147** (1989) 421–38.
22. Curtis, D. R., Duggan, A. W., Felix, D. & Johnston, G. A. R., GABA, bicuculline and central inhibition. *Nature* (*Lond.*), **226** (1970) 1222–4.
23. Scott, R. H. & Duce, I. R., Pharmacology of GABA receptors on skeletal muscle fibres of the locust (*Schistocerca gregaria*). *Comp. Biochem. Physiol.*, **86C** (1987) 305–11.
24. Sattelle D. B., Pinnock, R. D., Wafford, K. A. & David J. A., GABA receptors on the cell body membrane of an identified insect motoneurone. *Proc. Roy. Soc. Lond. B*, **232** (1988) 443–56.
25. Benson, J. A., Transmitter receptors on insect neuronal somata: GABAergic and cholinergic pharmacology. In *Molecular Basis of Drug and Pesticide Action*, ed. G.G. Lunt. Elsevier Science Publishers BV, Amsterdam, pp. 193–206.

26. Walker, R. J., Crossman, A. R., Woodruff, G. N. & Kerkut, G. A., The effect of bicuculline on the gamma-amino butyric acid (GABA) receptor of neurones of *Periplaneta americana* and *Helix aspersa. Brain Res.*, **33** (1971) 75–82.

27. Wafford, K. A. & Sattelle, D. B., Effects of amino acid neurotransmitter candidates on an identified insect motoneurone. *Neurosci. Lett.*, **63** (1987) 135–40.

28. Lees, G., Beadle, D. J., Neumann, R. & Benson, J. A., Responses to GABA by isolated insect neuronal somata: pharmacology and modulation by a benzodiazepine and a barbiturate. *Brain Res.*, **401** (1987) 267–78.

29. Murphy, V. F. & Wann, K. T., The action of GABA receptor agonists and antagonists on muscle membrane conductance in *Schistocerca gregaria. Br. J. Pharmac.*, **95** (1988) 713–22.

30. Usherwood, P. N. R. & Grundfest, H., Peripheral inhibition in skeletal muscle of insects. *J. Neurophysiol.*, **28** (1965) 497–518.

31. Duittoz, A. H. & Martin, R. J., SR95103 acts as a GABA antagonist in *Ascaris suum* muscle. *Br. J. Pharmac.*, **97** (1989) 490.

32. Martin, R. J., Pennington, A. J., Duittoz, A. H., Robertson, S. & Kusel, J. R., The physiology and pharmacology of neuromuscular transmission in the nematode parasite *Ascaris suum. Parasitol.*, **102** (1991) S41–S58.

33. Holden-Dye, L. & Walker, R. J., Avermectin and avermectin derivatives are antagonists at the 4-aminobutyric (GABA) receptor on the somatic muscle cells of *Ascaris*; is this the site of anthelmintic action? *Parasitol.*, **101** (1990) 265–71.

34. Schofield, P. R., Darlison, M. G., Fujita, N., Burt, D. R., Stephenson, F. A., Rodriquez, H., Rhee, L. M., Ramachandran, J., Reale, V., Glencorse, T. A., Seeburg, P. H. & Barnard, E. A., Sequence and functional expression of the GABA-A receptor shows a ligand-gated receptor super-family. *Nature Lond.*, **328** (1987) 211–27.

35. Zhang, Z.-W. & Feltz, P., Bicuculline blocks nicotinic acetylcholine response in isolated intermediate lobe cells of the pig. *Br. J. Pharmac.*, **102** (1991) 19–22.

36. Slater, N. T. & Carpenter, D. O., Blockade of acetylcholine induced inward current in *Aplysia* neurons by strychnine and desiprimine: Effect of membrane potential. *Cell. Mol. Neurobiol.*, **2** (1982) 53–8.

37. Schwartz, R. D. & Mindlin, M. C., Inhibition of the GABA receptor-gated chloride ion channel in brain by noncompetitive inhibitors of the nicotinic receptor-gated cation channel. *J. Pharmacol. exp. Therap.*, **244** (1988) 963–70.

38. Ramoa, A. S., Alkondon, M., Arcava, Y., Irons, J., Lunt, G. G., Deshpande, S. S., Wonnacott, S. S., Aronstram, R. S. & Albuquerque, E. X., The anticonvulsant MK-801 interacts with peripheral and central nicotinic acetylcholine receptor ion channels. *J. Pharmac. exp. Therap.*, **254** (1990) 71–82.

39. Davis, R. E. & Stretton, A. O. W., Passive membrane properties of motoneurons and their role in long distance signalling in the nematode *Ascaris. J. Neurosci.*, **9** (1989) 403–14.

40. Angstadt, J. D., Donmoyer, J. E. & Stetton, A. O. W., The retrovesicular ganglion of the nematode *Ascaris. J. Comp. Neurol.*, **284** (1989) 374–88.

41. Sithigorngul, P., Cowden, C., Guastella, J. & Stretton, A. O. W., Generation of monoclonal antibodies against a nematode peptide extract: another approach for identifying unknown peptides. *J. Comp. Neurol.*, **284** (1989) 389–97.

42. Kass, I. S., Wang, C. C., Walrond, J. P. & Stretton, A. O. W., Avermectin B_{1a}: A paralyzing anthelmintic that affects interneurons and inhibitory motorneurons in *Ascaris. Proc. Natl. Acad. Sci. USA*, **77** (1980) 6211–15.

43. Goh, S. L. & Davey, K. G., Localization and distribution of catecholaminergic structures in the nervous system of *Phocanema decipiens* (Nematoda). *Int. J. Parasitol.*, **6** (1976) 403–11.

44. Klemm, N., The distribution of biogenic monoamines in invertebrates. In *Neurobiology: Current Comparative Approaches*, ed. R. Gilles & J. Balthazart. Springer-Verlag, Berlin, 1985, pp. 280–96.

45. Stretton, A. O. W., Cowden, C., Sithigorngul, P. & Davis, R. E., Neuropeptides in the nematode *Ascaris suum. Parasitol.*, **102** (1991) S107–16.
46. Davenport, T. R. B., Lee, D. L. & Isaac, R. E., Immunocytochemical demonstration of a neuropeptide in *Ascaris suum* (Nematoda) using an antiserum to FMRFamide. *Parasitol.*, **97** (1988) 81–8.
47. Leach, L., Trudgell, D. L. & Gahan, P. B., Immunocytochemical localization of neurosecretory amines and peptides in the free living nematode *Godeyus ulmi. Histochem. J.*, **19** (1987) 471–5.
48. Cowden, C., Stretton, A. O. W. & Davis, R. E., AF1 a sequenced bioactive neuropeptide isolated from the nematode *Ascaris suum. Neuron,* **2** (1989) 1465–73.
49. Sithigorngul, P., Stretton, A. O. W. & Cowden, C., Neuropeptide diversity in *Ascaris*: an immunocytochemical study. *J. Comp. Neurol.*, **294** (1990) 362–76.

9

Chemical Intercellular Signalling Mechanisms in the Nervous System of the Nematode *Ascaris suum*: Potential Sites of Actions of New Generations of Anthelmintic Drugs

ANTONY O. W. STRETTON

Department of Zoology and Neurosciences Training Program, University of Wisconsin, Madison, Wisconsin 53706, USA

INTRODUCTION

The existence of genetic drug resistance to existing anthelmintics presents a serious challenge, and emphasizes the urgent need for new generations of drugs. Fortunately, the potential for developing new drugs, acting on so far untargeted systems in parasites, seems high.

In this article I will concentrate on the intercellular signalling systems of the nervous system of nematodes and suggest that these systems present many interesting opportunities for the development of novel drugs. That is not to say that there are not other important potential targets, such as molecules specific to development, cuticle formation, osmotic regulation, nutrient absorption, etc. However, at present the nervous system is a particularly appropriate center of attention because many of the intercellular signalling mechanisms are now beginning to be understood at several levels. These include the identification of the small molecules that carry the signals, the characterization of the receptors with which they interact, the electrophysiological consequences of the ligand–receptor interaction, and the localization to specific identified neurons (or to muscle cells) of both molecular partners of the signalling dyad.

In principle, each distinct signalling system presents several potential targets—the biosynthetic enzymes, receptors, mechanisms of termination of action (enzymatic breakdown or uptake), vesicular packaging, etc. It is hard to imagine any systematic pharmacological exploration, targeted to a particular signalling system, without knowledge of the chemical structure of the molecule that is released from one cell and acts (usually) on another. This phase of discovery is crucial, and is highly enabling for subsequent analyses of the basic mechanisms of each system.

NEUROTRANSMITTERS AND NEUROMODULATORS IN THE NEMATODE *ASCARIS*

I will discuss the progress that has been made in the identification of neurotransmitters and/or neuromodulators in the nematode *Ascaris suum,* a large intestinal parasite of pigs. *Ascaris* is a useful model for these studies because its large size permits a variety of experimental approaches to be applied to the study of the nervous system; these include anatomical studies that permit neurons to be individually identified,[1,2] electrophysiological experiments on neurons[3] and muscle cells[4] using both intracellular,[3,4,5] extracellular,[6] and (for muscle cells) patch electrodes,[7] immunocytochemical studies,[8-10] and the biochemical analyses of single neurons isolated by microdissection.[11] The nervous system has the further advantage that it comprises only 298 neurons, most of which are anatomically extremely simple in comparison with the elaborately branched neurons typical of almost all other organisms.[1] Taken together, these features make the *Ascaris* nervous system a very attractive nematode model system for the study of the basic mechanisms of neural activity at the cellular level[3,5,12] and for the analysis of the system as a whole.[13,14]

Conventional neurotransmitters

Acetylcholine

The cholinergic system is a classical target for anthelmintics. There is a large family of anti-cholinesterases which paralyze nematodes; another class of anthelmintics (e.g. levamisole, morantel, pyrantel) are cholinergic agonists in *Ascaris.*[15] Several recent papers have reviewed the pharmacological literature,[15-17] and it will not be repeated here. Suffice it to say that there is good evidence that acetylcholine is a neurotransmitter of excitatory motorneurons;[11] there is also some evidence suggesting that some of the interneurons that synapse onto the excitatory motorneurons are also cholinergic (Table 1). So far attempts to generate reagents that selectively label cholinergic neurons in anatomical preparations have not been successful—the assays for cholinergic function have involved microdissection of the processes of single motorneurons followed by biochemical assay for choline acetyltransferase (ChAT) activity.[11] Because it has yet not been possible to dissect out single identified interneurons for assay, the number of interneurons that are cholinergic is presently unknown.

One further intriguing finding is that in *Ascaris* there is a ChAT activity in hypodermal tissue.[13] In the hypodermis surrounding the commissures of the motorneurons the ChAT activity is relatively low, but in the head region it is much higher, and constitutes over 95% of the total ChAT activity in the worm. Furthermore the majority of this activity is localized anterior to the nerve ring, in a region which contains a swelling of the hypodermis anterior to the end of

TABLE 1
Distribution of Transmitters and Neuropeptides Among Neuronal Classes of
Ascaris suum

Classical transmitters	Motor neurons	Sensory neurons	Interneurons
Acetylcholine[11]	+		(+)[a]
GABA[8,22]	+		+
5HT[28]			+[b]
Dopamine[32]	−	+	−
Octopamine	?	?	?
Glutamate	?	?	?
Neuropeptides[9]			
FMRFamide	+	+	+
CCK		+	+
Gastrin			+
LHRH			+
α-MSH		+	+
NPY	+[c]	+	+
CRF		+	+
GCRP		+	+
L-11			+
12$_B$	+[c]		+
SCP$_B$		+	+
VIP	+		

[a] There is pharmacological evidence suggesting that ACh is released from some interneurons (Angstadt & Stretton, unpublished).
[b] The pharyngeal neurons containing 5HT-like immunoreactivity are probably neurosecretory.
[c] Present in DVB neurons which appear to innervate the anal depressor muscles.

he body wall musculature.[13] It seems likely that there is a second, non-neural, cholinergic signalling system in *Ascaris,* so it should not be immediately assumed that all cholinergic drugs that affect behavior are acting directly on he neuromuscular system.

Recently it was shown that *N*-methyl scopolamine, a classical muscarinic antagonist, opens channels in one class of dorsal excitatory motorneurons DE1) but not in a second class of dorsal excitatory neurons (DE2) nor in either class of inhibitory motorneurons (DI or VI).[18] Thus there may be receptors related to muscarinic receptors in *Ascaris,* although the identity of heir endogenous ligand is unknown.

GABAergic systems

GABA was first proposed to be an inhibitory neurotransmitter in *Ascaris* by del Castillo *et al.*[19] They also showed that the anthelmintic piperazine was a GABA agonist.[20] Subsequently it was shown that GABA and piperazine open

Cl⁻ channels in muscle cells.[7,21] The selective localization of GABA-immunoreactivity in inhibitory motorneurons, but not in excitatory motor-neurons, strongly supported the suggestion that GABA was an inhibitory neuromuscular transmitter.[8]

Further immunocytochemical studies showed that there is a total of 30 GABA-immunoreactive neurons in *Ascaris* (about 10% of the total neurons).[22] Of these, 19 are the inhibitory motor neurons that act on somatic muscle (6 dorsal inhibitors and 13 ventral inhibitors, innervating dorsal and ventral muscle, respectively).[8] The four RME neurons, which have cell bodies in the nerve ring, also contain GABA-like immunoreactivity; from their homologies to *C. elegans* neurons, they too are predicted to be motorneurons, innervating head muscles.[49] The single GABA-immunoreactive neuron in the tail is the DVB neuron, with its cell body in the preanal ganglion; again, by homology with its counterpart in *C. elegans*, it is probably a motor neuron innervating anal muscles. Thus 24 of the 30 GABA-containing neurons appear to be motorneurons. This leaves only 6 GABA-containing neurons in the head ganglia, 1 pair in the ventral ganglion, 1 in the amphid ganglion, and 1 in the deirid cluster; these are the only interneurons that contain GABA. We were surprised that this transmitter system is so sparsely represented among the interneurons.

A second aspect of GABAergic function that has recently been explored in *Ascaris* is a GABA uptake mechanism.[23] Exogenous GABA is taken up into muscle cells, which do not have detectable levels of GABA-immunoreactivity, so presumably GABA is rapidly metabolized in muscle. It is interesting that among the GABA-immunoreactive neurons, only some accumulate exogenous GABA, namely the four RME cells, and the six putative GABAergic interneurons. The dorsal and ventral inhibitory motorneurons (DI and VI) do not take up GABA despite their high levels of GABA-immunoreactivity (DVB has not yet been examined for GABA uptake). There is an additional class of ventral nerve cord neurons that, like muscle, take up GABA but do not contain detectable GABA-immunoreactivity. In light of these results, the question of which cells contain glutamic acid decarboxylase, the GABA biosynthetic enzyme, is very interesting: it is possible that *in vivo*, some cells accumulate GABA by uptake rather than by synthesis. A prediction from these results is that blockers of GABA uptake should lead to flaccid paralysis of the worm, and therefore be interesting candidates for anthelmintic activity.

Recently the pharmacology of *Ascaris* muscle GABA receptors has been explored. Tests with a series of agonists suggest similarities with vertebrate GABAa receptors,[24] but several GABAa antagonists, including bicuculline and picrotoxin, are inactive. However, certain arylaminopyridazine-GABA derivatives act as competitive GABA antagonists in *Ascaris*,[17,25] a discovery that has filled an important gap in nematode pharmacology.

Serotonin
The effects of serotonin on glycogen metabolism in *Ascaris* muscle,[26] and the ability of *Ascaris* to synthesize serotonin[27] suggested that serotonin may

mediate intracellular signalling in *Ascaris*. This was strengthened by two findings, firstly the demonstration that serotonin is localized to two neurons in the pharynx in females, and in five additional neurons in the ventral cord in the male tail,[28] and secondly, that high affinity serotonin receptors could be found in *Ascaris* membranes.[29] The anatomy of the pharyngeal neurons strongly suggests that they are neurosecretory: they are extensively branched (a highly atypical feature of *Ascaris* neurons) and have large numbers of varicosities at the surface of the pharynx. The recent finding that injection of serotonin into female worms that were ligatured posterior to the pharynx causes paralysis[30] suggests the existence of 5HT receptors which are another potential system that could be targetted by anthelmintics. The role that serotonin plays in egg-laying in *C. elegans*[31] suggests that there may be other important sites where serotonin acts, although there is no evidence at present whether serotonin influences egg-laying in *Ascaris*.

Dopamine
In female *Ascaris*, there are eight sensory neurons containing dopamine.[32] There are also suggestions that dopamine affects locomotory behavior.[30]

Octopamine and glutamate
The status of octopamine and glutamate as putative neurotransmitters in *Ascaris* is uncertain. No satisfactory demonstrations of cellular localization in subsets of neurons have been made. However, octopamine has effects on locomotory behavior,[30] and glutamate receptors have been detected on DE2 motorneurons. DE2 motorneurons are depolarized by glutamate and kainate, but not by NMDA (Davis, R. E. & Stretton, A. O. W., unpublished). Clearly more exploratory work is needed before either of these compounds can be taken as serious neurotransmitter candidates.

Neuropeptides

Neuropeptides are known to comprise the largest class of putative neurotransmitter substances in many organisms, both vertebrate and invertebrate.[33,34] Therefore, it is not surprising that nematodes have recently been shown to contain a wide variety of peptide-like immunoreactivities.[9,10,35] In some cases, peptides corresponding to these immunoreactivities have been isolated, sequenced, and shown to have potent biological activity.[36,37] However, much work remains to be done in this area. Considering the apparent richness of the neuropeptide signalling systems in nematodes, it seems safe to predict that they represent a large number of potential targets for drug action that are waiting to be explored. The characterization of these systems, starting with the chemical identification of the endogenous peptides, is an important goal.

Immunocytochemical screening for neuropeptides related to known peptides
One straightforward approach for the discovery of neuropeptides in *Ascaris* is to use antibodies, raised against peptides already identified in other organisms,

to test for the presence of immunoreactivity related to the original peptide in anatomical preparations. Such tests can be made on sections[10] or wholemounts,[9] and each method has its advantages: in sections, the identity of the immunoreactive neurons can be made more rigorous,[1] whereas in wholemounts, a much larger fraction of the nervous system can be screened rapidly.

This screening method has some important advantages compared to an alternative immunological approach—to screen tissue extracts by peptide-specific RIAs. These advantages stem from the fact that, besides scoring the presence or absence of the peptide-like immunoreactivity, they also give some indications about possible functional roles the peptides might play since they also identify those neurons which contain the putative peptide, and may have the potential to release it. There is considerable information available in *Ascaris* about the function of many of the neurons—this is one of the advantages of working with a system containing few neurons. This information has been derived either directly by electrophysiological techniques in the case of motorneurons and the interneurons that synapse onto them, or indirectly from anatomical correlates which suggest which neurons are sensory and which may be neurosecretory. This preliminary functional information has been very useful in allowing an informed choice to be made as to which of the putative peptides should be isolated for sequence determination. This is an important strategic choice since the isolations are intricate and time-consuming, and demand expertise in isolating very small quantities of peptide from a complex mixture.

A wholemount preparation of the head[8] and tail[35] regions of *Ascaris* has been developed in which the muscle cells are removed by digestion with collagenase; this leaves the nervous system attached to the hypodermis and cuticle, and readily accessible to antibodies. These two preparations together contain over 95% of the types of neurons present in the worm (the rest of the worm largely includes repeats of neuronal types that are present in the head, tail, or both). We routinely use the wholemount preparations for initial screening.

In such preparations, a set of 44 antisera, raised against 42 peptides (30 from mammals, nine from molluscs, and three from arthropods) showed that there were at least 12 different peptide-like immunoreactivities in *Ascaris* neurons[9] (Table 2). Each antiserum stained a different subset of neurons, and thus formally defined at least one distinct epitope that is selectively expressed in those neurons. The 12 positive immunoreactivities were seen with antibodies raised against luteinizing hormone releasing hormone (LHRH), small cardio-active peptide B (SCP$_B$), neuropeptide Y (NPY), FMRFamide, cholecystokinin (CCK), gastrin, α-melanocyte stimulating hormone (αMSH), corticotropin releasing factor (CRF), calcitonin-gene related peptide (CGRP), vasoactive intestinal peptide (VIP), and two peptides from *Aplysia* named L11 and 12$_B$. The number of neurons stained by different antisera varied widely, ranging from just one pair of neurons, in the case of the LHRH-like immunoreactivity,

TABLE 2
Immunocytochemical detection of peptide-like Immunoreactivity in *Ascaris* Neurons

Antiserum against	Subset of neurons stained	Blocked by peptide
Mammals		
ACTH	No	
Bombesin	No	
Calcitonin	No	
CCK8	Yes	Yes
CGRP	Yes	ND
CRF	Yes	Yes
β-endorphin	No	
FSH	No	
Gastrin	Yes	Yes
GIP	No	
Glucagon	No	
Growth Hormone	No	
GRF	No	
Insulin	No	
Leu-enkephalin	Yes	No
LH	No	
LHRH	Yes	Yes
Met-enkephalin	Yes	No
Motilin	No	
αMSH	Yes	Yes
Neurophysin	No	
Neurotensin	No	
NPY-1	Yes	ND
NPY-2	No	
Oxytocin	Yes	No
Prolactin	No	
Somatostatin	No	
Substance P	Inconsistent	
TSH	No	
Vasopressin	Inconsistent	
VIP-1	Yes	No
VIP-2	Yes	Yes
Molluscs		
BCP	Inconsistent	
ELH	No	
FMRFamide	Yes	Yes
L11	Yes	ND
SCP$_B$	Yes	Yes
Peptide A	No	
Peptide 9	No	
Peptide 16A	No	
Peptide 12$_B$	Yes	ND

TABLE 2—*contd.*

Antiserum against	Subset of neurons stained	Blocked by peptide
Arthropods		
LAKH	Inconsistent	
Proctolin	No	
RPCH	No	

Abbreviations. ACTH adrenocorticotropic hormone; αMSH, α-melanocyte stimulating hormone; BCP, bag cell peptide; CCK-8, cholecystokinin octapeptide; CGRP, calcitonin gene related peptide; CRF, corticotropin releasing factor; ELH, egg laying hormone; FMRFamide, phe-met-arg-phe-amide; FSH, follicle-stimulating hormone; GH, growth hormone; GIP, gastric inhibitory peptide; GRF, growth hormone releasing factor; LAKH, locust adipokinetic hormone; LH, luteinizing hormone; LHRH, luteinizing hormone releasing hormone; L11, peptide isolated from neuron L11 of *Aplysia* abdominal ganglion; M-ENK, met-enkephalin; ND, not done; NPY, neuropeptide Y; RPCH, red pigment concentrating hormone; SCP_B, small cardioactive peptide B; TSH, thyroid stimulating hormone; VIP, vasoactive intestinal peptide; peptides 9, 12_B, and 16_A, nona-, dodeca-, and hexadeca-peptide sequences occurring in the R3-R8 and R14 neurons of *Aplysia* abdominal ganglia.

to over half of the neurons in the worm, for FMRFamide-like immunoreactivity. There was no obvious relationship between the occurrence of staining and the phylogenetic relationships to the animal in which the peptide had originally been discovered (Table 2).

Suitable caution needs to be used in interpreting these immunocytochemical results. Firstly, an immunological control in which the serum is preabsorbed with the immunogenic peptide is necessary to show that the immunoreactivity is specific; this was carried out with eight of the above immunoreactivities (all except NPY, CGRP, L11 and 12_B), and in each case the immunoreactivity was blocked by the peptide. The results with several antisera which selectively stained subsets of neurons failed this control test (e.g. anti-leu-enkephalin, anti-met-enkephalin, and anti-oxytocin); in other cases an antiserum contained both specific and non-specific antibodies (e.g. anti-LHRH showed specific staining of one pair of neurons in the head, while the staining of three more pairs was not blocked by the peptide). Presumably this non-specific staining is due to the presence in the serum of antibodies to unknown epitopes that show specific cellular localization in some *Ascaris* neurons. Although they may be useful in general neuroanatomical studies for identifying neurons, there is no reason to believe that they provide any information about the presence or location of a neuropeptide. Secondly, even if the immunoreactivity is specific, it does not necessarily imply the existence of a neuropeptide—there might be selective cellular expression of an *Ascaris* protein which happens to include an epitope that is shared with the original peptide. Chemical studies are needed before the existence of a neuropeptide can be claimed. Thirdly, it is easy to be misled by the intensity of the immunocytochemical staining. Strong signals do not necessarily imply the presence of larger amounts of material than those giving rise to weak signals; the signals may be produced by epitopes with different degrees of cross-reactivity to the original peptide used for immunization. Similarly the complete absence of a signal does not necessarily imply the absence of a structurally or functionally related peptide. Because there is no necessary overlap between the sequence of an epitope and the sequence required for biological activity, the two sequences may well diverge during evolution. In the initial screening, two anti-NPY and two anti-VIP sera were used, and in each case one antiserum gave positive results and the other was negative.[9] As other antisera become available, it would be worthwhile to continue to re-screen with antibodies to the 30 peptides which gave negative results in the first round of screening. Finally, because of the possibility of convergent evolution, before any immunoreactivity can be claimed to be in the same structural or functional family as the parent peptide, the sequence of the peptide(s) must be determined so that direct comparison can be made. Fortunately, in many cases RIA-grade antisera are available so that an assay already exists for following the purification of the peptide leading to chemical isolation.

Among the cells that were specifically stained with the antibodies that were tested, there were many examples of neurons showing co-localization of different peptide-like immunoreactivities in the same cell.[9] For example, a pair of identified neurons called the RIG neurons contained six different immunoreactivities, and the ADE neurons (the pair of anterior deirid neurons which have a sensory dendrite in the lateral line) contain four different peptide-like immunoreactivities, as well as dopamine.[9,32] Whether some, or perhaps all, of the peptide-like immunoreactivities are due to coexistence of multiple epitopes on the same molecule, or whether they represent the expression of the products of several neuropeptide-encoding genes in the same cell, will not be clear until the peptide sequences are known so that their cross-reactivity with these antisera can be directly tested on synthetic peptides.

Isolation and biological activity of endogenous Ascaris peptides
The anatomical results obtained with the 12 positive antisera were indeed helpful in making a decision on which of the neuropeptides to attempt to purify (Table 1). Two of the antisera, anti-FMRFamide and anti-VIP, stained neurons innervating somatic muscles (anti-NPY and anti-12$_B$ stained the DVB neuron which appears to innervate anal muscles). Anti-FMRFamide stained many more neurons than anti-VIP; later studies on sectioned material showed that FMRFamide-like immunoreactivity was present in several interneurons in the ventral cord, as well as in a dorsal motorneuron called D-var which

synapses onto dorsal muscle and the dendrites of ventral inhibitory motor-neurons (Donmoyer, J. E., Desnoyers, P. A. & Stretton, A. O. W., unpublished). These results suggested that putative FMRFamide-like peptides might be involved in the activity of the motor nervous system.

Fractionation of an acid–methanol extract of 5000–10,000 *Ascaris* heads were therefore carried out,[36] and fractions were assayed with an RIA that used an antibody with broad specificity for FMRFamide-like peptides—it recognized peptides with *C*-terminal-RFamide.[38] The extract was first crudely fractionated on C18 cartridges, then subjected to between three and six steps of HPLC. The first HPLC step showed that there was considerable molecular heterogeneity among FMRFamide-like peptides: at least 10 peaks were found. After extensive further fractionation several peaks were obtained where immunoreactivity and UV absorbance at 214 nm corresponded. The first two to be sequenced were named AF1 and AF2 (AF = Ascaris FMRFamide-like peptide). Their sequences were Lys–Asn–Glu–Phe–Ile–Arg–Phe and Lys–His–Glu–Tyr–Leu–Arg–Phe, respectively.[36,37] Synthetic peptides were prepared in the terminally amidated form because of the specificity of the RIA, and in a second purification the natural peptides were isolated and sequenced, and their elution times in three HPLC steps were shown to correspond to those of the synthetic peptides. Finally the amino acid composition and the *C*-terminal amidation were confirmed by fast-atom bombardment mass spectroscopy. The sequence of AF1 is KNEFIRFamide, and that of AF2 is KHEYLRFamide.

Both AF1 and AF2 have potent biological activity on the neuromuscular system of *Ascaris*. Three levels of analysis of this activity have been explored so far. The first, and simplest, is to monitor the effects on locomotory behavior when solutions of the peptide are injected into intact *Ascaris*. Injection of AF1 or AF2 (0·1 ml of 10^{-6} M solution) anterior to the gonopore blocked the propagation of locomotory waveforms through the region near the injection site. The second preparation is a dorsal muscle strip in which the effects of peptides on muscle tension is measured. Both AF1 and AF2 produce multiple effects on muscle tension, including relaxation, contraction, and induction of rhythmic activity.[37,39]

The third, and more fundamental, analysis involves the study of the physiological effects of peptides on single nerve or muscle cells, through the use of intracellular recording techniques.[3,5] These techniques have allowed us to study five of the seven classes of *Ascaris* motorneurons, including their cellular electrical properties, their synapses to muscle and to other motor-neurons, and their spontaneous synaptic input, so we can investigate systematically many of the possible sites of action of chemicals on the motor nervous system. AF1, at 10^{-9}–10^{-7} M, abolished the slow oscillatory potentials that occur in dorsal and ventral inhibitory motorneurons.[36] The input resistance of these cells is drastically reduced by AF1; this effect is due, at least in large part, to the presence of receptors for AF1 on the inhibitory motorneuron, since when neurons are synaptically isolated by blocking synaptic transmission with cobalt ions, the AF1-induced effect on input resistance still occurs. Since *Ascaris* motorneurons depend on their unusually high membrane resistance for long-distance signalling (they do not produce classical action potentials),[3] AF1-induced reductions in membrane resistance would lead to their being effectively removed from the locomotory circuit. We believe that this is a possible explanation for the paralysis which occurs when worms are injected with AF1, although there may be additional sites of action we have not yet studied. A similar study of AF2 is in progress.

Besides AF1 and AF2, several other FMRFamide peptides have been isolated and sequenced (Cowden, C. & Stretton, A. O. W., unpublished). These sequences are unique and not related by post-translational modification or by proteolysis; they are also different from any other FMRFamide-like sequence from other organisms, whether obtained from isolated and sequenced peptides, or by deduction from the DNA sequences of genes. It is clear that there is a family of FMRFamide-like peptides in *Ascaris*. Whether they are members of an intragenic family or are the products of multiple genes will require cloning and sequencing of the *Ascaris* gene or genes that encode them. These sequences will be reported when we have characterized some of their physiological effects on *Ascaris* nerve and muscle cells, and have raised specific antibodies to study their cellular localization.

Other methods for peptide discovery
The approach for peptide discovery described above depends on the availability of antibodies against previously characterized peptides, and will only detect *Ascaris* peptides which share epitopes with known peptides. Clearly this approach is unacceptably limiting: there may well be *Ascaris* peptides which would not be detected by existing antibodies, either because they belong to so far undiscovered 'universal' families, or because they are structurally unique, and are found only in nematodes, or even only in *Ascaris suum*. There are several existing methods for detecting such peptides. The classical method is to use biological activity as an assay for purification. The isolation of leucosulfakinin-I and -II[40,41] and of leucopyrokinin[42] from the cockroach are examples of recent successes obtained by using this approach. A different strategy was used by Tatemoto and Mutt[43] who developed specific methods for detecting peptides with *C*-terminal amides, a common feature of many neuropeptides; this method led to the discovery of new members of the VIP-glucagon family and of the NPY family. More recent approaches have been based on molecular biology and recombinant DNA technology.[44] The isolation and sequencing of cDNA encoding known peptides led to an understanding of the proteolytic cleavage signals used in the processing of the precursor proteins[45] and hence to the discovery of other peptides (e.g. CGRP) also encoded by the same gene.[46] Another approach is to screen recombinant DNA libraries with cDNAs made from poly A^+ mRNA isolated from different cells or tissues. This approach has led to the isolation and sequence

determination of new neuropeptide genes specifically expressed in collections of cells or even in single identified neurons in *Aplysia*.[47]

We have recently devised an alternative method in which a crude peptide extract (from *Ascaris* or *C. elegans*) is conjugated to carrier protein, and used as an immunogen for generating monoclonal antibodies.[35] Screening of hybridoma-conditioned culture fluids is carried out by immunocytochemistry on *Ascaris* whole mounts. Clones were selected if they produce antibodies staining subsets of cells different from those stained by any previously tested antibody or antiserum. This criterion thus defines novel epitopes. This technology has recently been extended so that the monoclonal antibodies can be used to detect the unknown peptide during fractionation.[48] The method is a dot-blot procedure in which a BSA–peptide conjugate is formed from each fraction *in situ* on a nitrocellulose membrane, and antibody binding is detected with immunoperoxidase staining. The sensitivity is high (2–10 fmol peptide), and the method is of general applicability. Compared to RIA, there is the considerable advantage that a radiolabeled peptide is not required, an important consideration when the chemical nature of the peptide is still unknown!

CONCLUSIONS

In this survey I have tried to take a forward-looking view, realizing that the evidence that each of the 'signalling molecules' has been identified as such can be challenged in every case—the classical criteria for identifying transmitters have been met to varying extents, but in no case is the evidence complete. However, it is clear that these molecules include some that profoundly affect nematode behavior, and this can be sufficient information to identify a target site for a new anthelmintic.

To what extent is information gained by studying the *Ascaris* nervous system useful? Is it a good model system for other nematodes? The answer to this is not yet clear. There are some ways in which the nervous systems of nematodes appears to be very conservative. For example, the structure of the nervous system of *Ascaris* and *C. elegans* is remarkably similar at the cellular level. The morphology of individual cells, as well as their patterns of synaptic connectivity, are remarkably conserved.[1,2,49] This conservatism also applies to the expression of neurotransmitters such as GABA and dopamine.[8,22,32,50] It would be informative to extend such an analysis to many other species of nematodes, to see whether this structural conservatism is a feature of the entire phylum. If it is, we can speculate on the source of the behavioral diversity that enables the nematodes to occupy such a wide range of ecological niches. One of the recent discoveries to have emerged from the study of invertebrate nervous systems is that the activity of the system is critically dependent on the chemical context in which it operates. Neuromodulators can act as ganged switches which change the properties of voltage-gated and/or ligand-gated channels such that from a given network of neurons, many different operating circuits, controlling different behaviors, can be specified.[51,52] Perhaps this is a reason for the remarkable diversity of the neuropeptides; each one may encode a different behavior.

To someone interested in developing anthelmintics based on these signalling systems, it is strategically very important to know whether different species of nematodes have evolved different versions of peptide signalling mechanisms. Alternatively these mechanisms may also be conservative, and the behavioral diversity needed for speciation achieved by varying the cellular patterns of expression of neuropeptides and their receptors. These issues again emphasize the importance of comparative studies.

ACKNOWLEDGEMENTS

The author wishes to express his enormous debt and gratitude to the past and present members of his laboratory who have contributed to the ideas and experiments presented here. Judy Donmoyer, John Walrond, Ira Kass, Carl Johnson, Ralph Davis, Jim Angstadt, John Guastella, Marsha Segerberg, Paisarn Sithigorngul, Cindy Cowden, Jeff Meade, John Mulvihill, David Wright, and Catharine Buchanan are all gratefully acknowledged. The author also wishes to express his appreciation to the members of his laboratory for their forbearance while the neuropeptide work was being initiated, when some of them saw less of him than they should have.

REFERENCES

1. Stretton, A. O. W., Fishpool, R. M., Southgate, E., Donmoyer, J. E., Walrond, J. P., Moses, J. E. R. & Kass, I. S., Structure and physiological activity of the motoneurons of the nematode *Ascaris*. *Proceedings of the National Academy of Sciences, USA*, **75** (1978) 3493–7.

2. Angstadt, J. D., Donmoyer, J. E. & Stretton, A. O. W., Retrovesicular ganglion of the nematode *Ascaris*. *J. Comp. Neurol.*, **284** (1989) 374–88.

3. Davis, R. E. & Stretton, A. O. W., Passive membrane properties of motoneurons and their role in long-distance signalling in the nematode *Ascaris*. *J. Neurosci.*, **9** (1989) 403–14.

4. Jarman, M., Electrical activity in the muscle cells of *Ascaris lumbricoides*. *Nature, London*, **184** (1959) 1244.

5. Davis, R. E. & Stretton, A. O. W., Signalling properties of *Ascaris* motoneurons: graded active responses, graded synaptic transmission, and tonic transmitter release. *J. Neurosci.*, **9** (1989) 415–25.

6. Davis, R. E. & Stretton, A. O. W., Extracellular recording from the motornervous system of the nematode *Ascaris*. *Soc. Neurosci. Abstr.*, **16** (1990) 726.

7. Martin, R. J., γ–aminobutyric acid- and piperazine-activated single channel currents from *Ascaris suum* body muscle. *British Journal of Pharmacology*, **84** (1985) 445–61.

8. Johnson, C. D. & Stretton, A. O. W., GABA-immunoreactivity in inhibitory motor neurons of the nematode *Ascaris*. *J. Neurosci.*, **7** (1987) 223–35.

136   *A. O. W. Stretton*

9. Sithigorngul, P., Stretton, A. O. W. & Cowden, C., Neuropeptide diversity in *Ascaris*: an immunocytochemical study. *J. Comp. Neurol.*, **294** (1990) 362–76.
10. Davenport, T. R. B., Lee, D. L. & Isaac, R. E., Immunocytochemical demonstration of a neuropeptide in *Ascaris suum* (Nematoda) using antiserum to FMRFamide. *Parasitology*, **97** (1988) 81–8.
11. Johnson, C. D. & Stretton, A. O. W., Localization of choline acetyltransferase within identified motoneurons of the nematode *Ascaris*. *J. Neurosci.*, **5** (1985) 1984–92.
12. Angstadt, J. D. & Stretton, A. O. W., Rhythmic activity in inhibitory motorneurons in the nematode *Ascaris*. *J. Comp. Physiol. A.*, **166** (1989) 165–77.
13. Johnson, C. D. & Stretton, A. O. W., Neural control of locomotion in *Ascaris*: anatomy, electrophysiology and biochemistry. In *Nematodes as Biological Models*, Vol. 1. Academic Press, New York, pp. 159–95.
14. Stretton, A. O. W., Davis, R. E., Angstadt, J. D., Donmoyer, J. E. & Johnson, C. D., Neural control of behaviour in *Ascaris*. *Trends Neurosci.*, **8** (1985) 294–300.
15. Harrow, I. D. & Gration, K. A. F., Mode of action of the anthelmintics morantel, pyrantel and levamisole on the muscle cell membrane of the nematode *Ascaris suum*. *Pesticide Science*, **16** (1985) 662–72.
16. Colquhoun, L., Holden-Dye, L. & Walker, R. J., The pharmacology of cholinoceptors on the somatic muscle cells of the parasitic nematode *Ascaris suum*. *British Journal of Pharmacology*, **99** (1990) 253P.
17. Martin, R. J., Pennington, A. J., Buittox, A. H., Robertson, S. & Kusel, J. R., The physiology and pharmacology of neuromuscular transmission in the nematode parasite, *Ascaris suum*. *Parasitology*, **102** (1991) 541–58.
18. Segerberg, M. A. & Stretton, A. O. W., Two pharmacologically distinct classes of cholinergic receptors in nematodes. *Soc. Neurosci. Abstr.*, **12** (1986) 1022.
19. del Castillo, J., de Mello, W. C. & Morales, T., Inhibitory action of γ-aminobutyric acid (GABA) on *Ascaris* muscle. *Experientia*, **20** (1964) 141–3.
20. del Castillo, J., de Mello, W. C. & Morales, T., Mechanism of the paralysing action of piperazine on *Ascaris* muscle. *British Journal of Pharmacology*, **22** (1964) 463–77.
21. Martin, R. J., Electrophysiological effects of piperazine and diethylcarbamazine on *Ascaris suum* somatic muscle. *British Journal of Pharmacology*, **77** (1982) 255–65.
22. Guastella, J., Johnson, C. D. & Stretton, A. O. W., GABA-immunoreactive neurons in the nematode *Ascaris*. *J. Comp. Neurol.*, **307** (1991) 584–608.
23. Guastella, J. & Stretton, A. O. W., Distribution of ³H-GABA uptake sites in the nematode *Ascaris*. *J. Comp. Neurol.*, **307** (1991) 598–608.
24. Holden-Dye, L., Krogsgaard-Larsen, P., Nielsen, L. & Walker, R. J., GABA receptors on the somatic muscle cells of the parasitic nematode, *Ascaris suum*: stereoselectivity indicates similarity to a GABA$_a$-type agonist recognition site. *British Journal of Pharmacology*, **98** (1989) 841–50.
25. Duittoz, A. H., & Martin, R. J., Effects of the arylaminopyridazine-GABA receptor: the relative potency of the antagonists in *Ascaris* is different to that at vertebrate GABA$_a$ receptors. *Comp. Biochem. Physiol.*, **98C** (1991) 417–22.
26. Donahue, M. J., Yacoub, N. J., Michinoff, C. A., Masaracchia, R. A. & Harris, B. G., Serotonin (5-hydroxytryptamine): a possible regulator of glycogenolysis in perfused muscle segments of *Ascaris suum*. *Biophysics and Biochemical Research Communications*, **101** (1981) 112–17.
27. Martin, R. E., Chaudhuri, J. & Donahue, M. J., Serotonin (5-hydroxytryptamine) turnover in adult female *Ascaris suum* tissue. *Comp. Biochem. Physiol.*, **91C** (1988) 307–10.
28. Stretton, A. O. W. & Johnson, C. D., GABA and 5HT immunoreactive neurons in *Ascaris*. *Soc. Neurosci. Abstr.*, **11** (1985) 626.

29. Chaudhuri, J. & Donahue, M. J., Serotonin receptors in the tissues of adult *Ascaris suum. Molec. Biochem. Parasitol.*, **35** (1989) 191–8.
30. Buchanan, C. A. & Stretton, A. O. W., The effects of biogenic amines on *Ascaris* locomotion. *Soc. Neurosci. Abst.*, **17** (1991).
31. Desai, C., Garriga, G., McIntire, S. L. & Horvitz, H. R., A genetic pathway for the development of the *Caenorhabditis elegans* HSN motor neurons. *Nature*, **336** (1983) 293–4.
32. Sulston, J., Dew, M. & Brenner, S., Dopamine neurons in the nematode *Caenorhabditis elegans. J. Comp. Neurol.*, **163** (1975) 215–44.
33. Iversen, L. L., Neuropeptides—what next? *Trends Neurosci.*, **6** (1983) 293–4.
34. O'Shea, M. & Schaeffer, M., Neuropeptide function: the invertebrate contribution, *Ann. Rev. Neurosci.*, **8** (1985) 171–98.
35. Sithigorngul, P., Cowden, C., Guastella, J. & Stretton, A. O. W., Generation of monoclonal antibodies against a nematode peptide extract: another approach for identifying unknown neuropeptides. *J. Comp. Neurol.*, **284** (1989) 389–97.
36. Cowden, C., Stretton, A. O. W. & Davis, R. E., AF1, a sequenced bioactive neuropeptide isolated from the nematode *Ascaris suum. Neuron*, **2** (1989) 1465–73.
37. Cowden, C. & Stretton, A. O. W., AF2, a nematode neuropeptide. *Soc. Neurosci. Abstr.*, **16** (1990) 305.
38. Marder, E., Calabrese, R. L., Nusbaum, M. P. & Trimmer, B. A., Distribution and partial characterization of FMRFamide-like peptides in the stomatogastric ganglion of the rock crab, *Cancer borealis*, and the spiny lobster, *Panulirus interruptus. J. Comp. Neurol.*, **259** (1987) 150–63.
39. Cowden, C. & Stretton, A. O. W., Structure and physiological activity of AF2, an endogenous FMRFamide-like peptide from *Ascaris suum* (in preparation).
40. Nachman, R. J., Holman, G. M., Haddon, W. F. & Ling, N., Leucosulfakinin, a sulfated insect neuropeptide with homology to gastrin and cholecystokinin. *Science*, **234** (1986) 71–3.
41. Nachman, R. J., Holman, G. M., Cooke, B. J., Haddon, W. F. & Ling, N., Leucosulfakinin-II, a blocked sulfated insect neuropeptide with homology to cholecystokinin and gastrin. *Biochem. Biophys. Res. Commun.*, **140** (1986) 357–64.
42. Holman, G. M., Cook, B. J. & Nachman, R. J., Isolation, primary structure and synthesis of leucomyosuppressin, an insect neuropeptide that inhibits spontaneous contractions of the cockroach hindgut. *Comp. Biochem. Physiol.*, **85C** (1986) 329–33.
43. Tatemoto, K. & Mutt, V., Isolation and characterization of the intestinal peptide porcine PHI (PHI-27), a new member of the glucagon-secretin family. *Proc. Natl. Acad. Sci. USA*, **78** (1981) 6603–7.
44. Bloom, F. E., Identifying neuropeptides by gene cloning. *Psychopharmacol. Bull.*, **22** (1986) 701–7.
45. Nakanishi, S., Inoie, A., Kita, T., Nakamura, M., Chang, A. C., Cohen, S. N. & Numa, S., Nucleotide sequence of cloned *c*DNA for bovine corticotropin-B-lipotropin precursor. *Nature*, **278** (1979) 423–7.
46. Rosenfeld, M. G., Mermod, J. J., Amara, S. G., Swanson, L. W., Sawchenko, P. E., Rivier, J., Vale, W. W. & Evans, R. M., Production of a novel neuropeptide encoded by the calcitonin gene via tissue-specific RNA processing. *Nature*, **304** (1983) 129–35.
47. Nambu, J. R., Taussig, R., Mahon, A. C. & Scheller, R. J., Gene isolation with cDNA probes from identified *Aplysia* neurons: Neuropeptide modulators of cardiovascular physiology. *Cell*, **35** (1983) 47–56.
48. Sithigorngul, P., Stretton, A. O. W. & Cowden, C., A versatile dot-ELISA method with femtomole sensitivity for detecting small peptides. *J. Immunol. Methods*, **141** (1991) 23–32.

49. White, J. G., Southgate, E., Thomson, J. N. & Brenner, S., The structure of the nervous system of the nematode *Caenorhabditis elegans*. *Phil. Trans. Roy. Soc. Lond.* (*Biol.*), **314** (1986) 1–340.
50. McIntire, S. & Horvitz, H. R., Immunocytochemical reactivity of neurons in wildtype and mutant *C. elegans* to antisera against GABA, serotonin, and CCK. *Soc. Neurosci. Abstr.*, **11** (1985) 920.
51. Getting, P. & Dekin, M. S., Tritonia swimming: a model system for integration within rhythmic motor systems. In *Model Neural Networks and Behavior*, ed. A. I. Selverston. Plenum, New York, 1985, pp. 3–20.
52. Harris-Warrick, R. M. & Marder, E., Modulation of neural networks for behavior. *Ann. Rev. Neurosci.*, **14** (1991) 39–57.

10

The Peptidergic Neuron: Possible New Molecular Targets

MICHAEL O'SHEA & RICHARD C. RAYNE

Cell Biology Laboratory, Royal Holloway and Bedford New College (RHBNC), University of London, Egham, Surrey TW20 0EX, UK

INTRODUCTION

This is an essay for the non-specialist about neuropeptides, the peptidergic neuron and the potential of research in this area for the development of novel methods of pest control. Neuropeptides represent the largest and most diverse class of neurotransmitters and neurohormones in the nervous systems of both vertebrates and invertebrates.[1,2] They are responsible for regulating such diverse and essential biological processes as development and maturation, metabolism, behaviour, reproduction, excretion and water balance. Peptidergic systems would seem therefore to offer many opportunities for the development of novel methods for the control of invertebrate pests and parasites.[3,4]

In this brief essay we will review the basic biology of the peptidergic neuron. A number of features of the neuropeptidergic systems contrast with the classical neurotransmitter systems (aminergic and cholinergic, for example) and perhaps the greatest differences arise from the structural complexity of neuropeptides and their mode of biosynthesis. Neuropeptides are large and complex information carrying molecules which are derived from even larger and more complex precursor proteins. The classical transmitters, on the other hand, are simpler and more easily synthesized. They can, for example, be synthesized at the sites of transmitter release where they can be 'recycled' by the re-uptake of released material. Neuropeptides cannot be synthesized at the site of release because they are derived from proteins which are synthesized in the cell bodies of neurons and which may contain sequences of more than one neuropeptide. As a consequence of the mode of synthesis, peptidergic neurons often contain and release more than one biologically active neuropeptide.

An understanding of how neurons make neuropeptides provides the key to understanding how and perhaps why these neurons are able to generate multiple transmitter messages. We will therefore describe those processes

139

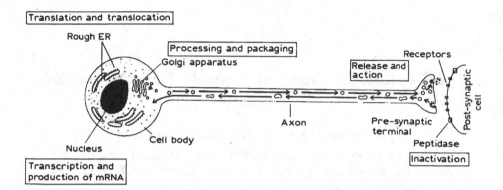

Fig. 1. A generalized peptidergic neuron. For further explanation, see the text.

involved in the generation of multiple peptide messengers. We shall also indicate how neuropeptides are released and how they interact with receptors, and finally how peptide signals are terminated. In order to cover this range of processes we have arbitrarily divided the peptidergic system into three parts: (1) the generation of multiple peptide messengers; (2) release and action; and (3) inactivation. By considering the peptidergic neuron in this way, we can perhaps identify new vulnerable sites for the development of novel control methods for insects and other invertebrates in which the neuropeptides perform such diverse and important functions.

To facilitate the description, peptidergic neurons have therefore been generalized in order to illustrate how in general terms they operate. The aim is to focus our thoughts on those features of peptidergic systems which may be potential targets. A diagram of a generalized peptidergic neuron is shown in Fig. 1.

THE GENERATION OF MULTIPLE PEPTIDE MESSENGERS

Transcription and production of neuropeptide encoding *m*RNAs

It seems unlikely that selective pesticides can be developed by interference with transcriptional mechanisms. However, the possibility that there may be an invertebrate specific splicing mechanism has arisen recently (see below) and this perhaps could be developed as a novel target for invertebrate-selective control methods.

Neuropeptide biosynthesis begins with the activation of genetic information which encodes the neuropeptide or neuropeptides. The structures of several genes which encode neuropeptides in a variety of both vertebrate and invertebrate systems are now known.[5] Not surprisingly, these genes do not differ fundamentally from others in eukaryotic systems in that they are constructed from exons containing information which will become part of the mRNA and intervening introns which will not. Within the nucleus, as in all eukaryotic systems, transcription results in the production of an RNA which is then edited by the removal of the introns and the splicing together of the exons. In a number of peptidergic systems, the phenomenon of alternative splicing has been identified which allows one neuropeptide encoding gene to generate more than one mRNA species.

A specific example of alternative splicing in invertebrates is provided by the gene which encodes the FMRFamide type peptides in the snail *Lymnaea stagnalis* (see article by J. Burke *et al.*, p. 179). This gene encodes the tetrapeptides FMRFamide and FLRFamide. The tetra- and heptapeptides, however, are encoded by different exons and this gene is subjected to sophisticated regulation at the level of RNA splicing. This results in the expression of either the heptapeptides or the tetrapeptides in subsets of neurons in the snail's CNS. This shows that the processes involved in regulating RNA splicing are highly specific and are important and regulated components of the individuality of identified neurons. A deeper understanding of how cell specific splicing is achieved would perhaps enable us to interfere with the expression of the appropriate peptidergic message in specific neurons. Such interference with peptide expression would doubtless disrupt the biological process controlled by the identified neuron concerned.

The exciting possibility that invertebrates may have a form of RNA splicing which is not found in vertebrates has recently arisen. This phenomenon has been called trans-splicing because the spliced mRNA has a short leader sequence derived from a different gene. Trans-splicing, to date, is known only in a few invertebrates, but it already offers a potential target for production of highly selective chemotherapy against nematodes. Should trans-splicing prove to be a feature unique to invertebrate species, it may provide a more general molecular target for the control of a wide variety of invertebrate pests.

Translation and translocation

In peptidergic neurons the translation of a neuropeptide mRNA produces the preprohormone, or prepropeptide. Protein synthesis begins in the cytosol when the mRNA becomes associated with ribosomes. For preprohormones (as for other proteins which are translocated across the ER membrane), the initial N-terminal amino acids (referred to as the signal sequence) are recognized by a protein complex on the ER called the signal recognition particle (SRP).[6,7] Interaction between the signal sequence and the SRP temporarily halts translation and therefore halts elongation of the preprohormone. The ribosomes with their partly translated preprohormones interact with the ER membrane through an association between the SRP and an SRP receptor. Once the ribosome is docked with the ER membrane the SRP is released and translation of the preprohormone is resumed. It is thought that the passage of

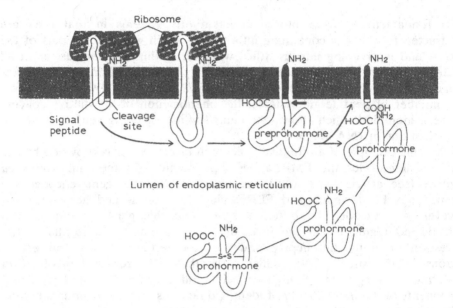

FIG. 2. A simplified diagram illustrating translocation and the transformation from the preprohormone to the prohormone in the ER lumen. A more detailed account of this process is found in the text.

the preprohormone through the ER membrane is facilitated by translocator proteins which form an aqueous pore.[8,9] The growing preprohormone is translocated through this aqueous pore and the signal sequence is cleaved from the prepropeptide by an enzyme, the signal peptidase, on the internal surface of the ER membrane. Elongation of the prohormone then proceeds through the aqueous pore in the ER membrane, enabling the synthesis of the prohormone and its translocation into the lumen of the ER to occur simultaneously. A simplified illustration of this process is shown in Fig. 2.

The peptide precursor protein, without its signal sequence, is released into the lumen of the ER and is now referred to as the prohormone or propeptide. The transformation from the prepro to prohormone represents the first post-translational modification of the preprohormone. Other post-translational modifications also occur in the ER, perhaps the most important being the formation of disulfide bridges.[10] These are usually intra-prohormone bridges, but inter-prohormone disulfide bridges are also known to occur. Other early post-translational processing events may also occur in the ER. For example, the N-terminal amino acid residues of the insect adipokinetic hormone (AKH) prohormones are modified to pyroglutamate prior to their transfer from the ER into the Golgi. Further post-translational processing events (see below) probably occur in the trans-Golgi or within the immature secretory vesicles.

Processing and packaging

The generation of biologically active peptides is achieved by the action of prohormone convertases and is usually initiated by endoproteolytic cleavage of the prohormone.[5] Such cleavages may liberate one or several peptides and their sequences may be identical, similar, or entirely different. Typically, basic amino acid residues (e.g. Arg, Lys) within the prohormone sequence serve as recognition sites for the endopeptidase(s) and most often a pair of basic amino acids (Lys–Arg, Arg–Lys, Arg–Arg) mark the sites of prohormone cleavage. After the endopeptidase has cleaved the prohormone C-terminal to the basic amino acid pair, the remaining basic residues are removed by carboxypeptidases. In many cases, a remaining C-terminal Gly residue is enzymatically converted to a C-terminal amide group and more than one enzyme may be necessary for this processing event.[11]

Neuropeptide diversity in neurons, therefore, can be produced by processing a prohormone containing more than one biologically active sequence. Diversity can also be generated by a phenomenon known as 'differential processing'. This refers to the fact that a single prohormone is not necessarily processed in the same way in different tissues or in different neurons.[12] Thus, a single prohormone (i.e. a single gene product) can generate a different cocktail of peptides, depending on the cell in which this precursor is expressed. Little is known about the mechanism of differential processing, but it may depend on differences in prohormone convertases contained in different neurons or on differences in precursor conformations in different neurons.

Many questions concerning the enzymes involved in the processing of prohormones remain unanswered. For example, how many enzymes are required? How similar are the enzymes in different organisms? How is the synthesis of the processing enzymes regulated? Do the processing enzymes involved in neuronal peptide biosynthesis resemble those which have been isolated from yeast[13] and which appear to have substrate specificities like those which must exist in neurons? Answers to such questions will help in the design of specific enzyme inhibitors which may interfere with the process of biosynthesis of essential neuropeptide transmitters and hormones.

As a first step toward isolating such enzymes, we are utilizing an in-vitro system to reconstitute neuropeptide processing of the locust AKH I prohormone, P1. AKH is synthesized by the neurosecretory cells of the corpora cardiaca (CC), major neurosecretory organs in insects. When incubated with homogenates of the CC, synthetic P1 is processed to yield the final peptide products found in the CC neurosecretory cells. In addition, we have isolated and sequenced intermediates of AKH I processing, the C-terminally extended peptides AKH–Gly–Lys–Arg, AKH–Gly–Lys, and AKH–Gly. The results thus indicate that AKH I is generated by sequential processing of P1 by endopeptidase (to remove AKH from the prohormone) carboxypeptidase (to remove Lys–Arg) and peptidyl-glycine mono-oxygenase (to form a C-terminal amide from gly) activities in the CC cells. Further studies using this system will enable us to identify the intracellular site(s) at which these processing steps occur and, we hope, to isolate and characterize the enzymes responsible.

Secretory vesicles produced by the Golgi apparatus contain the products of prohormone processing. For a prohormone which contains sequences of more than one biologically active peptide, this is illustrated diagrammatically in Fig. 3. Each vesicle may contain the full complement of peptide products. Since the stoichiometry of the peptides in the prohormone is fixed by the structure of the prohormone itself, it would be natural to assume that the ratios of the various peptides in a single secretory vesicle would reflect the stoichiometry of the prohormone. In fact this is not necessarily the case. Different vesicles containing peptides derived from a single prohormone can contain different ratios of peptides. This fact increases further the potential diversity in the chemical messages liberated by peptidergic neurons.

A specific example of this type of differential packaging of neuropeptides is found in the bag cell system of *Aplysia*.[14,15] The precursor of the egg-laying hormone (ELH) contains several peptides, the ELH itself and the so-called α,

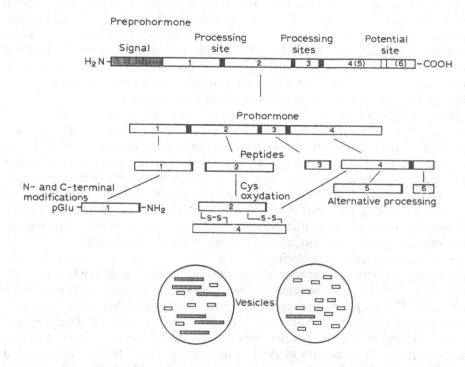

Fig. 3. A generalized diagram illustrating the processing of a preprohormone to the biologically active peptides. Alternative processing is also illustrated in this model. Peptides numbers 5 and 6 would be produced by the use of a potential processing site in peptide 4. Varying peptide stoichiometries in individual vesicles can result by a variety of means during processing (see text).

β and γ bag cell peptides, or BCPs. There is one copy of each of these peptides in the precursor. In the trans-Golgi the precursor is first cut into two segments, one containing ELH and the other the three BCPs. Somehow the two segments of the precursor are segregated by the Golgi into different vesicle types. This sorting process allows the cells to degrade differentially one of the vesicle types, resulting in the production of a peptide stoichiometry which would not be predicted from the precursor structure. Thus, in this example, whereas the precursor contains equal numbers of copies of the BCPs and ELH, the bag cells produce an ELH to BCP ratio of about 5 to 1. Some of the secretory vesicles in these cells contain predominantly ELH and others predominantly the BCPs.

This example and others suggest that the chemical signal derived from a peptidergic neuron cannot simply be understood in terms of the peptide content of a prohormone. Since these neurosecretory cells and others apparently regulate the peptide stoichiometry, it is possible that there is signalling information in the peptide stoichiometry. Some invertebrate pep-tidergic neurons produce peptide cocktails from more than one co-localized prohormone. An example of this is found in the CC system of the locust, the site of synthesis of the two adipokinetic hormones AKH I and AKH II.[16–18] The neurosecretory cells of the CC each produce two prohormones, pro-AKH I and pro-AKH II, from two different *m*RNAs. In the ER these prohormones form dimers prior to being processed. In this way three precursors are produced, the two homodimers and the one heterodimer of the two prohor-mones. Specific proteolysis of these dimeric precursors generates a cocktail of peptide products in the cell: AKHs I and II, and dimeric peptides called APRPs 1, 2 and 3 (for AKH precursor-related peptides 1, 2, and 3). The peptide stoichiometries are varied in this system simply by varying the relative rates of synthesis of the two prohormones.[19] When they are produced in equal amounts the random dimerization of the prohormones produces a binomial distribution of the three [dimer] precursors, 1:2:1. Predictable skewed binomial distributions of product peptide ratios are produced when pro-AKH I is produced in greater amounts than pro-AKH II. The cells producing the AKH peptides change the rates of synthesis of the two precursors systemati-cally through development, suggesting an important role for peptide stoichi-ometry in the generation of complex chemical signals from peptidergic neurons.[19]

RELEASE AND ACTION

Peptides, like other transmitters, are released when the vesicles containing them fuse with the cell membrane at the release sites (Fig. 4). Release is often a highly regulated process which is controlled ultimately by the concentration of intracellular free calcium at the release site.[20] Calcium concentration in the neurone or neuroendocrine cell can be affected by transmembrane voltage by

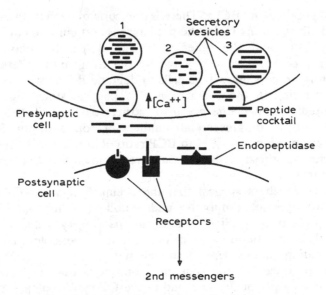

Fig. 4. Illustration of the release of peptide cocktails from neurosecretory cells. Different peptide stoichiometries can be contained in different vesicles. The endopeptidase on the postsynaptic cell is supposed to be a peptide inactivating protease. For further explanation see the text.

the activation of voltage-sensitive calcium channels. Alternatively, release can be caused by 'releasing factors' which may act through a chemical second messenger system to cause indirectly an increase in the concentration of calcium. Peptide releasing factors represent unexplored territory in the invertebrate nervous system, but they are clearly of great potential importance. An agent which would provoke unregulated or inappropriate peptide release would have a serious disruptive effect on the organism and could provide the basis for developing novel control compounds. An example of a peptide releasing factor in insects is octopamine which is involved in the release of the AKH peptides.[21]

Released peptides reach their targets (which can be very close to the release site for peptide transmitters, or at a distance for peptide hormones) and bind to a cell surface receptor linked to a signal-amplification system. There are essentially two ways in which signals from peptides (and other transmitters) are amplified by receptors.[22] The first involves receptors which change their ionic permeability when the ligand binds. Such receptors are known as ligand-gated ion channels. Since the cell membrane has a relatively high electrical resistance and capacitance, a small trickle of ions through the receptor channel produces significant voltage shifts across the membrane. Those shifts in voltage can then be amplified and used as signals for other events. Amplification is achieved because there are other ion channels in the membrane (voltage-gated) which sense the small changes in voltage caused by the ligand-gated channel and amplify them.

The second way to amplify the peptide signal is to involve an enzymatic activity. Receptors which do this are referred to as non-ion-channel-linked. Simply speaking there are two types of these, the G-protein-linked receptors in which the link to the enzymatic activity is indirect and the tyrosine kinase type, in which the receptor itself is also an enzyme. G-protein-linked receptors characteristically have three functional parts, the seven transmembrane spanning regions, an extracellular domain which may be involved in ligand binding and an intracellular domain which participates in linking the peptide signal to the enzymatic synthesis of second messengers (cAMP, IP_3 and DAG).[23] It is interesting to note that this class of receptor is quite well studied in vertebrates and there is a good deal of homology between different receptors of this type, especially in the transmembrane spanning regions. Perhaps a short-cut to obtaining structures for such receptors in invertebrates is to exploit this feature and to clone invertebrate receptors based on assumed homology with their vertebrate counterparts. This would, however, lead to obtaining structures of receptors prior to knowing their function and it is normally easier to proceed from function to structure than from structure to function.

The other class of non-ion-channel-linked receptors are the so-called tyrosine protein kinases.[24,25] When stimulated they phosphorylate tyrosine residues on a variety of proteins inside the cell—they can even phosphorylate their own tyrosines. It is through these types of receptor that many peptide growth factors and regulators of mitotic activity are thought to exert their influence. Unfortunately, very little is known about the tyrosine kinase type of receptor in invertebrate nervous systems. They are, however, probably very important with essential roles in growth and development.

Some peptide signals are known to mediate their actions through stimulating cGMP production but little is known about their class of receptor in invertebrates.

INACTIVATION

Chemical signals in the nervous system cannot be allowed to persist indefinitely and must be actively terminated. Non-peptide transmitters can be inactivated by high-affinity uptake systems. Soon after they are released they can be mopped up, internalized, and may be reused. Neuropeptides, on the other hand, generally are inactivated by degradative membrane-associated enzymes, the active sites of which are directed extracellularly.[26] Neuropeptides with protected *N*- and *C*-termini are rendered inactive by the action of endopeptidases which cleave the peptide at specific sites to yield fragments which have no biological activity (Fig. 4). Neuropeptides with free *N*- and *C*-termini may be inactivated by amino or carboxypeptidases. Examples of neuropeptide-degrading endopeptidase and aminopeptidase activities in the nervous system of insects have been documented.[27,28] This scheme appears to hold for insect neuropeptides with actions outside the nervous system, as well. We have recently demonstrated that circulating AKH I and II are inactivated by a membrane-associated endopeptidase located on the extracellular surfaces of several peripheral tissues in the desert locust.[29] Both the CNS and peripheral AKH-inactivating endopeptidases exhibit pH optima, inhibitor sensitivities and substrate specificities similar to those of endopeptidase 24.11, a mammalian enzyme implicated in peptide degradation.[26] Our work thus strengthens the notion put forward by Matsas *et al.*[30] that peptide-inactivating enzymes are limited in number, but relatively broad in their peptide specificity. Much work needs to be done in this area, particularly in light of the fact that one could use these enzymatic activities to develop assays for finding potentially useful enzyme inhibitors. If these peptide inactivating enzymes are located at an accessible target in the parasite or pest, inhibitors may be expected to disrupt the normal functioning of the nervous system or of other peptide target tissues.

SUMMARY AND CONCLUSIONS

The involvement of a diversity of neuropeptides and peptide hormones in the regulation of a wide variety of essential processes in the invertebrates suggests that neuropeptidergic systems offer a fertile area for the development of pesticidal and insecticidal compounds. Here, we suggest that there are three aspects of peptidergic systems which may be vulnerable to interference: (i) neuropeptide synthesis; (ii) neuropeptide action and (iii) neuropeptide inactivation.

Interference with neuropeptide synthesis could probably be most readily achieved by the development of specific inhibitors of the enzymes involved in processing of peptide precursor proteins. Interference with neuropeptide action requires information about the interaction between the peptide ligand and its receptor. Currently very little is known about the pharmacology or structure of neuropeptide receptors in invertebrates. While this information may soon be forthcoming, it seems unlikely that peptides themselves could be used directly as pesticides or insecticides. Peptide mimetics, however, might be useful if they could be designed with the appropriate characteristics allowing them, for example, to be ingested. Finally, interference with neuropeptide inactivation, as with biosynthesis, would depend upon the development of specific enzyme inhibitors.

It is of course easy to underestimate the difficulties encountered in applying neurobiological knowledge to the development of new leads for novel pesticides of novel targets. Certainly the views of Drs Voss and Neumann in their prologue served to focus the mind on the real problems faced at the cutting edge of R&D in this industry. Neuroscience, according to the thesis of Drs Voss and Neumann, has little utility in the development of new lead compounds but does have an important role to play in explaining the modes of action, resistance and selectivity of existing compounds. From this point of view neuroscience has a secondary role to play, supporting for example the development and maintenance of pesticides which are already in the market-place, or close to the market. Views expressed in this essay disagree quite fundamentally with the conclusion. Unfortunately, however, we cannot cite the specific examples of the primary utility of neuroscience in developing lead compounds. This is because neuroscience has not yet had sufficient time to prove itself. It would be unfortunate indeed if the agrochemical industry relegated neuroscience to a secondary role at this early stage. That surely would ensure that it certainly will not have a primary role to play in developing leads.

Perhaps the agrochemical industry could benefit in the future from developments in the pharmaceutical industry in which a far more important role is played by neuroscience in developing drugs for the future. In this regard it is interesting to note that the pharmaceutical industry is very actively involved in basic research on peptidergic systems.[31] The growing interest in peptide pharmaceuticals will undoubtedly lead to the development of sophisticated methods and strategies for developing peptide mimetics and delivery systems for interacting with on peptidergic systems in mammals. Moreover, major advances are currently being made in the design of protease inhibitors, driven largely by the need to produce anti-viral drugs. We feel sure that these developments in the pharmaceutical industry could be very instructive and may well provide the signposts for future developments in agrochemical R&D. The impact of such developments in the pharmaceutical industry would be facilitated by far closer ties between scientists designing new drugs (ligands and enzyme inhibitors) and neuroscientists in the agrochemical industry. No such benefits will derive unless neuroscience is given a primary role to play in the development of novel pesticides.

REFERENCES

1. O'Shea, M., Insect neuropeptides: pure and applied. In: *Proceedings of USDA International Conference on Insect Neurochemistry and Neurophysiology*. Humana Press, Clifton, New Jersey, 1986, pp. 3–27.
2. O'Shea, M. & Schaeffer, M., Neuropeptide function: The invertebrate contribution. *Ann. Rev. Neurosci.*, 8 (1985) 171–98.
3. O'Shea, M., Neuropeptides in insects: possible leads to new control methods. In: *New Approaches to New Leads for Insecticides*, ed. H. von Keyserlingk. Springer-Verlag, Berlin, 1985, pp. 133–51.
4. O'Shea, M., Introduction to neuropeptides: perspectives for the parasitologist. *Parasitology*, 102 (1991) 871–5.
5. Sossin, W. S., Fisher, J. M. & Scheller, R. H., Cellular and molecular biology of neuropeptide processing and packaging. *Neuron*, 2 (1989) 1407–17.
6. Walter, P. & Blobel, G., Translocation of proteins across the endoplasmic reticulum. II. Signal recognition protein (SRP) mediates the selective binding to microsomal membranes of *in vitro*-assembled polysomes synthesizing secretory proteins. *J. Cell Biol.*, 91 (1981) 551–6.
7. Rapoport, T. A., Protein transport across the ER membrane. *Trends Biochem. Sci.*, 15 (1990) 355–8.

8. Simon, S. M. & Blobel, G., A protein-conducting channel in the endoplasmic reticulum. *Cell,* **65** (1991) 371–80.
9. Lingappa, V. R., More than just a channel: provocative new features of protein traffic across the ER membrane. *Cell,* **65** (1991) 527–30.
10. Wold, F., *In vivo* chemical modification of proteins (post-tranlational modification). *Ann. Rev. Biochem.,* **53** (1981) 783–814.
11. Bradbury, A. F. & Smyth, D. G., Peptide amidation. *Trends Biochem. Sci.,* **16** (1991) 112–15.
12. Douglass, J., Civelli, O. & Herbert, E., Polyprotein gene expression: generation of diversity of neuroendocrine peptides. *Ann. Rev. Biochem.,* **53** (1984) 665–715.
13. Fuller, R. S., Sterne, R. E. & Thorner, J., Enzymes required for yeast prohormone processing. *Ann. Rev. Physiol.,* **50** (1988) 345–62.
14. Fisher, J. M., Sossin, W., Newcomb, R. & Scheller, R. H., Multiple neuropeptides derived from a common precursor are differentially packaged and transported. *Cell,* **54** (1988) 813–22.
15. Sossin, W. S., Fisher, J. M. & Scheller, R. H., Sorting within the regulated secretory pathway occurs in the trans-Golgi network. *J. Cell. Biol.,* **110** (1990) 1–12.
16. Hekimi, S., Burkhart, W., Moyer, M., Fowler, E. & O'Shea, M., Dimer structure of a neuropeptide precursor established: consequences for processing. *Neuron,* **2** (1989) 1363–8.
17. Schultz-Aellen, M.-F., Roulet, E., Fischer-Lougheed, J. & O'Shea, M., Synthesis of a homodimer neurohormone precursor of locust adipokinetic hormone studied by *in vitro* translation and cDNA cloning. *Neuron,* **2** (1989) 1369–73.
18. O'Shea, M., Hekimi, S. & Schulz-Aellen, M.-F., Locust adipokinetic hormones: Molecular biology of biosynthesis. In *Molecular Insect Science,* ed. H. H. Hagedorn, J. G. Hildebrand, M. G. Kidwell & J. H. Law. Plenum Press, New York, 1990, pp. 189–97.
19. Hekimi, S., Fischer-Lougheed, J. & O'Shea, M., Regulation of neuropeptide stoichiometry in neurosecretory cells. *Journal of Neuroscience,* **11** (10) (1991) 3246–56.
20. Smith, S. J. & Augustine, G. J., Calcium ions, active zones and synaptic transmitter release. *Trends Neurosci.,* **11** (1988) 458–64.
21. Orchard, I., Adipokinetic hormones—an update. *J. Insect Physiol.,* **33** (1987) 451–63.
22. Hollenberg, M. D., Mechanisms of receptor-mediated transmembrane signalling. *Experientia,* **42** (1986) 718–27.
23. Weiss, E. R., Kelleher, D. J., Woon, C. W., Sopakkar, S., Osawa, S., Heasley, L. E. & Johnson, G. L., Receptor activation of G-proteins. *Faseb J.,* **2** (1988) 2841–8.
24. Hanks, S. K., Quinn, A. M. & Hunter, T., The protein kinase family: conserved features and deduced phylogeny of the catalytic domains. *Science,* **241** (1988) 42–52.
25. Yarden, Y. & Ullrich, A., Growth factor receptor tyrosine linases. *Ann. Rev. Biochem.,* **57** (1988) 443–78.
26. Turner, A. J., Neuropeptide signalling and cell-surface peptidases. In *The Biology and Medicine of Signal Transduction,* ed. Y. Nishizuka *et al.* Raven Press, New York, 1990, pp. 467–71.
27. Isaac, R. E., Proctolin degradation by membrane peptidases from nervous tissues of the desert locust (*Schistocerca gregaria*). *Biochem. J.,* **245** (1987) 365–70.
28. Isaac, R. E., Neuropeptide-degrading endopeptidase activity of locust (*Schistocerca gregaria*) synaptic membranes. *Biochem. J.,* **255** (1988) 843–7.
29. Rayne, R. C. & O'Shea, M., Inactivation of neuropeptide hormones (AKHI and AKHII) studied *in vivo* and *in vitro*. *Insect Biochem. Mol. Biol.,* **22** (1992) 25–34.

0. Matsas, R., Fulcher, I. S., Kenny, A. J. & Turner, A. J., Substance P and [Leu]enkephalin are hydrolysed by an enzyme in pig caudate synaptic membranes that is identical with the endopeptidase of kidney microvilli. *Proc. Natl. Acad. Sci. USA,* **80** (1983) 3111–15.

1. Ward, F. (ed.), *Peptide Pharmaceuticals.* Open University Press, Buckingham, 1991.

SECTION 3

Applications of Molecular Biology

11

Cloning Nematode Acetylcholine Receptor Genes

JAMES A. LEWIS

Division of Life Sciences, The University of Texas at San Antonio, San Antonio, Texas 78249, USA

JOHN T. FLEMING

The MRC Laboratory of Molecular Biology, Hills Road, Cambridge CB2 2QH, UK

&

DAVID BIRD

Department of Nematology, University of California, Riverside, California 92521, USA

INTRODUCTION

Recombinant DNA technology has made it possible to clone receptors from many organisms by cross-hybridization or by the polymerase chain reaction. It may be difficult, though, to establish the functional importance of any clone obtained. We describe the cloning of nematode acetylcholine receptor genes by selection for resistance to levamisole, a scheme providing assurance that the clones obtained are functionally related.

Levamisole kills nematodes by causing toxic muscle hypercontraction. Brenner[1] found that mutants of the nematode resistant to levamisole could easily be isolated and he suggested that levamisole acted by stimulating a cholinergic pathway. Study of the comparative pharmacological responses of normal, wild-type worms and such levamisole-resistant mutants indicated that levamisole and the related nematocide pyrantel were both nicotine analogs that acted on an acetylcholine receptor present on nematode muscle.[2] These conclusions were supported by electrophysiological observations made on *Ascaris* muscle.[3] Binding assays performed with a tritiated derivative of levamisole detected a receptor with a nicotinic pharmacological profile in wild-type extracts.[4] Mutants of the six genes conferring the greatest resistance to levamisole were found to be altered or deficient in levamisole receptor binding.[5] Surprisingly, the loss of receptor function only moderately impairs the motor behavior of receptor mutants as adults or older juveniles but strongly incapacitates the mutants as young juveniles.[6] Additionally, receptor mutants, although completely unresponsive to nicotine analogs, still have moderate

155

response to direct acetylcholine analogs and cholinesterase inhibitors.[2,5] Thus, the levamisole receptor, although present throughout life, appears to be of greatest functional importance to young juvenile nematodes and there appears to be more than one kind of acetylcholine receptor in nematodes. Receptor mutants should not only be useful for studying the levamisole receptor itself, as we describe in this report, but by lack of levamisole receptor function, the appropriate mutant should facilitate the pharmacological study of other acetylcholine receptor types in the nematode.

We have used the dispensability of levamisole receptor function and the ease of isolating levamisole receptor mutants by drug-resistant selection to obtain both transposon insertion mutants and chromosomal aberration mutants useful for cloning receptor genes by restriction fragment polymorphism analysis. Transposons inserts causing inactivation of a receptor gene were identified by comparing the pattern of Tc1 transposon-containing restriction fragments in mutants with the pattern found in unmutated control constructs bearing the same chromosomal region from the parent mutator strain. Chromosomal aberrations found in independently generated *gamma*-ray mutants were used to confirm the association of a particular restriction fragment with a receptor gene. By this approach, mutants useful for cloning have been isolated in six genes known to be important to levamisole receptor expression and two genes, *unc-29* and *unc-38* likely to encode subunits of this apparent nicotinic acetylcholine receptor,[5] have been cloned.

METHODS

Nematode strains

The wild-type strain of *C. elegans* used was the Bristol strain.[1] The Bergerac BO transposon mutator strain[7] was obtained from the Caenorhabditis Genetics Center and TR679 mutator strain was kindly provided by P. Anderson.[8]

Mutant isolation

For the isolation of spontaneous transposon-induced mutations, 30 (BO strain) or 40 (TR679 strain) adult hermaphrodites were placed on a 100 mm Petri plate spread with bacteria. After 4–5 days at 20°C, progeny worms were washed off the plate and placed to one side of a plate containing 1 mM levamisole, as previously described.[6] Drug plates were then screened at daily intervals for extremely levamisole-resistant mutants. To isolate γ-ray-induced mutants, worms were first irradiated over 6 minutes with 1500 rads from a ^{60}Co source and 20 mutagenized adults instead put out on each 100 mm plate.

Identification of restriction fragment length polymorphisms in mutants

The mutator strains used contain high copy numbers of the transposon Tc1. To be able to identify a novel Tc1 element possibly causing a levamisole resistance mutation by genomic probing with Tc1 DNA and to largely eliminate extraneous background Tc1 elements, the mutation was balanced against a Bristol chromosome containing left- and right-flanking markers and back-crossed 12 times into a strain homozygous for the same flanking Bristol genetic markers. The mutation was then recombined with left- and right-flanking markers and, if necessary to achieve adequate viability, separated by further recombination from one or both markers, before isolating a strain homozygous for the backcrossed mutation. Control constructs homozygous for approximately the same high-copy Tc1 chromosomal region but otherwise having a Bristol low-copy Tc1 genetic background were generated by backcrossing the wild-type allele present in the parent mutator strain into a Bristol strain having an ethylmethane sulfonate-induced mutation in the gene of interest usually flanked by the same left and right genetic markers used in the mutant constructions. For producing *unc-38* mutant constructs, *unc-57(e406)* and *dpy-5(e61)* marker mutations, 0·5 map units to either side of *unc-38* were used and for control constructs, *unc-11(e47)* was used instead of *unc-57*; for *unc-29* constructs, *unc-13(e450)* and *lin-11(n389)*, 1·1 and 1·6 map units to either side of *unc-29*, were used.

Recombinant DNA techniques

Standard recombinant DNA techniques were employed.[9] The Tc1 transposable element used as a genomic probe was an EcoRV fragment prepared by D. Bird from an isolate supplied by S. Emmons. The flanking genomic DNA associated with a novel Tc1-induced mutation was separated from the Tc1 DNA in any genomic subclone by excision with EcoRV and gel purification. An aliquot of an EMBL4 genomic library was generously supplied by C. Link and a Lambda ZAP total nematode cDNA library by R. Barstead.

RESULTS AND DISCUSSION

Mutant isolation

Mutations representing putative spontaneous transposon insertions and putative γ-ray-induced chromosomal aberrations could be readily isolated (Table). Southern blot analysis (below) showed a substantial fraction of *unc-38* and *unc-29* mutants tested contained inserts or chromosomal aberrations (from -rays) in the mutated genes. Tom Barnes (pers. comm., 8 August 1989) and Michael Hengartner (per. comm., 5 February 1991) have obtained similar results upon analyzing the *lev-1* and *unc-50* mutations, respectively. These

TABLE 1

The Isolation of Levamisole Receptor Mutants

Method of isolation	Number of extremely drug-resistant isolates					
	unc-29	unc-38	unc-63	unc-74	unc-50	lev-1
EMS[a]	74	44	59	27	5	2[b]
BO mutator	2	11	10	2	1	2
TR679 mutator	6	3	12	2	0	5
γ-rays	5	7	8	8	2	4

[a] EMS = ethylmethane sulfonate. Data taken from Ref. 6 for comparison.
[b] For *lev-1*, only partially resistant isolates were obtained, except for EMS mutagenesis, where partially resistant isolates were discarded and rare extremely resistant mutants were counted.[6]

results are consistent with the previous conclusion based on ethylmethane sulfonate-induced forward mutation rate that most, if not all, receptor genes identified by levamisole resistance are not essential to viability and can be freely mutated.[6] The frequency of isolating mutants with the TR679 mutator strain was at least fourfold higher than with the BO mutator strain, as previously observed,[8] and the relative frequency with which *unc-38* mutations were isolated in a TR679 background appears decreased. The BO mutants in Table 1 were isolated from 163 selection plates whilst the same number of TR679 mutants was found on only 36 selection plates. γ-ray-induced mutants were also found at high frequency (34 mutants on 40 plates).

We first chose to clone *unc-38* and *unc-29,* two genes likely to encode levamisole receptor structural subunits, both to confirm that the genes affected by levamisole resistance mutation encoded an acetylcholine receptor and to obtain the molecular tools to analyze the effects of other resistance mutations on receptor *m*RNA and protein expression.

Cloning *unc-38*

The ability to easily obtain multiple independent putative transposon-induced mutations in *unc-38* made the gene very easy to clone. We reasoned that if the gene were contained within one to a few restriction fragments, that mutating the gene independently over and over again by Tc1 transposon insertion ought to lead to the creation of one of a very few novel Tc1-containing restriction fragments in each of the backcrossed mutants examined, but that none of these novel restriction fragments should ever be observed when the same unmutated *unc-38* chromosomal region from the parent mutator strain was backcrossed into a Tc1 low-copy number wild-type strain. This expectation proved correct (Fig. 1, lanes 4–6). Four of five *unc-38* BO strains contained the same novel 4·7 kilobase HindIII restriction fragment created by the insertion of Tc1 into a 3·1 kilobase genomic fragment. Further indication that this restriction frag-

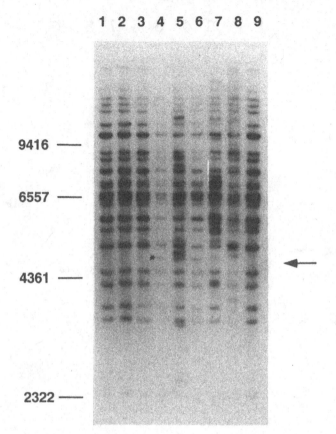

FIG. 1. Identification of Tc1 insertions into the *unc-38* receptor gene. A Southern blot of HindIII-digested genomic DNA from three backcrossed *unc-38* mutants, five unmutated backcrossed control constructs, and the Bristol wild-type strain of *C. elegans* was probed with nick-translated Tc1. The asterisk (*, indicated by arrow) indicates a 4·7 kilobase Tc1-containing HindIII fragment found in all three independent *unc-38* mutants (lanes 4–6) and a fourth *unc-38* mutant (not shown) but absent in all five control constructs (lanes 2, 3, and 7–9) and the wild type (lane 1). The *unc-38* mutants, respectively, are strains ZZ1034, ZZ1007, and ZZ1021, and the control constructs ZZ1026, ZZ1030, ZZ1009, ZZ1032, and ZZ1035.

ment arose *de novo* concurrently with the mutation of *unc-38* was obtained by probing a genomic Southern blot of the parent BO mutator strain, the *unc-38* mutants, the normal Bristol wild type, and several control constructs with the genomic DNA flanking one of the mutant Tc1 inserts (Fig. 2). The size of the restriction fragment is 3·1 kilobases in all but the four *unc-38* mutants in which the novel Tc1 insert was identified. The fragment size in these mutants is 4·7 kilobases, larger by the size of Tc1 (lanes 2, 4, 6 and 7).

Tc1 transposon-induced mutations often spontaneously revert.[7] The loss of mutant phenotype is usually accompanied by corresponding reversion to the

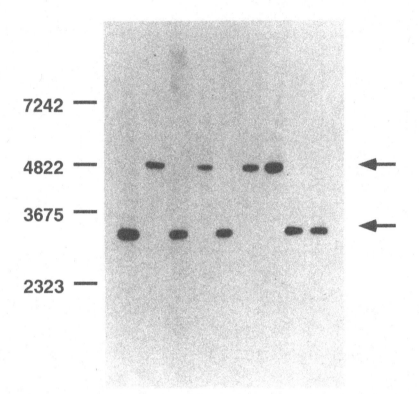

Fɪɢ. 2. Confirmation of restriction fragment length differences between *unc-38* mutants and the parent mutator strain, BO. A Southern blot of HindIII-digested genomic DNA from five spontaneous *unc-38* mutants, two control constructs, the Bergerac BO mutator strain, and the Bristol wild-type strain of *C. elegans* was probed with subcloned genomic DNA flanking the Tc1 insertion identified by the asterisk in Fig. 1, lane 6. Four of the five *unc-38* mutants (lanes 2, 4, 6, 7) show a fragment 1·6 kilobases larger (top arrow) than the 3·1 kilobase HindIII fragment (bottom arrow) found in the BO Bergerac parent mutator strain (lane 3), the Bristol wild type (lane 1) or two control constructs ZZ1009 and ZZ1026 (lanes 8 and 9). Although the control construct ZZ1009 shown in lane 7 of Fig. 1 has a Tc1 fragment very similar in size to that associated with *unc-38* mutation (see asterisk, Fig. 1), the *unc-38* region of that same control construct is shown to be 3·1 kb (kilobase) in size and unmutated by Tc1 (lane 8, this figure). The *unc-38* mutants are ZZ1021, ZZ1034, ZZ1037, ZZ1028, and ZZ1007 (lanes 2, 4–7).

wild-type restriction fragment size, confirming an association between fragment size and mutant state. Since we failed to obtain any *unc-38* revertants, we isolated homozygously viable γ-induced *unc-38* mutations to obtain independent confirmation that the HindIII fragment identified was part of the *unc-38* gene. Since most genes are essential to nematode viability but levamisole receptor genes do not appear to be, selecting for homozygous viability in a γ-ray mutant helps limit how far a deletion containing *unc-38* can extend to the left or right along the chromosome beyond the vicinity of *unc-38*. Figure 3 shows that two of five *unc-38* γ-ray mutants suffer alteration of the same HindIII fragment first identified in association with *unc-38* Tc1 mutations (lanes 4 and 6).

Cloning *unc-29*

A different approach was taken to clone *unc-29* because few spontaneous mutants were initially available. Mapping experiments by J. Thomas and ourselves identified two Tc1 elements closely linked to a spontaneous *unc-29*

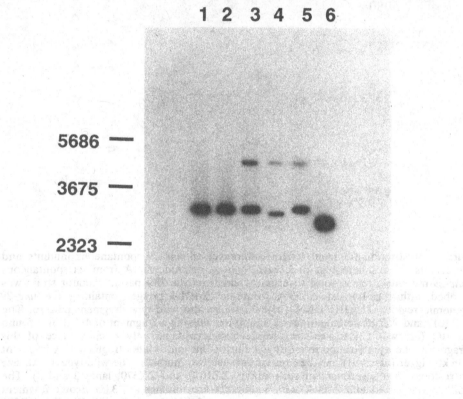

FIG. 3. Restriction fragment length differences in *unc-38* γ-ray-induced mutants. A Southern blot of HindIII-digested genomic DNA from five *unc-38* γ-ray-induced mutants and from the Bristol wild-type strain of *C. elegans* was probed with subcloned genomic DNA flanking the Tc1 insertion identified by the asterisk in Fig. 1, lane 6. There was some difficulty with partial cutting but at least two of the five γ-ray-induced mutants show restriction fragment patterns different from the 3·1 kb HindIII fragment seen in the wild type (lane 1). The fragment in the ZZ411 mutant (lane 4) is about 60 base pairs smaller and a doublet of fragments about 400 base pairs smaller is seen in the ZZ419 mutant (lane 6), indicating some more complex rearrangement of the *unc-38* gene for this mutant.

mutation but only one of the Tc1 elements (equivalent to the fragment identified by arrow D, strain ZZ568, lane 10, Fig. 4) was truly novel and not found in the vicinity of *unc-29* in an unmutated control construct. As for *unc-38* (Fig. 2), genomic DNA flanking the novel Tc1 element identified restriction fragment alterations associated with *unc-29* mutation, but for only two of six *unc-29* mutants tested. An EMBL4 genomic phage library was

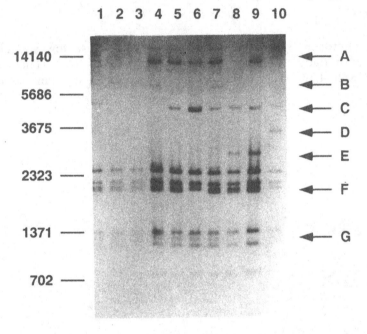

FIG. 4. Restriction fragment length differences in *unc-29* spontaneous mutants and revertants. A Southern blot of EcoRI-digested genomic DNA from six spontaneous *unc-29* mutants, from several revertants, and from the BO parent mutator strain was probed with nick-translated DNA from an EMBL4 phage containing the *unc-29* genomic region. The BO DNA (lane 1) shows the wild-type fragment pattern. The ZZ1017 and ZZ581 mutants (lanes 2 and 4) are missing a fragment of 4·7 kb size found in BO (arrow C). Because the phage chosen contains only a small piece of this fragment, the hybridization intensity is faint for this and related fragments. A fragment 1·6 kb bigger (arrow B) may be present in these two mutants. The wild-type 4·7 kb size is restored in a revertant of each mutant (ZZ1022 and ZZ570, lanes 3 and 5). The ZZ586 (lane 8) and ZZ569 (lane 9) mutants are missing a 1·3 kb EcoRI fragment (arrow G) and each has a new fragment about 1·6 kb bigger (arrow E). The ZZ578 (lane 6) and ZZ568 (lane 10) mutants are missing a 2·0 kb EcoRI fragment (arrow F), with new fragments 2·5 kb and 1·6 kb appearing at arrows C and D, respectively. New restriction fragments (vicinity of arrow A) involving DNA contained in the phage probe are seen in three of the four TR679-derived *unc-29* mutants (lanes 4, 6, and 9) and their associated revertants (lanes 5 and 7) but not in the two BO-derived mutants (lanes 2 and 10), an associated BO revertant (lane 3), or a single TR679 mutant (lane 8). The source of this anomaly may be the extremely high spontaneous mutation rate of the TR679 strain.

screened with the flanking genomic DNA probe and one phage isolate was used (Fig. 4) to reprobe the six *unc-29* mutants, several spontaneous revertants, and one of the parent mutator strains (BO). Two of the *unc-29* mutants show alteration of a 4·7 kilobase EcoRI fragment (lanes 2 and 4), two other mutants are altered in a neighboring 1·3 kilobase EcoRI fragment (lanes 8 and 9), and yet two other mutants mutated in another neighboring 2·0 kilobase EcoRI fragment (lanes 6 and 10). The cloning of *unc-29* was confirmed by observing restoration of wild-type fragment size in revertants of insertion mutations into the 4·7 kilobase (Fig. 4, lanes 3 and 5) and 2·0 kilobase fragments (lane 7) and by alteration of the 4·7 kilobase fragment in one of five homozygously viable *unc-29* γ-ray mutants (data not shown).

Conclusion

Because the 1·3 kilobase *unc-29* EcoRI fragment was flanked to each side by two other fragments associated with *unc-29* mutation (the 4·7 and 2·0 kilobase EcoRI fragments), the 1·3 kilobase fragment appeared to be well-contained within the *unc-29* gene. It was used to probe a Lambda ZAP total nematode cDNA library and cDNA clones hybridizing to the central 1·3 kilobase genomic fragment were found at a frequency of 1/100 000–1/200 000, indicating that it will probably not be hard to obtain cDNA clones of levamisole receptor genes for determination of the protein sequences encoded. In a parallel approach, one of us (J. Fleming) has used a probe derived from the *Drosophila* ARD acetylcholine receptor subunit gene[10] to screen a nematode genomic library and has found clones that by cross-hybridization, by restriction fragment pattern similarity, and by detection of restriction fragment pattern polymorphisms in our mutants correspond to the *unc-29, unc-38,* and *lev-1* receptor genes (Fleming, J., Squires, M. & Barnes, T., pers. comm., 28 June 1988, 26 July 1988 and 8 August 1989, respectively). These results further suggest that these nematode genes may encode functionally related subunits of one type of nematode acetylcholine receptor.

ACKNOWLEDGMENT

We are grateful to the Caenorhabditis Genetics Center, which is funded by the NIH National Center for Research Resources (NCRR), for providing marker mutations and relevant map data. We thank Jim Thomas, Michael Hengartner, and Tom Barnes for sharing their results with us. This investigation has been aided by NIGMS grant GM 08194.

REFERENCES

1. Brenner, S., The genetics of *Caenorhabditis elegans*. *Genetics,* **77** (1974) 71–94.
2. Lewis, J. A., Wu, C.-H., Levine, J. H. & Berg, H., Levamisole-resistant mutants of the nematode *Caenorhabditis elegans* appear to lack pharmacological acetylcholine receptors. *Neuroscience,* **5** (1980) 967–89.
3. Harrow, I. D. & Gration, K. A. F., Mode of action of the anthelmintics morantel, pyrantel and levamisole on muscle cell membrane of the nematode *Ascaris suum*. *Pestic. Sci.,* **16** (1985) 662–72.
4. Lewis, J. A., Fleming, J. T., McLafferty, S., Murphy, H. & Wu, C., The levamisole receptor, a cholinergic receptor of the nematode *Caenorhabditis elegans*. *Mol. Pharm.,* **31** (1987) 185–93.
5. Lewis, J. A., Elmer, J. S., Skimming, J., McLafferty, S., Fleming, J. & McGee, T., Cholinergic receptor mutants of the nematode *Caenorhabditis elegans*. *Journal of Neuroscience,* **7** (1987) 3059–71.
6. Lewis, J. A., Wu, C.-H., Berg, H. & Levine, J. H., The genetics of levamisole resistance in the nematode *Caenorhabditis elegans*. *Genetics,* **95** (1980) 905–28.
7. Moerman, D. G. & Waterston, R. H., Spontaneous unstable *unc-22 IV* mutations in *Caenorhabditis elegans* var. Bergerac. *Genetics,* **108** (1984) 859–77.
8. Collins, J., Saari, B. & Anderson, P., Activation of a transposable element in the germ line but not the soma of *Caenorhabditis elegans*. *Nature,* **328** (1987) 726–8.
9. Sambrook, J., Fritsch, E. F. & Maniatis, T., *Molecular Cloning: A Laboratory Manual,* 2nd edn. Cold Spring Harbor Laboratory Press, Cold Spring Harbor, New York, 1989.
10. Sawruk, E., Hermans-Borgmeyer, I., Betz, H. & Gundelfinger, E. D., Characterization of an invertebrate nicotinic acetylcholine receptor gene: the *ard* gene of *Drosophila melanogaster*. *FEBS Letters,* **235** (1988) 40–6.

12

The Molecular Biology of Potassium Channels and Mutations that Alter the Selectivity of the Pore

THOMAS L. SCHWARZ & ANDREA J. YOOL

Department of Molecular and Cellular Physiology, Beckman Center, Stanford University Medical Center, Stanford, California 94305-5426, USA

INTRODUCTION

Molecular biology and *Drosophila* genetics have combined to give us an enormous amount of information about K^+ channels. Work in many labs has resulted in the cloning of a large family of these channels. Originally cloned in *Drosophila*, homologs in other species have been isolated as well. Diversity within the family is generated by several mechanisms: six different classes of K^+ channel genes, multiple closely related genes within a single class, and alternative RNA splicing within a single gene. Easy and abundant expression of the cloned channels in *Xenopus* oocytes has promoted structure/function studies. Our lab has concentrated on identifying the pore of the channel and characterizing mutations that alter ion selectivity. Starting with the ShB splicing variant from the Shaker locus, we have identified mutations within a region called H5 that greatly enhance the ability of Rb^+ and NH_4^+ to pass through the channel. This H5 region is highly conserved in all the K^+ channel sequences of the family and has an intermediate hydrophobicity. One model is proposed, involving an eight-strand beta-barrel, that could account for the physiological data now in hand. Because of the intrinsic interest of the mechanism of ion permeation, and because of the importance of the pore as a site of pharmacological intervention, elucidating the detailed nature of the pore will be an important area for future study.

THE CLONING OF THE SHAKER LOCUS

In the molecular biology of K^+ channels, *Drosophila* has led the way. Each of the six major classes that have been discovered was found first in the fly. The first of these is represented by the *Shaker* locus. At a time when no biochemical purification for a K^+ channel could be seen on the horizon, this fly mutation offered an opportunity to learn the sequence of a channel and proved to be the key to many more. The *Shaker* fly, when lightly anesthetized with ether, rapidly shakes its legs and scissors its wings and twitches its abdomen.[1] The groundwork for the molecular biology was laid by physiological studies that indicated that the phenotype of these flies could be traced to a defect in a particular type of K^+ channel: a voltage-dependent, inactivating K^+ channel.[2,3] Once the cytological location of the locus was established,[4] three labs set out to clone the genomic DNA from this region.[5-7]

The structure of a K^+ channel

The first complete cDNA from the locus that was sequenced[8] indicated that it was indeed a channel gene and taught us the basic design that would be found in each subsequent addition to the family (Fig. 1). The hallmark of these and each of the other voltage-gated ion channels cloned to date is the so-called S4 region. Positive charges (arginine or lysine residues) are regularly spaced at every third position, with more hydrophobic residues in between. This structure is generally believed to span the membrane, despite its numerous charges and to represent the voltage-sensor. Mutagenesis of this region of Na^+ channels[9] and K^+ channels[10] has been consistent with this hypothesis; mutations altered the voltage sensitivity of the gating of the channel. A clear picture of how this structure operates has not yet emerged; how it is folded and how it moves in response to changes in the membrane potential is still unknown.

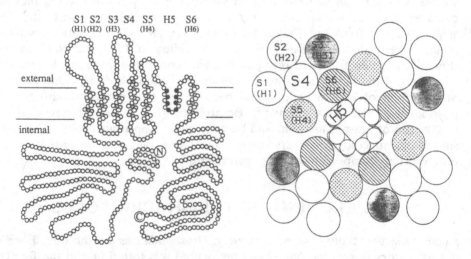

FIG. 1. Left: A diagram of the ShB potassium channel showing the most likely disposition of the channel in the membrane. Right: A cartoon of a plausible relationship of the transmembrane domains of a potassium channel, with the pore-lining H5 sequence at the center of the four subunit channel.

In addition to the putative membrane-spanning S4 segment, five highly hydrophobic regions are also predicted to cross the membrane. These were originally designated H1, H2, etc., to indicate that they are hydrophobic and to avoid presuming a particular disposition in the membrane,[8] but they are most frequently now termed S1–S6, by analogy to the presumed membrane-spanning segments of the Na^+ and Ca^{2+} channels. The long and hydrophilic amino and carboxy terminal regions are likely to lie in the cytoplasm of the cell. Originally a speculation based on comparison with Na^+ and Ca^{2+} channel sequence, this now has experimental confirmation for K^+ channels, at least with regard to the amino terminus; functional deficits due to the deletion of amino-terminal residues can be remedied by the application of a peptide mimicking this sequence to the cytoplasmic side of the channel.[11]

A region of intermediate hydrophobicity precedes the last hydrophobic segment. It has been termed H5,[8] or alternatively SS1 & SS2,[12] or the S5–S6 linker (the latter term does not clearly differentiate the hydrophobic portion of the linker from the surrounding portions). As will be described below, this region is now thought to be the lining of the pore and it is exceptionally highly conserved among all the classes of K^+ channels.

Four *Shaker* proteins are thought to come together to form a functional channel. Originally this hypothesis was based on analogy to Na^+ and Ca^{2+} channels, which look like four *Shaker* proteins attached to one another in series. Recently a clever physiological counting experiment that used mutations that alter sensitivity to a channel-blocking toxin has supported this hypothesis.[13] Figure 1 shows a model of how the putative transmembrane domains may be arranged in the protein. In constructing this model we have taken into account the following factors: (1) the H5 region probably lines the pore (see below); (2) the positively charged S4 is unlikely to be in contact with the lipid and is probably surrounded by other transmembrane regions that can provide negative charges to form ion pairs; (3) transmembrane helices are seldom arranged so that their connecting loops cross over one another (and many of these loops are very short in K^+ channels; and (4) the first three hydrophobic domains show the least sequence conservation, particularly along a single face of a helical net, and this may represent the portion in contact with membrane lipids.[14] A model with the positions of S5 and S6 reversed from the positions shown would be equally plausible.

SIX CLASSES OF K⁺ CHANNEL

A seemingly endless parade of new channels has followed the cloning of the *Shaker* channel. Four methods have been used to find them. Most were found by using the *Shaker* sequence or one of its derivatives to screen for homologs. Both PCR based methods[15] and low-stringency hybridization methods[14,16] have been used with success. The third method, expression cloning in *Xenopus* oocytes, could have yielded something unrelated to *Shaker*, but the clone proved to be a homolog of one of *Shaker's* cousins.[17] Recently, the genetic approach that was used for *Shaker* has resulted in the identification of two new classes that are too distantly related to the other channels to have been found by homology.[18,19] Mention should also be made of an unusual protein that is only 130 amino acids in length and contains only a single transmembrane domain. This protein, when expressed in *Xenopus* oocytes, produces a slowly activating voltage-dependent K^+ current.[20] Because it is so dissimilar from other channel proteins, there is still some question as to whether it is the channel itself or an inducer of channels latent in the oocyte. Mutagenesis and reconstitution into lipid bilayers may eventually resolve this issue, but the so-called I_{SK} protein will not be discussed further here.

Six classes of K^+ channels can now be distinguished, and each is represented by a single *Drosophila* gene. The mammalian homologs of these genes are much closer to the fly genes than they are to members of other classes and so the evolutionary divergence into these classes is clearly older than the vertebrate/invertebrate split. The *Drosophila* genes will be used here to name these classes for two reasons: (1) the *Drosophila* genes, without exception, came first and (2) with only a single gene for each class, they are much simpler to keep track of than their mammalian homologs. These genes are *Shaker*, *Shab*, *Shaw*, *Shal*, *eag*, and *slo*. Other nomenclatures are used and we can only hope that some consensus is reached quickly, since the literature is becoming tangled.

Shab, *Shaw*, and *Shal* were each cloned in the Salkoff lab.[16,21] They share approximately 40% homology to one another and to *Shaker*. Each resembles *Shaker* in its basic design: an S4 sequence and an H5 are found in the same relative positions among the hydrophobic domains. The sequence and the length of the terminal cytoplasmic domains are the most variable regions; in *Shab*, for example, the amino terminus is over 400 amino acids in length, twice the size of the comparable part of *Shaker*. Like *Shaker*, they are each sufficient to produce a K^+ channel in *Xenopus* oocytes and they all make voltage-dependent channels.[21] Where they differ most is in their kinetics and voltage dependence. Inactivation is much slower in *Shab* and is essentially absent in *Shaw*. *Shaw* is activated at much lower voltages. *Shaker* is the fastest channel, both in its activation and inactivation.[21]

The two newest classes, *eag* and *slo*, are named after the mutations *ether-a-gogo* and *slowpoke*. Very distant cousins of the other four classes, they have their highest homology to the others in the H5 region. They also contain an S4 sequence, though in *slo* it is very short. Neither has yet been expressed in oocytes and the function of the *eag* product remains mysterious. Physiological studies of muscles in *slo* flies, however, have already implicated *slo* in a fast inactivating K^+ current that is dependent on Ca^{2+} influx.[22] Thus this gene is likely to supply the first example for a Ca^{2+}-activated K^+ channel. Vertebrate homologs and perhaps additional fly homologs will no doubt be appearing in the wake of the fly genes.

GENERATING MORE DIVERSITY

The diversity of K$^+$ channels is augmented by two mechanisms. In vertebrates, a single class (e.g. *Shaw* or *Shaker*) may be represented by six or more separate and closely related genes. In flies, such close homologs are not known, but multiple protein products are known to be made by alternative splicing of RNA in at least two of the genes, *Shaker*[23] and *Shal*.[21] Alternative splicing is known to occur in some mammalian K$^+$ channel genes as well.[24]

Shaker is the best studied of the alternatively spliced loci;[23] it shows variation in the region that codes for the amino terminus and in the region that codes for the last transmembrane domain through the carboxy terminus. At least five amino terminals and two carboxy terminals have been found. A single amino terminus may occur with either of the two carboxy termini. It is not known, however, whether all 10 of the possible permutations arise in nature. Physiologically, these variations can show large variations from one another in their kinetics.[25] Different amino termini provide different inactivation rates. Channels with different carboxy terminals also differ in their inactivation rates but, more dramatically, differ as well in their ability to recover from inactivation. When considering these splicing variants and the large gene families in the vertebrates, however, we must realize that we do not know how these channels function in their normal milieu. Fascinating though the kinetic differences may be, they are sometimes very subtle. It is quite possible that the most important differences between the variants have eluded us so far. They could differ in their levels of expression due to promoter differences; they could differ in their turnover rates, in their signals for targeting to subcellular regions, or in their ability to be modulated by kinases or G-proteins.

A final potential source of diversity comes from the finding that different splicing variants or the products of different vertebrate homologs can coassemble in the oocyte to make a heteromultimeric channel.[26] The possible permutations of products from the *Shaker* locus now go into the thousands. The limit so far seems to be that no two products of different classes can coassemble. When *Shab, Shaw, Shal,* and *Shaker* were tested in a meticulous series of coinjection experiments,[27] no evidence for heteromultimeric channels could be found. Although within a class, coassembly of variants can occur, it should be stressed that no evidence for this 'mix and match' mechanism has been found yet in nature; it may be a contrived (but interesting) phenomenon of the oocyte expression system.

MUTATING THE PORE OF THE CHANNEL

In trying to understand the functioning of K$^+$ channels, we have concentrated on identifying the pore. We targeted the H5 region for mutagenesis because it was extremely well conserved among all the K$^+$ channels and because its intermediate hydrophobicity seemed appropriate for a region that would lie between a hydrophilic conduction pathway and the more hydrophobic core of the protein. Other topological constraints made it unlikely that it would form a transmembrane helix (this would cause the carboxy terminal to lie outside the cell), and it seemed implausible that a large extracellular loop would be hydrophobic and conserved.

Our hypothesis, therefore was that this domain would line the pore, perhaps as a beta hairpin, and we tested this hypothesis with site-directed mutagenesis.[28] Several mutations in this region, despite their conservative nature, prevented us from recording any detectable currents. Three mutations, however, have proven to be very informative: phenylalanine 433 to tyrosine (F433Y), phenylalanine 433 to serine (F433S) and threonine 441 to serine (T441S).

We examined the selectivity of the normal ShB channel and these three mutants in single channel recordings from cell-attached and pulled-off patches from *Xenopus* oocytes. To determine the relative selectivity of a test ion (X^+) to K^+, we used a method introduced by Hille.[29] In 'biionic conditions', that is when one side of the membrane has only KCl and the other has only XCl, the reversal potential that is measured is a function of the relative permeabilities of the two cations. It can be thought of as a competition between the two cations, the only charge carriers available to the channel, and the reversal potential is the energy needed to offset the relative advantage of the more permeant ion. The equation

$$P_X/P_K = \exp[(F/RT)(E_R)]$$

expresses the relationship of the permeability ratio to the reversal potential (E_R). For the wild-type channel, the selectivity sequence and the ratios of permeability relative to K^+ were: $K(1\cdot0) > Rb(0\cdot55) > NH_4(0\cdot11) \gg Na(0\cdot03)$. The permeability ratio for Na^+ is likely to be an overestimate. The single channel I/V relationships from which this sequence was obtained are shown in Fig. 2. Reversal potentials were interpolated for each ion except Na^+; no inward Na^+ currents were detectable.

The selectivity of the mutant channels was substantially altered. F433S and T433S showed dramatically increased inward currents carried by either Rb^+ or NH_4^+. The reversal potential in NH_4^+ shifted dramatically as well (Fig. 2). In F433Y, similar but smaller changes in permeation were observed. For NH_4^+ versus K^+, for example, the permeability ratio of $0\cdot11$ in the wild-type channel was shifted to $1\cdot3$ in F433S and to $0\cdot85$ in T441S, but only to $0\cdot23$ in F433Y. The two most dramatic mutations, F433S and T441S, have decreased the size of the amino acid side chains and decreased their hydrophobicity. F433Y has increased the size of the side chain while decreasing the hydrophobicity. Thus one interpretation of the effect of the mutations is that we improved the passage of the larger ions, Rb^+ and NH_4^+, by making the lining of the pore wider and more hydrophilic.

Interestingly, currents in symmetrical K^+ were unaltered by these mutations

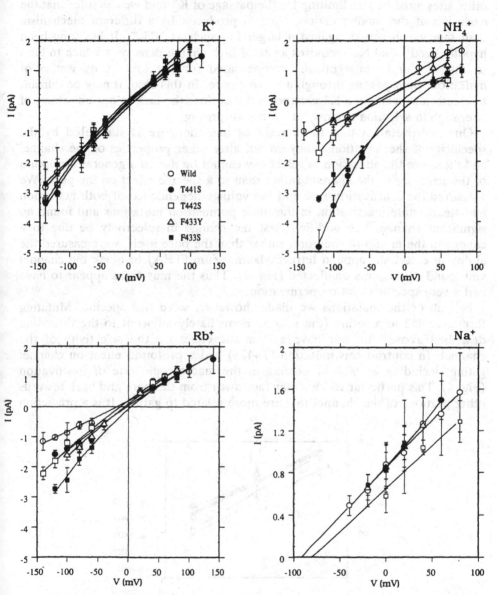

FIG. 2. Current/voltage relationships for wild-type and mutant ShB channels derived from single channel recordings from *Xenopus* oocytes. In each case the cytoplasmic surface of the patch was bathed in 100 mM KCl and the external surface was exposed to a pipette solution containing 100 mM of the chloride salt of the ion listed above. For details of the recording conditions, see Ref. 28.

(Fig. 2), and Na^+ ions continued to appear impermeable. We can infer that other sites must be rate limiting for the passage of K^+ and we can infer that the exclusion of the smaller cation, Na^+, is produced by a different mechanism than that which selects against a larger ion such as NH_4^+. It has long been hypothesized[30] that Na^+ requires a site of high charge density to which to bind so that it can be energetically compensated for the loss of its waters of hydration when passing through a narrow pore. In this case, it may be difficult to create such a site in a K^+ channel and therefore the lack in our mutations of alteration in selection against Na^+ is not surprising.

Our interpretation that these residues line the pore is supported by the specificity of the mutations; they do not alter other properties of the channel and therefore the alteration in selectivity cannot be due to a general disruption of the structure of the channel rather than to a specific effect on the pore. We measured the inactivation rate and the voltage dependence of both activation and steady state inactivation in the three permeation mutations and found no significant change.[28] In addition, lest the change in selectivity be due to a change in the mouth of the pore, rather than the pore itself, we measured the ability of externally applied tetraethylammonium (TEA) to block the channel and found it, too, was unaffected (Fig. 3). Thus the mutations appear to have had a very specific effect on permeation.

Not all of the mutations we made, however, were this specific. Mutating threonine 442 to a serine (the residue immediately adjacent to the threonine described above) did not have a dramatic effect on the selectivity of the channel. In contrast this mutation (T442S) had a profound effect on channel gating, including a 100-fold slowing in the macroscopic rate of inactivation (Fig. 4). This particular residue may face away from the pore and back towards other portions of the channel that are more related to gating. It is a prediction

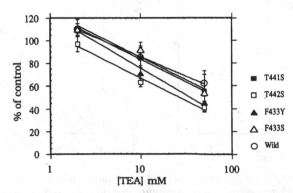

FIG. 3. The sensitivity of the ShB channel to externally applied tetraethylammonium (TEA) is not altered by the mutations. The response was measured as the amplitude of the peak outward current at +40 mV recorded by two electrode voltage clamps from *Xenopus* oocytes. The size of the currents in the presence of TEA has been normalized to the current size before TEA application (mean ± S.D.). For details of the recording conditions, see Ref. 28.

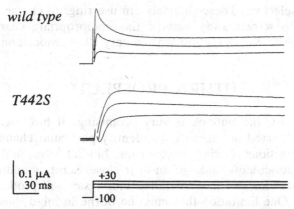

wild type

T442S

0.1 µA
30 ms
+30
-100

Fig. 4. Mutation of threonine 442 to a serine slows the rate of inactivation.[28] Currents recorded in two electrode voltage clamp of *Xenopus* oocytes are compared.

of the beta-barrel model that half of the residues in H5 would point away from the pore.

Our evidence that H5 lines the pore is corroborated by two independent lines of experimentation in two other labs. One group has mapped residues that alter the binding of pore-blocking compounds, TEA and a component of charybdotoxin. These residues are adjacent to the H5 sequence on either side.[31] These flanking regions, which are less well conserved and more hydrophilic, are likely to form the external vestibule. In addition, the mutation T441S, the same mutation that altered selectivity, was shown to have a decreased affinity for TEA when it is applied to the inside surface of the patch.[32] This residue falls approximately midway through the H5 and, if the pore is indeed a beta barrel, might be at the opening of the pore onto the internal vestibule.

Another laboratory[33] constructed a chimeric channel; they switched a 21 amino acid region that contains H5 between two members of the *Shaw* family that differed in their unit conductance and sensitivities to internal and external TEA. All of these properties were changed as the H5 region was changed; channel properties followed the nature of the donor of the H5 region.

Taken together, the changes in selectivity, conductance, and pharmacology argue strongly that H5 is the pore-lining region. While some of the effects might be explained if H5 was an extracellular domain near the outer vestibule (changes in toxin binding and conductance), other properties point to models in which the H5 is folded into the plane of the membrane, extending at least partway across. These are the effect on internal TEA binding and the selectivity changes. Selectivity of highly selective channels such as these has long been thought to depend on the lining of the pore itself. To distinguish ions of the same charge and similar size, a very close association of the ion with the channel must occur at the selectivity filter.[30] In this regard it is very different from acetylcholine receptors and other ligand-gated channels which are only charge selective. These channels can use rings of charge at either end of the channel to screen away ions of the inappropriate charge.[34] But to discriminate between K^+, Na^+, and Rb^+, a very close association is needed.

FUTURE PROSPECTS

For the biophysicist, the outlook is very promising. It has been remarkably easy to use site-directed mutagenesis to identify particular channel sequences with particular functions. Gating mechanisms, binding sites, and conductance properties will undoubtedly undergo an ever closer scrutiny and the combination of more mutations and more detailed biophysics will produce ever more detailed models. One limitation that must be borne in mind, however, is that only the side chains of the amino acids can be altered in this way: we cannot, for example, remove one of the carbonyl groups from the peptide backbone to see if it is essential to the conductance pathway. Hard structural information, from crystallography or NMR, will be very much in demand but no one can predict how quickly those answers will come. In the meantime, the challenge is great and the functional information derived from biophysical analysis of mutations will provide a framework with which to interpret future structural data.

With regard to K^+ channel permeation, the need for more mutations is clear. How many residues in the H5 region will alter selectivity? Which will affect K^+ permeation, or the selection against Na^+, or, like those reported above, the permeation of the larger cations? Will the beta-barrel model hold up? The most testable prediction of this structure is that the side chain of every other residue will project into the pore, while the others will point back towards the center of the subunit. The biophysical analysis of each mutation can also be expanded to look at binding sites for blocking ions and the permeation of small organic cations. The latter were effectively used to get an estimate of the dimensions of Na^+ channels.[35] If the pore of K^+ channels can truly be widened, these same ions, though impermeant in the wild-type channel, may now become permeant.

The hardest property to study may be the gating of the pore: what is the conformational difference between the conducting and the closed state? How is the voltage sensor coupled to the pore? If large-scale changes in protein conformation occur, mutations in many regions may alter the opening step, and other properties as well. Perhaps mutations such as T442S will provide information about this process.

The extension of what we have learned in K^+ channels to their cousins, the Na^+ and Ca^{2+} channels will offer new insights. Comparable regions of intermediate hydrophobicity and high sequence conservation can be identified in the S5–S6 linkers of these channels and the overall similarity of the structure of these channels makes it very likely that these regions will also be forming the pore. Permeation studies on Ca^{2+} channels have made very specific predictions about multiple Ca^{2+}-binding sites within the pore. Can these now be identified? These channels can conduct Na^+ and K^+ if Ca^{2+} is absent and so they provide an interesting starting point from which to study selectivity: can a mutation interfere with the permeation of one of these ions without disrupting passage of the others?

The study of K^+ channels is on a vastly different plane than it was just 5 years ago. Can the availability of clones and expression systems and our new structural insights be put to use? For the design of pesticides, the picture is uncertain. The more we know about the structure of *Shaker* and the other K^+ channels in flies, the easier it will be to design and test potential blockers. The great similarity of *Drosophila* channels to mammalian homologs has been a boon to the cloners but may be the bane of those who wish to target insects or worms for destruction, and not vertebrates. Can the subtle differences be exploited to give specificity?

The pharmacologist has great cause for optimism. Though the enormous array of K^+ channel subtypes in a single species is bewildering to the scientist, it holds out the promise of very selective targets for drugs. The variability of the regions that flank H5, i.e. the amino acids that are thought to form the vestibule region, are sufficiently varied that one can hope to design drugs to target a particular class of channels in a particular tissue while minimizing side effects on other channels and tissues. Perhaps we can look forward to channel blockers that strengthen the heartbeat without major neurological side effects. Other agents may assist neuronal spike propagation for multiple sclerosis patients, without cardiovascular or intestinal complications.

REFERENCES

1. Novitski, E., Shaker—report of a new mutant. *Drosophila Information Service*, **23** (1949) 61–2.
2. Jan, Y. N., Jan, L. Y. & Dennis, M., Two mutations of synaptic transmission in Drosophila. *Proc. Roy. Soc. London, B Biol. Sci.*, **198** (1977) 87–108.
3. Salkoff, L., Genetic and voltage clamp analysis of a Drosophila potassium channel. *Cold Spring Harbor Symp. Quant. Biol.*, **48** (1983) 221–31.
4. Tanouye, M., Ferrus, A. & Fujita, S., Abnormal action potentials associated with the Shaker complex locus of Drosophila. *Proc. Natl. Acad. Sci. USA*, **78** (1981) 6548–52.
5. Papazian, D., Schwarz, T., Tempel, B., Jan, Y. N. & Jan, L. Y., Cloning of genomic and complementary DNA from *Shaker*, a putative potassium channel gene from Drosophila. *Science*, **237** (1987) 749–53.
6. Kamb, A., Iverson, L. & Tanouye, M., Molecular characterization of *Shaker*, a gene that encodes a potassium channel. *Cell*, **50** (1987) 405–13.
7. Baumann, A., Krah-Jentgens, I., Mueller, R., Mueller-Holtkamp, F., Seidel, R., Kecskemethy, N., Casal, J., Ferrus, A. & Pongs, O., Molecular organization of the maternal effect region of the *Shaker* complex of *Drosophila*: characterization of an I_A channel transcript with homology to vertebrate Na^+ channel. *EMBO J.*, **6** (1987) 3419–29.

8. Tempel, B., Papazian, D., Schwarz, T., Jan, Y. N. & Jan, L., Sequence of a probable potassium channel component encoded at the *Shaker* locus of *Drosophila. Science,* **237** (1987) 770–75.
9. Stuhmer, W., Conti, F., Suzuki, H., Wang, X., Noda, M., Yahagi, N., Kubo, H. & Numa, S., Structural parts involved in activation and inactivation of the sodium channel. *Nature,* **339** (1989) 597–603.
10. Papazian, D., Timpe, L., Jan, Y. N. & Jan, L. Y., Alteration of voltage-dependence of *Shaker* potassium channel by mutations in S4 sequence. *Nature,* **349** (1991) 305–10.
11. Zagotta, W., Hoshi, T. & Aldrich, R., Restoration of inactivation in mutants of *Shaker* potassium channels by a peptide derived from ShB. *Science,* **250** (1990) 568–71.
12. Guy, H. & Seetharamulu, P., Molecular model of the action potential sodium channel. *Proc. Natl. Acad. Sci. USA,* **83** (1986) 508–12.
13. MacKinnon, R., Determination of the subunit stoichiometry of a voltage-activated potassium channel. *Nature,* **350** (1991) 232–5.
14. Tempel, B., Jan, Y. N. & Jan, L. Y., Cloning of a probable potassium channel gene from mouse brain. *Nature,* **232** (1988) 837–9.
15. Kamb, A., Weir, M., Rudy, B., Varmus, H. & Kenyon, C., Identification of genes from pattern formation, tyrosine kinase, and potassium channel families by DNA amplification. *Proc. Natl. Acad. Sci. USA,* **87** (1989) 4372–6.
16. Butler, A., Wei, A., Baker, K. & Salkoff, L., A family of putative potassium channel genes in Drosophila. *Science,* **243** (1989) 943–7.
17. Frech, G., VanDongen, A., Schuster, G., Brown, A. & Joho, R., A novel potassium channel with delayed rectifier properties isolated from rat brain by expression cloning. *Nature,* **340** (1989) 642–5.
18. Warmke, J. W., Drysdale, R. & Ganetzky, B., A distinct potassium channel polypeptide encoded by the Drosophila *eag* locus. *Science,* **252** (1991) 1560–2.
19. Atkinson, N., Robertson, G. & Ganetzky, B. A component of calcium–activated potassium channels encoded by the Drosophila *slo* locus. *Science,* **253** (1991) 551–5.
20. Takumi, T., Ohkubo, H. & Nakanishi, S., Cloning of a membrane protein that induces a slow voltage-gated potassium current. *Science,* **424** (1988) 1042–5.
21. Wei, A., Covarrubias, M., Butler, A., Baker, K., Pak, M. & Salkoff, L., K^+ current diversity is produced by an extended gene family conserved in *Drosophila* and mouse. *Science,* **248** (1990) 599–603.
22. Elkins, T., Ganetzky, B. & Wu, C.-F., A *Drosophila* mutation that eliminates a calcium-dependent potassium current. *Proc. Natl. Acad. Sci. USA,* **83** (1986) 8415–9.
23. Schwarz, T., Tempel, B., Papazian, D., Jan, Y. N. & Jan, L. Y., Multiple potassium channel components are produced by alternative splicing at the *Shaker* locus in Drosophila. *Nature,* **331** (1988) 137–42.
24. Luneau, C., Smith, J. & Williams, J. Molecular cloning of a rat brain K^+ channel family with homology to Shaw. *Biophysical J.,* **59** (1991) 451a.
25. Timpe, L., Schwarz, T., Tempel, B., Papazian, D., Jan, Y. N. & Jan, L. Y., Expression of functional potassium channels from *Shaker* cDNA in Xenopus oocytes. *Nature,* **331** (1988) 143–5.
26. Isacoff, E., Jan, Y. N. & Jan, L. Y., Evidence for the formation of heteromultimeric potassium channels in Xenopus oocytes. *Nature,* **345** (1990) 530–4.
27. Covarrubias, M., Wei, A. & Salkoff, L., Four cloned K^+ channel subfamilies are independent current systems. *Biophysical J.,* **59** (1991) 3a.
28. Yool, A. & Schwarz, T., Alteration of ionic selectivity of a K^+ channel by mutation of the H5 region. *Nature,* **349** (1991) 700–704.

29. Hille, B., Potassium channels in myelinated nerve. Selective permeability to small cations. *J. Gen. Physiol.*, **61** (1973) 669–86.
30. Eisenmann, G. & Horn, R., Ionic selectivity revisited: The role of kinetic and equilibrium processes in ion permeation through channels. *J. Memb. Biol.*, **76** (1983) 197–225.
31. MacKinnon, R. & Yellen, G., Mutations affecting blockade and ion permeation in voltage-activated potassium channels. *Science*, **250** (1990) 276–80.
32. Yellen, G., Jurman, M., Abramson, T. & MacKinnon, R., Mutations affecting internal TEA blockade identify the probable pore-forming region of a K⁺ channel. *Science*, **251** (1991) 939–42.
33. Hartmann, H., Kirsch, G., Drewe, J., Taglialatela, M., Joho, R. & Brown, A., Exchange of conduction pathways between two related K⁺ channels. *Science*, **251** (1991) 942–4.
34. Imoto, K., Busch, B., Sakmann, B., Mishina, M., Konno, T., Nakai, J., Bujo, H., Mori, Y., Fukuda, K. & Numa, S., Rings of negatively charged amino acids determine the acetylcholine receptor channel conductance. *Nature*, **313** (1988) 645–8.
35. Hille, B., The permeability of the sodium channel to organic cations in myelinated nerve. *J. Gen. Physiol.*, **58** (1971) 599–619.

13

FMRFamide-Related Peptides: Organisation and Expression of the Gene in the Snail *Lymnaea*

JULIAN F. BURKE, KERRIS BRIGHT, SUSAN E. SAUNDERS, ELAINE KELLETT &
PAUL R. BENJAMIN

*Sussex Centre for Neuroscience, Biological Sciences, University
of Sussex, Brighton BN1 9QG, UK*

INTRODUCTION

A neuropeptide that has aroused a great deal of interest in the last few years is the tetrapeptide, FMRFamide (pronounced 'Femerfamide'), named after the single letter abbreviations for its constituent amino acids (Phe–Met–Arg–Phe–NH₂). FMRFamide was originally isolated from the sunray Venus clam *Macrocallista nimbosa* by Price and Greenberg in 1977,[1] and since then attempts have been made to detect FMRFamide in many diverse phyla. Authentic FMRFamide appeared to be restricted to the molluscs. However, Krajniak and Price[2] recently demonstrated the presence of FMRFamide in the polychaete annelid worm *Nereis virens*. This was not entirely unexpected, since the annelid worms are believed to share a common ancestry with molluscs and thus should contain authentic FMRFamide or a closely related peptide. Although FMRFamide is restricted to molluscs and polychaetes, FMRFamide immunoreactivity shows a broad phylogenetic distribution with FMRFamide-related peptides (FaRPs) having been detected in every organism studied including the mammalian CNS.[3] These observations suggest that a family of FMRFamide-related peptides exists within the animal kingdom. To date, over 50 FMRFamide–related peptides have been isolated and their structures characterised.

The molluscan FMRFamide–related peptides have generally been identified on the basis of biological activity such as their excitatory action on hearts of *Mercenaria mercenaria* or ability to cause contraction of the radula protractor muscle (RPM) of *Busycon contrarium*.[4] However, the FMRFamide–related peptides of other phyla have usually been identified by immunoassays. Most of these substances react very weakly, if at all, in standard radioimmunoassays (RIA) or the bioassays described above.[5] FMRFamide antisera are relatively unselective, and the Arg–Phe–NH₂ C–terminus is generally regarded as the minimum structural requirement for immunoreactivity, but some anti-

179

FMRFamide antisera have been found to recognise Arg–Tyr–ANM$_2$.[6] Thus the total set of immunoreactive FMRFamide–related peptides may not share common biological function; it includes the biologically related FMRFamide-like peptides as a subset, but also contains a collection of peptides showing some degree of sequence homology to FMRFamide, but little biological cross-reactivity. It has been argued that the observed similarities in structure such as the C–terminal amide group, and repeated appearance of two aromatic residues separated by two intervening non-aromatic residues stem from processing mechanisms or from common physical characteristics of peptides that interact with receptors and mediate intercellular communication.[4,7] More recently, related peptides have been identified by molecular cloning techniques.[8]

FMRFamide–RELATED PEPTIDES IN MAMMALIAN SYSTEMS

FMRFamide and related peptides exhibit a diverse range of physiological actions. In mammalian systems evidence has been presented that FMRFamide or the mammalian FMRFamide–related peptide Phe–Leu–Phe–Gln–Pro–Gln–Arg–Phe–NH$_2$ (F–8–Famide) may function as endogenous antagonists of opioid peptides. For example, intracerebroventricular injections of FMRFamide into mice blocked morphine–induced analgesia,[9] while intrathecal injection of antibodies directed against F–8–amide produced long-lasting analgesia in rats, which was reversible by the opioid agonist, nalaxone.[10] It has been argued that these effects are mediated through specific receptors which differ from opioid receptors, thus indicating a physiological antagonism of FMRFamide–related peptides and opioids.[10] More recent data suggest that F–8–Famide and also CCK8, may be specific for the μ opioid receptor.[11] In addition to anti-opioid actions, FMRFamide, the chicken peptide Leu–Pro–Leu–Arg–Phe-amide (LPLRFamide) and F–8–Famide have been shown to exhibit cardioactive properties. For example, exogenously applied FMRFamide increased blood pressure in rats by a naloxone-sensitive mechanism and F–8–F amide has been found to elevate mean arterial pressure.[12] In addition, treatment of rats with 15 μg of F–8–Famide induces a morphine-withdrawal-like behavioural syndrome.[13]

FMRFamide–RELATED PEPTIDES IN MOLLUSCAN SYSTEMS

The biological roles of FMRFamide–related peptides have been most extensively studied in molluscs and in particular the pulmonates and opisthobranchs. Immunocytochemical and radioimmunoassay studies have led to the identification of FMRFamide in all molluscan species examined and although

FLRFamide has been definitely reported in only a few species, it may also be a ubiquitous, although quantitatively minor, molluscan peptide.[14] In addition to these tetrapeptides the pulmonates also contain at least one (and usually several) heptapeptide with the formula XDPFLRFamide (X = Glycine, Serine, Aspartate or pGlutamate) for example *Lymnaea* has been shown to contain at least two heptapeptides, GDPFLRFamide and SDPFLRFamide.[15,16]

PERIPHERAL ACTIONS OF FMRFamide–RELATED PEPTIDES

The FMRFamide–related peptides exhibit potent cardioexcitatory actions in several species of mollusc.[17] Experiments that compare the relative pharmacological potencies of different members of this family have shown differential effects of the tetrapeptides when compared with heptapeptides. In some species such as *Lymnaea stagnalis* and *Rapana thomasiana*,[18] the tetrapeptides are more potent than the *N*–terminally extended peptides, HPLC also showed that FMRFamide was the main type of peptide present in the *Lymnaea* heart.

In *Lymnaea* a pair of excitatory cardioactive motoneurons called E_{he} (E group, heart excitatory) have recently been discovered[19] whose neuronal effects of increasing the frequency of heart beat are closely mimicked by the application of FMRFamide.[20–22] Given that the neurons are immunoreactive for antibodies raised against FMRFamide (immunocytochemistry and RIA) it appears likely that FMRFamide is the excitatory neurotransmitter for the E_{he} cells. Two cardioexcitory monoamines (5–HT and dopamine) are also candidate co-transmitters for the E_{he} cells but have been discounted because application of potent monoamine antagonists to the snail heart does not block the ability of E_{he} cells to excite the heart. Furthermore mapping of dopamine[23] and serotonin cells[24] using immunocytochemistry confirms that E_{he} cells do not contain these two transmitters. In contrast, the heart of *Helix aspersa* is more sensitive to the extended forms than FMRFamide itself.[25]

Non-cardiac muscles have also been shown to be susceptible to the actions of these peptides. For example, contractions of the radula protractor muscle of *Busycon* and gill muscles of *Aplysia* are both enhanced by FMRFamide, whereas contractions of the gut in *Aplysia* are inhibited by the peptide.[26,27] The tentacle retractor muscle of *Helix* has been an interesting system of study as it is innervated by the acetylcholine–(ACh) and FMRFamide-containing C3 neuron. ACh has been found to be the major transmitter involved in mediating contraction of the tentacle muscle, but exogenously applied FMRFamide also contracts this muscle; the heptapeptide pQDPFLRFamide, however, causes relaxation when applied in the same way.[28] Bulloch *et al.* have also reported evidence for differential action of tetrapeptides as compared to heptapeptides at peripheral synapses in *Helisoma*.[29]

CENTRAL ACTIONS OF FMRFamide–RELATED PEPTIDES

FMRFamide has been shown to have effects on a variety of ionic currents in central molluscan neurons, including changes in sodium (Na^+), potassium (K^+), calcium (Ca^{2+}), and chloride (Cl^-) conductances.[30–37] In *Helix* the different ion channel targets of the endogenous FMRFamide–related peptides on central neurons have been studied in some detail and have shown marked differences in their effects. For example, the extended peptides activate a fast increase in K^+ permeability, not seen with FMRFamide itself which only activates a slow increase in K^+ permeability. Interestingly, although all the endogenous FMRFamide–related peptides can activate increases in K^+ conductance, only the tetrapeptides have been shown to activate an increase in Na^+ conductance.[28]

The differential actions of tetra- and heptapeptides both in central and peripheral systems suggests that both tetrapeptide and heptapeptide receptors occur and function separately. The evidence for this is particularly strong in *Helix* where in addition to pharmacological evidence structure–activity relation (SAR) studies show that the heptapeptides do not interact at FMRFamide receptors.[38,39] Similar studies show this is also the case for the *Lymnaea* neuron RPeD1 (unpublished observations).

Data relating to the ionic actions of FMRFamide–related peptides in *Lymnaea* are limited in comparison to those available for *Helix*. However, it has been shown recently that the dual action of FMRFamide in neurosecretory caudodorsal cells (CDCs) is mediated by one receptor type.[40]

In addition to the transmitter/modulatory actions of molluscan FMRFamide–related peptides at neuro–muscular junctions and between central neurons, they also appear to be able to 'act at a distance', i.e. they have the properties of neurohormones. The presence of 'FMRFamide' immunoreactivity in the blood of all molluscan species examined, at concentrations ranging from 1 nM in bivalve species to 10-fold that in gastropods such as *Lymnaea* and *Helix* is in itself indicative of a hormonal function for these peptides.[25] More specific evidence for this mode of action is provided by a study of the effect of FMRFamide on neuron R14 in *Aplysia*. This neuron, located in the abdominal ganglion, is thought to be excited primarily by substances in the body fluid.[41] Ichinose and McAdoo[42] showed that application of FMRFamide–related peptides thus probably function as transmitters, modulators and neurohormones.

The existence of two classes of peptide and the differential actions of the tetra– and heptapeptides raise several important questions. Do, for example, the peptides coexist within individual neurons? Do particular neurons exclusively express either the tetrapeptides or the heptapeptides and if so, is expression of either type associated with a particular neural system? Furthermore, can either of the transmitter, modulatory or neurohormonal roles be specifically ascribed to either the tetrapeptides or heptapeptides? And how are these peptides encoded in the genome?

An experimental approach that endeavours to answer these questions has to fulfill two main criteria. Firstly, a molecular probe must be available that can specifically distinguish between tetrapeptide and heptapeptide expression and secondly, neural systems that are known to contain FMRFamide–related peptides are required in which to study the expression. Antibody probes have been raised to FMRFamide, which is shared by both the tetra– and heptapeptides and have so far proved to be unsuitable probes at least for immunocytochemical mapping. The second criterion is more easily fulfilled because the CNS of *Lymnaea* is an ideal model system and contains several neural networks and identified cell groups in some of which there is evidence of FMRFamide–related peptide expression.

MOLECULAR APPROACH TO THE STUDY OF FMRFamide–RELATED PEPTIDES IN *Lymnaea*

The widespread distribution of FMRFamide–related peptides and the sugges-tion of their association with defined neural systems in *Lymnaea*[19] presents an ideal experimental model in which to study the expression of these peptides. In particular, biochemical data showed[15] that at least three classes of related peptides existed in the *Lymnaea* nervous system. The question therefore arose as to whether two classes of peptides, the tetrapeptide (FMRFamide) and heptapeptides (GDPFLRFamide and SDPFLRFamide) were encoded by the same gene, different exons of the same gene or different genes entirely. Furthermore, using defined neuronal systems it should be possible to determine at the level of a single cell how expression is regulated. In order to do this a molecular approach is required, particularly as both classes of peptide react with the same anti-sera raised against FMRFamide.[4]

Isolation of a *Lymnaea* FMRFamide cDNA

A cDNA library was hybridised with synthetic oligonucleotides containing all possible codons for the sequence Phe–Met–Arg–Phe–Gly–Lys. From this two positive clones were isolated and sequenced. An analysis of the sequence revealed that FMRFamide was encoded nine times throughout the cDNAs. In addition, sequences encoding a number of novel potential peptides were also found. The organisation of this sequence is shown in Fig. 1.

From comparison of the *Lymnaea*[8] and *Aplysia*[16] sequences based upon the FASTP programme, it is observed that homology is not uniform throughout the region. The most conserved region between the genes is in the CRF (corticotrophin-release factor) sequence. Interestingly, this sequence is not present in the *Drosophila*[43,44] sequence and it is not yet known whether it is cleaved from the precursor as a peptide with a physiological function, although the conservation between *Lymnaea* and *Aplysia* makes this likely.

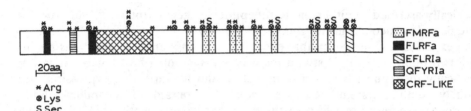

FIG. 1. Representation of the peptides encoded by the FMRFamide precursor of *Lymnaea*.[8] The first 18 amino acids make up a hydrophobic leader sequence encoded by the genomic DNA in the same reading frame as the FMRFamide peptides.

Genomic cloning of the *Lymnaea* FMRFamide gene

To isolate genomic sequences containing FMRFamide, a λEMBL3 library was screened with FMRFamide cDNA clones. This resulted in the isolation of three independent phage. One (λ4·5) was digested with *Eco*RI sub-cloned into a pUC19 vector screened again with FMRFamide cDNA and a single colony isolated. This was sequenced using the chain termination method and universal primers, as well as synthetic oligonucleotides. No difference was observed between the cDNA and genomic sequences. In contrast to *Aplysia,* the sequence showed that in *Lymnaea* there are no splicing events between the FMRFamide repeats. In *Aplysia,* on the basis of various cDNA sequences it has been proposed that splicing within the repeats may occur. Unfortunately, in the *Aplysia* case the possibility of instability in the cDNA, in the M13 sub-clones cannot be ruled out.[45]

It is clear that neither the *Aplysia* nor *Lymnaea* FMRFamide cDNAs, although encoding multiple peptides, encode the extended heptapeptides GDPFLRFamide and SDPFLRFamide. As members of a gene family are sometimes located in close proximity to each other,[46] the λ4·5 *Lymnaea* genomic clone was hybridised with sequences specific to GDPFLRFamide. As shown in Fig. 2, bands are observed that are clearly different from those hybridising to a FMRFamide-specific probe. This suggested that indeed genomic sequences encoding GDPFLRFamide and FMRFamide are co-linear in the genome. Sub-cloning screening and sequencing using a combination of *Exo*III and mung bean nuclease together with synthetic primers for double stranded sequencing[47] revealed a genomic sequence encoding both GDPFLRFamide as well as SDPFLRFamide. The organisation of this sequence is shown in Fig. 3.

The region of DNA encoding GDP/SDPFLRFamide sequences lacks an in-frame methionine start but is flanked by 3' and 5' splice sites. This suggests that this sequence makes up an exon.[48] As most, if not all, FMRFamide transcripts are spliced then it is probably best to consider the two neuropeptide precursor sequences as exons rather than separate genes. It is likely that they are both exons of the same gene.[50]

The organisation of the peptides within each precursor differs. In the

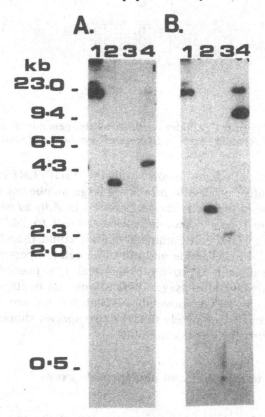

Fig. 2. Restriction enzyme digest of λ4·5 encoding FMRFamide and GDPFLRFamide. The phage DNA was digested as indicated, run on a 0·8% agarose gel, transferred to Hybond-N and hybridised first (A) with a FMRFamide specific probe, and then (B) washed and hybridised with a GDPFLRFamide specific probe. The figure shows the autoradiograph after each hybridisation. The DNA was treated as follows: (1) undigested, (2) *Eco*RI, (3) *Eco*RI/*Sal*I, (4) *Sal*I.

FMRFamide precursor then the peptides are separated from each other by other amino acids whereas the heptapeptide precursor contains evenly spaced peptides flanked by the presumptive protein cleavage sites. The exception to this is the peptide that is most *N*-terminal. This peptide (EFFPLamide) is a pentapeptide and is separated from the GDP/SDPFLRFamide peptides by a length of 21 amino acids. In this respect the precursor environment of the peptide is more similar to those encoded by the tetrapeptide precursor than the heptapeptide precursor.

It is interesting to speculate how these two peptide precursors arose. One possibility is that originally there was only a single gene comprising one exon (the tetrapeptide exon). This exon may have encoded the pentapeptide (EFFPLamide) now found on the heptapeptide exon and also GDPFLR-Famide. Over time two independent events occurred, the original gene was

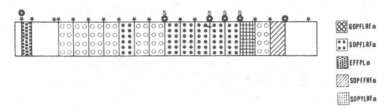

F‌ɪɢ. 3. Representation of the peptides encoded by the genomic sequences making up the GDP/SDPFLRFamide exon in *Lymnaea*,[48] symbols and scale as in Fig. 1.

split by the insertion of an intron, and the GDPFLRFamide sequences underwent a reiterative process to produce a large number of copies.[48]

There is a precedent for the reiterative process in *Aplysia* where the number of the FMRFamide repeat copies increased from 9 to 28.[16,25] In *Lymnaea* mutations within the GDPFLRFamide sequence could then have resulted in the formation of SDPFLRFamide and the other related peptides in this exon. Unfortunately, it is difficult to prove such a model. One possible source of data could come by studying codon usage. It is curious that heptapeptides have not been identified in *Aplysia*. One possibility is that they too are encoded by an as yet unidentified exon. It is unlikely that the two species should be so different with respect to their neuropeptide content.

Expression of the tetrapeptide and heptapeptide exons

In order to determine the pattern of expression of the two individual exons radioactive probes were generated using exon-specific genomic DNA and hybridised to alternative sections of the *Lymnaea* brain. The autoradiograph in Fig. 4 shows an example from one part of the brain. It can be seen that the FMRFamide probe hybridises to one cluster of cells, whereas the heptapeptide probe hybridises strongly to different cells in adjacent ganglia. This suggests that expression of the two exons is cell-specific and individual cells in the region of the brain shown here express either the heptapeptides or tetrapeptides but not both. Furthermore the level of regulation is either at the level of transcription or RNA processing. By sequencing the region of DNA between the two exons to determine whether any consensus transcription start sites are present and by the use of intron specific probes for *in-situ* hybridisation these two mechanisms can be distinguished.

The FMRFamide genes from a number of organisms have now been sequenced, *Lymnaea*,[8,48] *Aplysia*,[16] *Drosophila*[43,44] and *Anthozoa*.[49] As yet it is not clear whether any generalities can be made with respect to their organisation and expression. The genes in *Lymnaea*[50] and *Drosophila* are spliced, but the 5′ exon in *Drosophila* is non-coding,[43,44] furthermore the *Drosophila* gene only encodes *N*-terminally extended peptides.[43,44] In *Aplysia* only the FMRFamide encoding sequence has been found but it is possible that

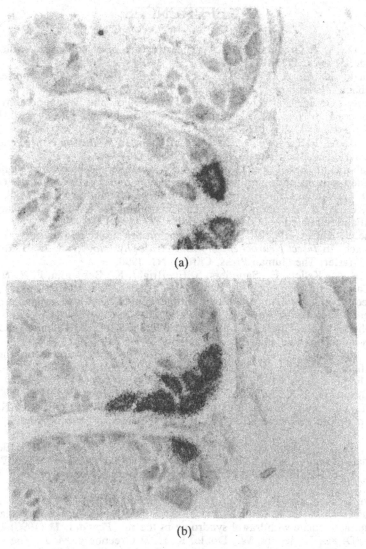

(a)

(b)

FIG. 4. In-situ hybridisation of two alternate sections of the *Lymnaea* brain hybridised with (A) a FMRFamide specific probe[8] or (B) a probe specific for GDPFLRFamide.[48]

another exon encoding heptapeptide sequence may exist. *Helix* clearly contains both FMRFamide and extended peptides but the organisation has not yet been fully resolved, although it is likely to be similar to *Lymnaea*. What is now required is to match the physiological role(s) of these peptides with the molecular genetics in individual neuronal systems.

REFERENCES

1. Price, D. A. & Greenberg, M. J., The structure of a molluscan cardioexcitatory peptide. *Science,* **197** (1977) 670.
2. Krajniak, K. G. & Price, D. A., Authentic FMRFamide is present in the polychaete *Nereis virens. Peptides,* **11** (1990) 75–7.
3. Dockray, G. J., Characterisation of FMRFamide-like immunoreactivity in rat spinal cord by region-specific antibodies in radioimmunoassay and HPLC. *J. Neurochem.,* **45** (1985) 152–8.
4. Price, D. A., FMRFamide: Assays and artifacts. In *Molluscan Neurobiology,* ed. J. Lever & H. H. Boer. North-Holland, 1982, pp. 184–9.
5. Greenberg, M. J. & Price, D. A., The phylogenetic and biomedical significance of extended neuropeptide families. In *Biomedical Importance of Marine Organisms,* ed. D. G. Fautin. California Academy of Sciences, San Francisco, 1988.
6. Price, D. A. & Greenberg, M. J., The hunting of the FaRPs: The distribution of FMRFamide related peptides. *Biol. Bull.,* **177** (1989) 198–205.
7. Joosse, J. & Geraerts, W. P. M., Neuropeptides: unity and diversity a molecular approach. In *Insect Neurochemistry and Neurophysiology,* ed. A. B. Borkovec & E. P. Masler. The Human Press, Clifton, NJ, 1990.
8. Linacre, A., Kellett, E., Saunders, S. E., Bright, K., Benjamin, P. R. & Burke, J. F., Cardioactive neuropeptide Phe–Met–Arg–Phe–NH$_2$ (FMRFamide) and novel related peptides are encoded in multiple copies by a single gene in the snail *Lymnaea stagnalis. J. Neurosci.,* **10** (1990) 412–19.
9. Kavaliers, M. & Hirst, M., Inhibitory influences of FMRFamide and PLG on stress-induced opioid analgesia and activity. *Brain Res.,* **372** (1986) 370–4.
10. Tang, J., Yang, H.-Y. T. & Costa, E., Inhibition of spontaneous and opiate-modified nociception by an endogenous neuropeptide with Phe–Met–Arg–Phe–NH$_2$-like immunoreactivity. *Proc. Natl. Acad. Sci. USA,* **81** (1983) 6680–83.
11. Magnuson, D. S. K., Sullivan, A. F., Simonnet, G., Roques, B. P. & Dickenson, A. H., Differential interactions of cholecystokinin and FLFQPFQRF-NH$_2$ with μ and σ opioid antinociception in the rat spinal cord. *Neuropeptides,* **16** (1990) 213–18.
12. Roth, B. L., Disimone, J., Majane, E. A. & Yang, H.-Y. T., Elevation of arterial pressure in rats by two new vertebrate peptides FLFQPQRF-NH$_2$ and AGEGLSSPFWSLAAPQRF-NH$_2$ antiserum. *Neuropeptides,* **10** (1987) 37–42.
13. Malin, D. H., Lake, R., Fowler, D. E. Hammond, M. W., Brown, S. L., Leyva, J. E., Prasco, P. E. & Dougherty, T. M., FMRF-NH$_2$-like mammalian peptide precipitates opiate-withdrawal syndrome in the rat. *Peptides,* **11** (1990) 277–80.
14. Price, D. A., Davies, N. W., Doble, K. E. & Greenberg, M. J., The variety and distribution of FMRFamide-related peptides in molluscs. *Zoological Science,* **4** (1987) 395–410.
15. Ebberink, R. H. M., Price, D. A., Loenhout, H. van, Doble, K. D. E., Riehm, J. P., Geraerts, W. P. M. & Greenberg, M. J., The brain of *Lymnaea* contains a family of FMRFamide-like peptides. *Peptides,* **8** (1987) 515–22.
16. Schaefer, M., Picciotto, M. R., Kreiner, T., Kaldany, R.-R., Taussig, R. & Scheller, R. H., *Aplysia* neurons express a gene encoding multiple FMRFamide neuropeptides. *Cell,* **41** (1985) 457–467.
17. Painter, S. D. & Greenberg, M. J., A survey of the responses of bivalve hearts to the molluscan neuropeptide FMRFamide and to S-hydroxy pyptamine. *Biol. Bull.,* **162** (1982) 311–32.
18. Kobayashi, N. & Muneoka, Y., Structural requirements for FMRFamide-like activity on the heart of the prosobranch *Rapana thomasiana. Comp. Biochem. Physiol.,* **84** (1986) 349–52.

19. Benjamin, P. R., Buckett, K. J. & Peters, M., Neurons containing FMRFamide-like peptides in the model invertebrate system *Lymnaea. Symposia Biologica Hungarica*, **3** (1988) 247–59.
20. Buckett, K. J., Dockray, G. J., Osborne, N. N. & Benjamin, P. R., Pharmacology of the myogenic heart of the pond snail *Lymnaea stagnalis. J. Neurophysiol.*, **63** (1990) 1413–25.
21. Buckett, K. J., Peters, M., Dockray, G. J., Van Minnen, J. & Benjamin, P. R., Regulation of heartbeat in *Lymnaea* by motoneurons containing FMRFamide-like peptides. *J. Neurophysiol.*, **63** (1990) 1426–35.
22. Buckett, K. J., Peters, M. & Benjamin, P. R., Excitation and inhibition of the heart of the pond snail *Lymnaea* by non-FMRFamidergic motoneurons. *J. Neurophysiol.*, **63** (1990) 1436–77.
23. Elekes, K., Kemenes, G., Hiripi, L., Geffard, M. & Benjamin, P. R., Dopamine-immunoreactive neurons in the central nervous system of the pond snail *Lymnaea stagnalis. J. Comp. Neurol.*, **307** (1991) 1–11.
24. Kemenes, G., Elekes, K., Hiripi, L. & Benjamin, P. R., A comparison of four techniques for mapping the distribution of serotonin and serotonin-containing neurons in fixed and living ganglia of the snail *Lymnaea. J. Neurocyto.*, **18** (1989) 193–208.
25. Price, D. A., Evolution of a molluscan cardioregulatory neuropeptide. *American Zoologist*, **26** (1986) 1007–15.
26. Price, D. A., Cottrell, G. A., Doble, K. E., Greenberg, M. J., Jorenby, W., Lehman, H. K. & Reihm, J. P., A novel FMRF-amide-related peptide in *Helix*: pQDPFLRFamide. *Biol. Bull.*, **169** (1985) 256–66.
27. Austin, T., Weiss, S. & Lukoviak, K., FMRFamide effects on spontaneous and induced contractions of the anterior gizzard in *Aplysia. Can. J. Physiol. Pharmacol.*, **61** (1982) 949.
28. Cottrell, G. A., The biology of the FMRFamide-series of peptides in molluscs with special reference to *Helix. Comp. Biochem. Physiol.*, **93** (1989) 41–5.
29. Bulloch, A. G. M., Price, D. A., Murphy, A. D., Lee, T. D. & Bowers, H. N., FMRFamide peptides in *Helisoma*: identification and physiological actions at a peripheral synapse. *J. Neurosci.*, **8** (1988) 3459–69.
30. Cottrell, G. A., FMRFamide neuropeptides simultaneously increase and decrease K^+ currents in an identified neurone. *Nature*, **296** (1982) 87–9.
31. Cottrell, G. A., Davies, N. W. & Green, K. A., Multiple actions of a mulluscon cardioexcitatory neuropeptide and related peptide on identified *Helix* neurons. *J. Physiol. (Lond).*, **356** (1984) 315–33.
32. Boyd, P. J. & Walker, R. J., Actions of the molluscan neuropeptide FMRFamide on neurons in the suboesophageal ganglia of the snail *Helix aspersa. Comp. Biochem. Physiol.*, **81** (1985) 379–86.
33. Colombiaoni, L., Paupardin-Tritsch, D., Vidal, P. P. & Gerschenfeld, H. M., The neuropeptide FMRFamide decreases both the Ca^{2+} conductance and a cyclic $3',5'$-adenosine monophosphate-dependent K^+ conductance in identified molluscan neurons. *J. Neurosci.*, **5** (1985) 2533–8.
34. Rubin, P., Johnson, J. W. & Thompson, S., Analysis of FMRFamide effects on *Aplysia* bursting neurons. *J. Neurosci.*, **6** (1986) 252–9.
35. Belardetti, F., Kandel, E. R. & Siegelbaum, S. A., Neuronal inhibition by the peptide FMRFamide involves opening of $S-K^+$ channels. *Nature*, **325** (1987) 1530–156.
36. Thompson, S. & Ruben, P., Inward rectification in response to FMRFamide in *Aplysia* neuron L2: Summation with transient K^+ current. *J. Neurosci.*, **8** (1988) 3200–207.
37. Taussig, R., Sweet-Cordero, A. & Scheller, R., Modulation of ionic currents in *Aplysia* motor neuron B15 by serotonin, neuropeptides and second messengers. *J. Neurosci.*, **9** (1989) 3218–29.

38. Payza, K., FMRFamide receptors in *Helix aspersa. Peptides,* **8** (1987) 1065–75.
39. Payza, K., Greenberg, M. J. & Price, D. A., Further characterisation of *Helix* FMRFamide receptors kinetics, tissue distribution, and interactions with the endogenous heptapeptides. *Peptides,* **10** (1989) 657–61.
40. Brussaard, A. B., Kits, K. S. & Ter Maat, A., One receptor type mediates two independent effects of FMRFamide on neurosecretory cells of *Lymnaea. Peptides,* **10** (1989) 289–97.
41. Coggleshall, R. E., Kandel, E. R., Kupfermann, I. & Waziri, R. A., Morphological and functional study on a cluster of identifiable neurosecretory cells in the abdominal ganglion of *Aplysia californica. J. Cell Biol.,* **31** (1966) 363–8.
42. Ichinose, M. & McAdoo, D. J., The voltage-dependent, slow inward current induced by the neuropeptide FMRFamide in *Aplysia* neuron R14. *J. Neurosci.,* **8** (1988) 3891–900.
43. Chin, A. C., Reynolds, E. R. & Scheller, R. H., Organization and expression of the *Drosophila* FMRFamide-related prohormone gene. *DNA Cell. Biol.,* **9** (1990) 263–71.
44. Schneider, L. E. & Taghert, P. H., Isolation and characterization of a *Drosophila* gene that encodes multiple neuropeptides related to Phe–Met–Arg–Phe–NH$_2$ (FMRFamide). *Proc. Natl. Acad. Sci. USA,* **85** (1988) 1993–7.
45. Taussig, R. & Scheller, R. H., The *Aplysia* FMRFamide gene encodes sequences related to mammalian brain peptides. *DNA,* **5** (1986) 453–61.
46. Geraerts, W. P. M., Smit, A. B., Li, K. W., Vreugdenhil, E. & Van-Heerikhuizen, H., Neuropeptide gene families that control reproductive behaviour and growth in molluscs. *Current Aspects of Neurosci.,* **3** (1991) 255–304.
47. Saunders, S. E. & Burke, J. F., Rapid isolation of miniprep DNA for double strand sequencing. *Nucl. Acid. Res.,* **18** (1990) 4948.
48. Saunders, S. E., Bright, K., Kellett, E., Benjamin, P. R. & Burke, J. F., Neuropeptides GDPFLRFamide and SDPFLRFamide are encoded by an exon 3′ to FMRFamide in the snail *Lymnaea stagnalis. J. Neurosci.,* **11** (1991) 740–5.
49. Darmer, D., Schmutzler, C., Diekhoff, D. & Grimmelikhuijzen, C. J. P., Primary structure of the precursor for the sea anemone neuropeptide Antho-RFamide (<Glu–Gly–Arg–Phe–NH$_2$). *Proc. Natl Acad. Sci. USA,* **88** (1991) 2555–9.
50. Saunders, S. E., Kellett, E., Bright, K., Benjamin, P. R. & Burke, J. F., Cell specific RNA splicing of a FMRFamide gene transcript in the brain. *J. of Neurosci.* **12** (1992) 1033–9.

14

Expression of Neurotransmitter Receptors in Insect Cells Using Baculovirus Vectors

LINDA A. KING, ALLAN E. ATKINSON, LOUIS, A. OBOSI, ISABEL BERMUDEZ & DAVID J. BEADLE

School of Biological and Molecular Sciences, Oxford Polytechnic, Gipsy Lane, Headington, Oxford OX3 0BP, UK

INTRODUCTION

Successful expression of neurotransmitter receptors in heterologous systems *in vitro* requires that the host cell can synthesise adequate amounts of correctly processed and targeted protein. To date the most popular expression systems for receptors have been the injection of DNA or mRNA into *Xenopus* oocytes,[1,2] or the use of vectors to transform *Escherichia coli*,[3] yeast[4] or mammalian cells, e.g. fibroblasts.[5,6] One of the main disadvantages of these systems is that often only very small quantities of protein are synthesised, and in the case of *E. coli,* the proteins are often not correctly processed to allow functional analyses.

The expression of foreign genes in insect cells using baculovirus vectors has gained increasing popularity in recent years and is now a well-established technology in many laboratories.[7–10] There are several advantages to using the baculovirus system for the expression of receptor proteins: (1) the vectors are not pathogenic to other invertebrates, vertebrates or plants; (2) the vectors do not depend on a helper virus system; (3) large quantities of protein can be synthesised using the hyper-expressed polyhedrin and p10 promotors; (4) insect cells can carry out most, if not all, of the necessary co- and post-translational modifications necessary for authentic protein function; (5) several different proteins may be expressed together by making use of the multiple expression vectors that are now available,[11] or by co-infecting cells with two or more recombinant viruses; (6) insect cells in culture are amenable to analysis by electrophysiology;[12] (7) recombinant viruses can be prepared in a relatively short period of time (3–4 weeks). Possible disadvantages include: (1) glycosylation is different in insect cells, being predominantly of the high-mannose type;[7,10] (2) synthesis of proteins in the very late phase of virus replication (using polyhedrin or p10 promoters) occurs when the cell is dying from virus infection and, therefore, may not be as capable as a non-infected cell at processing complex proteins. The latter problem may be overcome by utilising virus promoters that are active earlier in the virus replication cycle, such as the basic protein promoter.[13]

This paper provides an introduction to the baculoviruses for those unfamiliar with these viruses, describes how they are used as vectors for foreign gene expression and discusses how this system may be used to analyse neurotransmitter receptors.

BACULOVIRUS BIOLOGY

The family of viruses comprising the Baculoviridae consists of a single genus, baculovirus, that is divided into the three subgroups: subgroup A, the nuclear polyhedrosis viruses (NPV) in which the virions are occluded into a protein matrix, the polyhedron; subgroup B, the granulosis viruses (GV) in which single virions are occluded into a protein matrix, the granule; and subgroup C, the non-occluded viruses.[14] Over 800 different baculoviruses have been isolated and these have been named according to the original host insect from which they were isolated.[14] Relatively few baculoviruses have been studied in detail, probably because only a few have been adapted to growth in insect cells *in vitro*. The two viruses that have been developed as expression vectors are the *Autographa californica* (Ac) NPV, originally isolated from the alfalfa looper (*A. californica*) and cultured *in vitro* in cells derived from *Spodoptera frugiperda,*[15] and the *Bombyx mori* (Bm) NPV, which is propagated in the silkworm (*B. mori*).[16] The silkworm system has, however, received little attention outside of Japan and most workers will be more familiar with the AcNPV system, which forms the subject of this paper.

The baculovirus genome consists of a covalently closed, circular molecule of double-stranded DNA, ranging from 80 to 220 kilobase pairs (kbp). The DNA is packaged into rod-shaped nucleocapsids (about 40×350 nm) which are enveloped by a lipid membrane to form the virus particle. The NPVs may package a single (S) or multiple (M) nucleocapsids per virus particle; AcNPV is an example of a multiple nucleocapsid NPV. The virus particles are further packaged into large proteinaceous occlusion bodies, termed 'polyhedra' because of their polyhedral morphology. The crystalline matrix of polyhedra consists of a single, 29 kDa polypeptide (polyhedrin).[17]

Within infected insects, the NPVs have a bi-phasic replication cycle that generates two forms of infectious virus: extracellular or budded virus (ECV) and occlusion bodies or polyhedra (OB). Polyhedra are an important part of the natural life cycle of the virus, providing protection for the embedded virions, during horizontal transmission of infection. When an infected larva dies, millions of polyhedra remain in decomposing tissue and on leaf surfaces, where they may be ingested by other feeding larvae. Following ingestion, the polyhedra dissolve in the alkaline conditions of the insect mid-gut, releasing the infectious virus particles. Virus particles subsequently attach to and enter the epithelial cells lining the mid-gut. Following fusion, the nucleocapsids enter the cell cytoplasm. Baculovirus replication takes place in the nucleus, although the processes by which the nucleocapsids migrate to the nucleus and how uncoating occurs are unknown. Secondary infection of other tissues within the insect is accomplished by the release of ECV from the plasma membrane of gut cells into the haemolymph. ECV collect an integral membrane protein, gp64, as they pass through the plasma membrane, and this glycoprotein is thought to be involved in attachment of the virus to other cells (including haemocytes, fat body, nerve and ovary). In lepidopteran species, polyhedra production in the midgut cells is severely restricted; favouring ECV production to spread the virus infection. Later, the other infected tissues produce many polyhedra per nucleus. Nucleocapsids that are destined to be occluded derive their lipid membrane *de novo* in the nucleus and thus do not contain the gp64 surface protein; genetically, ECV and polyhedra-derived virus are identical. Large numbers of polyhedra, up to 10^8, are produced in each larva.

BACULOVIRUS REPLICATION AND GENE EXPRESSION

The replication cycle of AcNPV has been studied in the greatest detail using virus infection of cultured *S. frugiperda* cells. Virus replication was originally established in cell culture by using ECV harvested from infected larval haemolymph or by alkaline dissolution of polyhedra to release occluded virus particles; the former (ECV form) are 1000-fold more infectious for cells in culture due to the presence of gp64 in the outer membrane. Virus-infected cells *in vitro* release ECV into the culture medium, which can subsequently be harvested and used to inoculate further samples of cells.

During virus replication in cell culture there is a complex, sequential and often transient appearance of virus-induced polypeptides. The production of proteins in AcNPV-infected cells has been divided into four phases (see Fig.

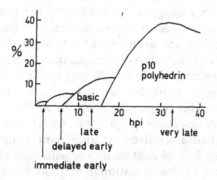

FIG. 1. Schematic representation of the four phases of baculovirus gene expression.

1): immediate early (α); delayed early (β); late (γ); and very late (δ). The early classes of gene expression occur before DNA synthesis occurs and are not inhibited by treatment of cells with cycloheximide. After DNA replication, during the late phase of gene expression, the virus structural proteins are synthesised, e.g. the basic, arginine-rich protein that associates with viral DNA. Virus particles are then released from the plasma membrane by budding, acquiring gp64 as they do so. At about 15–18 h post-infection (hpi), the very late phase of gene expression begins, during which time the virus particles within infected nuclei are occluded into polyhedra. This last phase may continue until 72 hpi and is characterised by the hyper-expression of two virus genes, polyhedrin and p10. Polyhedrin, as mentioned above, forms the protein matrix of polyhedra and p10 forms fibrillar structures in the nucleus, which are thought to play a role in polyhedra formation. Both genes have been characterised and hyper-expression can be attributed to very efficient promoters. Neither p10 nor polyhedrin are necessary for virus (ECV) production *in vitro*, and therefore, their coding sequences may be replaced by foreign genes; this forms the basis of the baculovirus expression vector system.

BACULOVIRUS EXPRESSION VECTORS

The methodology required to derive recombinant baculoviruses for foreign gene expression is outlined in Fig. 2. The AcNPV genome is too large to permit direct insertion of foreign genes, therefore, a portion of the genome encoding the polyhedrin (or p10) gene is first transferred into a plasmid vector, usually referred to as the transfer vector. Within the transfer vector, the polyhedrin coding sequences have largely been deleted and a restriction enzyme site engineered to facilitate insertion of the foreign gene coding sequences, downstream of the polyhedrin gene promoter.[18] The natural polyhedrin transcriptional termination signals are also utilised. After insertion of the foreign gene coding sequences, the transfer vector is co-transfected into insect cells with wild-type AcNPV DNA. Within the transfected cell homologous recombination, between the DNA flanking the foreign gene in the transfer vector and the homologous DNA flanking the polyhedrin gene in the virus genome, results in production of recombinant virus that is polyhedrin-negative. Various methods are available to introduce the DNAs into insect cells, including calcium phosphate co-precipitation, electroporation and lipofection. The latter is probably the most efficient method. Recombinant virus is separated from wild-type virus by harvesting the co-transfection culture supernatant and titrating this by plaque-assay. Recombinant virus gives rise to plaques with a polyhedra-negative phenotype, whilst wild-type virus will give plaques that are polyhedra-positive. Recombinant plaques are picked and re-titrated until they are free of contaminating wild-type virus.

One of the problems of the traditional method for making recombinant viruses, as described above, is that the rate of recombination is low, typically

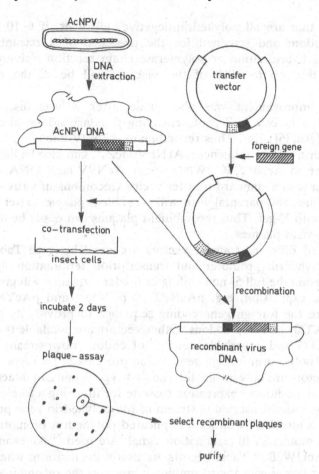

FIG. 2. An outline of the methodology required to derive recombinant baculoviruses. The open areas represent polyhedrin gene flanking sequences; the shaded area, the polyhedrin promoter; the stippled area, the polyhedrin transcription termination signals; the hatched area, the foreign gene coding sequences; and the cross-hatched area, the polyhedrin gene coding sequences.

1% or less, and screening for recombinant viruses can be very time-consuming. Recently, a considerable improvement has been made in the way that recombinant viruses can be made and selected for. This involves the use of linearised AcNPV DNA in the co-transfection of insect cells with the transfer vector.[19] Wild-type AcNPV was engineered to contain a unique *Sau*I (*Bsu*36I) restriction enzyme site within the polyhedrin gene coding sequences. When AcNPV.SC (for Single Cut) is linearised and used in a co-transfection experiment, the parental virus DNA is unable to replicate; only DNA that has recircularised, following homologous recombination with the transfer vector, can replicate.[19] Titration of the co-transfection supernatant by plaque-assay gives plaques that are all polyhedrin-negative, however, if 6–10 plaques are picked at random and screened for the presence of a recombinant virus genome (using hybridisation or polymerase chain reaction techniques), it has been shown that up to 40% of the plaques will be of the recombinant phenotype.[19]

A further improvement was also made after it was discovered that, fortuitously, the *E. coli lacZ* gene, encoding β-galactosidase, also contains a unique *Sau*I (*Bsu*36I) site. Thus recombinant AcNPV encoding *lacZ* in place of the polyhedrin coding sequences, AcNPV.*lacZ*,[20] can also be linearised, in a similar manner to AcNPV.SC. When linear AcNPV.*lacZ* DNA is used in a co-transfection with a standard transfer vector, recombinant virus will give rise to 'white' plaques and parental virus will give 'blue' plaques, after staining the plaque-assay with X-gal. Thus recombinant plaques can easily be distinguished from parental-virus plaques.

A number of different transfer vectors are available (see Table 1), most utilise the polyhedrin promoter and transcription termination signals. Only vectors that retain the full 5′ non-coding or leader sequence will give maximum levels of gene expression, e.g. pAcRP23,[20] pEV55[21] and pAcYM1.[22] These vectors require the foreign gene coding sequence to provide its own translational start (ATG) and stop codons. Other vectors are available that retain the polyhedrin ATG and a variable number of codons downstream and can be used to make polyhedrin-foreign gene fusion proteins.[23] Two types of multiple expression vectors are now available. The first type retains an intact polyhedrin gene and has an additional expression cassette for inserting a foreign gene. The foreign gene is usually inserted upstream of the polyhedrin gene promoter and expression is controlled by either a duplicated polyhedrin promoter or a copy of the p10 promoter; SV40 termination signals are used.[24] An example of such a vector is pAcUW2B.[24] These vectors are useful in situations where you need to infect insect larvae with the recombinant virus; as the recombinant virus will be polyhedrin-positive and thus can infect the insect through the normal, oral route. The second type of multiple expression vector contains two expression casettes, one based on the polyhedrin promoter and one on the p10 promoter, so that two foreign genes can be expressed simultaneously.[11] An example of such a vector is pAcUW3.[11] Co-expression of heterologous proteins can also be achieved by co-infection of cells with singly expressing recombinant viruses.

One perceived disadvantage of utilising the polyhedrin and p10 promoters for foreign gene expression, is the fact that they are only activated very late in the virus replication cycle, when the host cell is starting to die. Thus the cell may not be in the best condition to process complex proteins through the secretory pathways. New vectors are just becoming available that may overcome this problem. These vectors use the same basic polyhedrin transfer vector, as described above, but the polyhedrin promoter sequences have been deleted and replaced with copies of other baculovirus virus gene promoters that are activated earlier in the virus replication cycle, e.g. the basic protein promoter (pAcMP1).[13]

TABLE 1

Baculovirus Vectors for Heterologous Gene Expression in Insect Cells

Vector	Promoter utilised	Reference number	Comments
pAcYM1	Polyhedrin	22	Complete 5′ non-coding region; gives maximal expression. *Bam*HI insertion site.
pAcRP23	Polyhedrin	20	Complete 5′ non-coding region; gives maximal expression. *Bam*HI insertion site.
pVL941	Polyhedrin	31	Complete 5′ non-coding region and first 34 bases of polyhedrin gene; ATG removed by conversion to ATT.
pEV55	Polyhedrin	21	Complete 5′ non-coding region. Polylinker insertion site incorporating an ATG, in case foreign gene does not have its own start codon.
pAcCL29	Polyhedrin	34	Derived from pAcYM1, but plasmid vector incorporates a portion of the M13 genome allowing single-stranded DNA to be produced.
pAcAL1-3	Polyhedrin	34	Derived from pAcRP23 but the plasmid backbone is pT7T3, allowing single-stranded DNA to be produced. The vectors are also much smaller allowing the insertion of larger foreign genes.
pAcUW1	p10	11	Standard vector using p10 promoter; *Bgl*II and *Hind*III insertion sites.
pAcUW2B	p10	11	Foreign gene is expressed under p10 control; recombinant virus is polyhedrin-positive.
pAcUW3	p10 and polyhedrin	11	Dual expression vector for the simultaneous expression of two foreign genes; *Bgl*II and *Bam*HI insertion sites.
pAcMP1	Basic protein	13	Used as a standard transfer vector (derived from pAcRP23); allows synthesis of proteins earlier in the virus replication cycle.

EXPRESSION OF FOREIGN GENES

A wide variety of prokaryotic, eukaryotic and viral genes have now been expressed in insect cells using baculovirus vectors.[7–10,25] With the large, and diverse, numbers of recombinant proteins that have been synthesised in insect cells, it has generally been found that most are very similar to their authentic counterparts, as determined by enzymatic, immunological, structural and functional assays. Levels of expression obtained vary widely, from levels equivalent to the polyhedrin/p10 proteins (approximately 30% of total cell protein at 72 hpi), to very low levels that require detection using immunological techniques. In general, it appears that secreted and membrane-targeted glycoproteins are the most poorly expressed, although the levels observed are still usually better than are obtained with other expression systems. The insect cell can carry out most of the necessary co- and post-translational modifications required for normal functioning of proteins, including: glycosylation, phosphorylation, fatty acid acylation, ADP-ribosylation, signal peptide cleavage, disulphide bond formation, insertion into membranes, tertiary and quaternary structure formation.[7–10,25] Glycosylation in insect cells is significantly different to that found in mammalian cells, although this does not appear to affect the biological activity of the foreign protein. Insect cells add predominantly short, unbranched side chains, consisting mainly of mannose; recombinant glycoproteins may be readily identified by labelling the insect cells with [14C]mannose.[26,27] As with mammalian cells, the most common type of glycosylation is N-linked through asparagine residues, which can be inhibited by treatment of cells with tunicamycin.[26,27] More recently, O-linked glycosylation has also been reported to occur.[28]

EXPRESSION OF NEUROTRANSMITTER RECEPTORS

To date relatively few neurotransmitter receptors have been expressed in insect cells. However, from those that have been the results look quite promising. Klaiber *et al.* obtained functional expression of the *Drosophilia Shaker* K$^+$ channel.[12] In this study, a cDNA for the K$^+$ channel was placed in a conventional transfer vector, downstream of the polyhedrin promoter. Recombinant viruses were selected and used to infect *Spodoptera frugiperda* cells (Sf9). One advantage of using these insect cells is that they do not have detectable endogenous voltage-dependent ion channels. They therefore provide a very low electrophysiological background when examining functional expression. The insect cells are also very amenable to analysis by patch clamp and whole cell recording techniques. When the Sf9 cells were infected with recombinant virus, voltage-dependent currents were recorded that were characteristic of *Shaker* A-type K$^+$ currents. The voltage dependence of the peak current was similar to that seen for *Shaker* currents naturally expressed in *Drosophila* muscle and for cDNA expressed in *Xenopus* oocytes. Armstrong and Miller[29] also expressed a *Drosophila Shaker* cDNA in Sf9 cells and used patch clamp and whole cell recording techniques to show that Ca^{2+} ions are necessary for proper functioning of voltage-dependent K$^+$ channels and for maintenance of the native conformation of these membrane proteins.

Birnbaumer *et al.*[30] have reported high-level expression (25–30% total cell proteins) of two G-protein receptor alpha subunits, G-α_i-3 and G-α_s, using the polyhedrin promoter in the transfer vector pVL941.[31] They reported the use of a single chromatography step to isolate G-α_i-3 from Sf9 cells, obtaining purity of about 80%. The purified subunit protein was shown to interact with $\beta\gamma$ dimers to form trimers susceptible to ADP-ribosylation by pertussis toxin.

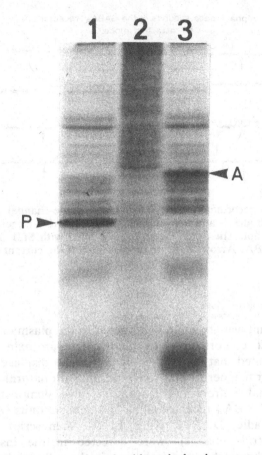

Fig. 3. Autoradiograph of a polyacrylamide gel showing expression of the alpha subunit of the chick nicotine acetylcholine receptor, Sf21 cells were infected with wild-type virus (track 1), recombinant virus (track 3) or were mock-infected (track 2) and were pulse-labelled with [^{35}S]methionine at 22–24 hpi. The polyhedrin (P) and alpha subunit (A) proteins are indicated.

Expression of the β_2-adrenergic receptor in Sf9 cells has recently been reported.[32] In this study a cDNA for the receptor was inserted into the transfer vector pAc373[31] and recombinant viruses obtained in the normal fashion, as described above. Binding studies using [^{125}I]iodocyanopindolol showed that Sf9 cells infected with the recombinant virus expressed about 1×10^6 receptors on the cell surface. Photoaffinity labelling of whole cells and membranes demonstrated that the receptors had a molecular weight of 46 000. Following purification from insect cells by alprenolol affinity chromatography, the receptor was shown to activate isolated G_s-protein.

The alpha subunit of the ligand-gated nicotinic acetylcholine receptor (chick muscle) has also been synthesised in insect cells (Sf21),[33] as shown in Fig. 3.

Alpha + beta subunits of the GABA$_A$ receptor
Cell-attached mode

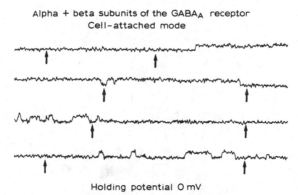

Holding potential 0 mV

FIG. 4. Patch-clamp recordings of GABA-induced single-channel activity from Sf21 cells co-infected with alpha- and beta-subunit (human GABA$_A$ receptor) recombinant baculoviruses, at 20 hpi. The patch electrode was filled with Sf21 physiological saline containing 1 μM GABA. Arrows indicate the baseline of the current trace. Calibration: 1 pA, 40 ms.

The receptor subunit was shown to be targeted to the plasma membrane and to be in the correct conformation to bind α-bungarotoxin. Ligand-binding analyses demonstrated that cholinergic ligands could displace α-bungarotoxin binding in a similar manner to that which occurs with natural receptors.

More recent studies from our laboratory have demonstrated functional expression of the GABA$_A$ receptor alpha and beta subunits (Atkinson, A. E., Bermudez, I., Beadle, D. J. & King, L. A., manuscript in preparation). GABA-gated chloride channels were detected in the insect cell plasma membrane, using the patch clamp technique in the cell-attached configuration, when the cells were infected with recombinant baculovirus expressing either the beta subunit alone or both subunits together (Fig. 4). No recordings were obtained when cells were infected with a recombinant virus expressing only the alpha subunit. Neither were the chloride channels active in the absence of GABA or in non-infected, wild-type virus-infected cells.

ACKNOWLEDGEMENTS

We thank Professor E. A. Barnard and Dr M. G. Darlison, Cambridge for providing us with the GABA$_A$ receptor cDNA clones and Dr R. D. Possee, Oxford for providing the baculovirus transfer vectors. A.E.A. and L.A.O. are supported by grants from ICI Agrochemicals and the Polytechnic and Colleges Funding Council.

REFERENCES

1. Marshall, J., Buckingham, S. D., Shingai, R., Lunt, G. G., Goosey, M. W., Darlison, M. G., Sattelle, D. B. & Barnard, E. A., Sequence and functional expression of a single α subunit of an insect nicotinic acetylcholine receptor. *EMBO J.*, **9** (1990) 4391–8.
2. Schofield, P. R., Darlison, M. G., Fujita, N., Burt, D. R., Stephenson, F. A., Rodriguez, H., Rhee, L. M., Ramachandran, J., Reale, V., Glencorse, T. A., Seeburg, P. H. & Barnard, E. A., Sequence and functional expression of the GABA_A receptor shows a ligand-gated receptor super-family. *Nature*, **328** (1987) 221–7.
3. Gershoni, J. M., Expression of the α-bungarotoxin binding site of the nicotinic acetylcholine receptor by *Escherichia coli* transformants. *Proc. Natl Acad. Sci. USA*, **84** (1987) 4318–21.
4. Fujita, N., Nelson, N., Fox, T. D., Claudio, T., Lindstrom, J., Riezman, H. & Hess, G. P., Biosynthesis of the *Torpedo californica* acetylcholine receptor α subunit in yeast. *Science*, **231** (1986) 1284–7.
5. Blount, P., Smith, M. M. & Merlie, J. P., Assembly intermediates of the mouse muscle nicotinic acetylcholine receptor in stably transfected fibroblasts. *J. Cell. Biol.*, **111** (1990) 2601–11.
6. Claudio, T., Green, W. N., Hartman, D. S., Hayden, D., Paulson, H. L., Sigworth, F. J., Sine, S. M. & Swedlund, A., Genetic reconstitution of functional acetylcholine receptor channels in mouse fibroblasts. *Science*, **238** (1987) 1688–94.
7. Miller, L. K., Insect baculoviruses: powerful gene expression vectors. *BioEssays*, **4** (1989) 91–5.
8. Cameron, I. R., Possee, R. D. & Bishop, D. H. L., Insect cell culture technology in baculovirus expression systems. *TIBTech.*, **7** (1989) 66–70.
9. Atkinson, A. E., Weitzman, M. D., Obosi, L. A., Beadle, D. J. & King, L. A., Baculoviruses as vectors for foreign gene expression in insect cells. *Pestic. Sci.*, **28** (1990) 215–24.
10. Bishop, D. H. L. & Possee, R. D., Baculovirus expression vectors. *Adv. Gene Technol.*, **1** (1990) 55–72.
11. Weyer, U., Knight, S. & Possee, R. D., Analysis of very late gene expression by *Autographa californica* nuclear polyhedrosis virus and the further development of multiple expression vectors. *J. Gen. Virol.*, **71** (1990) 1525–34.
12. Klaiber, K., Williams, N., Roberts, T. M., Papazian, D. M., Jan, L. Y. & Miller, C., Functional expression of *Shaker* K$^+$ channels in baculovirus-infected insect cell line. *Neuron*, **5** (1990) 221–6.
13. Hill-Perkins, M. S, & Possee, R. D., A baculovirus expression vector derived from the basic protein promoter of *Autographa californica* nuclear polyhedrosis virus. *J. Gen. Virol.*, **71** (1990) 971–6.
14. Martignoni, M. E. & Iwai, P. J., A catalogue of viral diseases of insects, mites and ticks. USDA Forest Service PNW-195, Washington, DC: USGPO.
15. Vaughan, J. L., Goodwin, R. H., Tompkins, G. J. & McCawley, P., The establishment of two insect cell lines from the insect *Spodoptera frugiperda* (Lepidoptera: *Noctuidae*). *In Vitro*, **13** (1977) 213–17.
16. Maeda, S., Expression of foreign genes in insects using baculovirus vectors. *Ann. Entomol.*, **34** (1989) 351–72.
17. Rohrmann, G. F., Polyhedrin structure. *J. Gen. Virol.*, **67** (1986) 1499–513.
18. Smith, G. E., Fraser, M. J. & Summers, M. D., Molecular engineering of the *Autographa californica* nuclear polyhedrosis virus genome: deletion mutants within the polyhedrin gene. *J. Virol.*, **46** (1983) 584–93.

19. Kitts, P. A., Ayres, M. D. & Possee, R. D., Linearization of baculovirus DNA enhances the recovery of recombinant virus expression vectors. *Nucleic Acid. Res.*, **18** (1990) 5667–72.

20. Possee, R. D. & Howard, S. C., Analysis of the polyhedrin gene promoter of the *Autographa californica* nuclear polyhedrosis virus. *Nucleic Acid Res.*, **15** (1987) 10233–48.

21. Miller, D. W., Safer, P. & Miller, L. K., An insect baculovirus host-vector system for high-level expression of foreign genes. In *Genetic Engineering*, Vol. 8, ed. J. K. Setlow & A. Hollaender. Plenum, New York, 1986, pp. 277–98.

22. Matsuura, Y., Possee, R. D. & Bishop, D. H. L., Expression of the S-coded genes of lymphocytic choriomeningitis arenavirus using a baculovirus vector. *J. Gen. Virol.*, **67** (1986) 1515–29.

23. Carbonell, L. F., Hodge, M. R., Tomalski, M. D. & Miller, L. K., Synthesis of a gene coding for an insect-specific scorpian neurotoxin and attempts to express it using baculovirus vectors. *Gene*, **73** (1988) 409–18.

24. Weyer, U. & Possee, R. D., Analysis of the promoter of the *Autographa californica* nuclear polyhedrosis virus p10 gene. *J. Gen. Virol.*, **70** (1989) 203–8.

25. Luckow, V. A., Cloning and expression of heterologous genes in insect cells with baculovirus vectors. In *Recombinant DNA Technology and Applications*, ed. C. Ho, A. Prokop & R. Bajpai. McGraw-Hill, New York, 1990, pp. 97–152.

26. Butters, T. D. & Hughes, R. C., Isolation and characterization of mosquito cell membrane glycoproteins. *Biochim. Biophys. Acta*, **640** (1981) 655–71.

27. Butters, T. D., Hughes, R. C. & Visher, R. C., Steps in the biosynthesis of mosquito cell membrane glycoproteins and the effects of tunicamycin. *Biochim. Biophys. Acta*, **640** (1981) 672–86.

28. Jarvis, D. L., Oker-Blom, C. & Summers, M. D., Role of glycosylation in the transport of recombinant glycoproteins through the secretory pathways of lepidop-terian insect cells. *J. Cell. Biochem.*, **42** (1990) 181–91.

29. Armstrong, C. M. & Miller, C., Do voltage-dependent K^+ channels require Ca^{2+}? A critical test employing a heterologous expression system. *Proc. Natl Acad. Sci. USA*, **87** (1990) 7579–82.

30. Birnbaumer, L., Abramowitz, J. & Brown, A. M., Receptor-effector coupling by G proteins. *Biochim. Biophys. Acta*, **1031** (1990) 163–224.

31. Luckow, V. A. & Summers, M. D., Trends in the development of baculovirus expression vectors. *Bio/Technol.*, **6** (1988) 47–55.

32. Reilander, H., Boege, F., Vasudevan, S., Maul, G., Hekman, M., Dees, C., Hampe, W., Halmreich, E. J. M. & Hartmut, M., Purification and functional expression of the human β_2–adrenergic receptor produced in baculovirus-infected insect cells. *FEBS*, **282** (1991) 441–4.

33. Atkinson, A. E., Earley, F. G. P., Beadle, D. J. & King, L. A., Expression and characterization of the chick nicotinic acetylcholine receptor α-subunit in insect cells using a baculovirus vector. *Eur. J. Biochem.*, **192** (1990) 451–8.

34. King, L. A. & Possee, R. D., *The Baculovirus Expression Vector System: a Laboratory Guide*. Chapman and Hall, London, 1992.

15

Functional Expression in *Xenopus* Oocytes of Invertebrate Ligand-Gated Ion Channels

DAVID B. SATTELLE,[a] SARAH C. R. LUMMIS,[b] HOWARD A. RIINA,[a] JOHN T. FLEMING,[c] NICOLA M. ANTHONY[a] & JOHN MARSHALL[c,d]

[a] *AFRC Laboratory of Molecular Signalling, Department of Zoology, University of Cambridge, Downing Street, Cambridge CB2 3EJ, UK*
[b] *Department of Zoology, University of Cambridge, Downing Street, Cambridge CB2 3EJ, UK*
[c] *Laboratory of Molecular Biology, MRC Centre, Hills Road, Cambridge CB2 2QH, UK*
[d] *Department of Pharmacology, Yale University, New Haven, Connecticut, 06510-8066, USA*

INTRODUCTION

Ligand-gated ion channels are hetero– or homo–oligomeric proteins exhibiting pseudosymmetry, with polypeptide subunits arranged around a central hydrophilic pore.[1] The channels directly operated by neurotransmitters belong to this class of membrane proteins in which the channel is an integral part of the receptor protein.[2] Ligand-gated ion channels have been particularly well studied in mammals, and targets for several chemically distinct classes of medicinal drugs have been recognised. Investigations of invertebrate ligand-gated ion channels have revealed target sites for insecticides and anthelmintics.[3] DNA cloning of polypeptide subunits of ligand-gated ion channel molecules has been achieved in the case of skeletal muscle, electric tissue[4] and nervous system nicotinic acetylcholine receptors[5] of vertebrates. Subunits of vertebrate $GABA_A$,[6] glycine[7] and L–glutamate receptors[8] have also been cloned recently, and the cells most widely used to test for functional expression of putative subunits are the oocytes of *Xenopus laevis*.[9] These cells permit expression in the plasma membrane of functional neurotransmitter-operated ion channels following cytoplasmic injection of crude RNA, poly-adenylated (messenger) RNA (poly $(A)^+$ mRNA) and cDNA-derived mRNA. Functional channels of this type can also be expressed as a result of nuclear injection of cDNA and an appropriate promotor.[10]

Recently, several laboratories have reported DNA cloning of putative invertebrate ligand-gated ion channel subunits,[11–18] and in order to establish the functional role of such putative receptors it is essential to demonstrate functional expression. Here we describe some recent expression studies from our laboratory aimed at addressing three particular problems: (1) is it possible to express functional ligand-gated ion channels using invertebrate tissue-specific (e.g. nervous system) preparations; (2) can the capacity to express function be demonstrated for a cloned putative invertebrate ligand-gated ion channel subunit; (3) can expression techniques permit investigation for the first time of functional invertebrate ligand-gated ion channels that have not so far proved accessible by any other means. For example, to date physiological study of neurotransmitter-gated ion channels of the nematode *Caenorhabditis elegans*, the only known source of viable nicotinic acetylcholine receptor mutants,[19] has not been possible. As a prelude to examining in detail these particular questions, we first consider the potential of the *Xenopus* oocyte as an expression system for invertebrate ligand-gated ion channels.

THE OOCYTE OF *XENOPUS LAEVIS* AS AN EXPRESSION SYSTEM FOR INVERTEBRATE LIGAND-GATED ION CHANNELS

The use of *Xenopus laevis* as a vehicle for protein production from messenger RNA was initiated by John Gurdon and his colleagues in the early 1970s,[20] but it was not until after the 1980s that *Xenopus* oocytes were routinely used to express ligand-gated ion channels.[9,21] The functions of such membrane proteins are not amenable to investigation by the conventional molecular biology approach of expression which employs a cell-free translation system. An intact cell expression system is required because many ligand-gated ion channels have been shown to be hetero-oligomers. The subunit composition of these molecules may vary between cell types and at different stages of development (see Ref. 7 for review). Also, the correct assembly of individual channels appears to require specific recognition between the subunits.[22,23] Recent studies on muscle nicotinic acetylcholine receptors have shown that specific extracellular domains mediate subunit interactions.[24]

It became apparent from the early 1980s that *Xenopus* oocytes injected with the appropriate messenger RNAs will correctly synthesise and assemble a variety of neurotransmitter-gated ion channels.[25,26] These can be characterised with relative ease using a variety several electrophysiological methods, and are also amenable to biochemical studies.[9,21,27,28]

There are a number of transient and stable expression systems available to molecular biologists, but the *Xenopus* oocyte system remains one of the most versatile for the study of ligand-gated ion channels. Its uses include: (1) biosynthesis/assembly of multisubunit protein molecules (an approach that is particularly useful when determining which combinations of subunits are functionally active, and which confer specific properties on the protein); (2) electrophysiological studies on receptor/channel molecules in tissues that are difficult to study *in situ* (examples include the *Torpedo* electroplax nicotinic acetylcholine receptor, a rich source of this channel protein which is not readily studied *in situ* by path-clamp electrophysiology as the molecules are too dense for a precise analysis, and some invertebrate ligand-gated ion channels where in-situ electrophysiology has hitherto proved impossible); (3) assay of messenger RNA fractions prior to cDNA cloning (thereby determining a suitable source and an appropriate developmental stage of messenger RNA production for subsequent cloning work); (4) screening cDNA libraries by expression has been applied successfully in the case of a number of ligand-gated channels; (5) use as a 'test-tube' for mutagenesis, enabling specific functions to be assigned to individual amino acids or amino acid segments.

Several comprehensive reviews have summarised in detail the physiological properties of *Xenopus* oocytes, including endogenous channels.[27,28] The native oocyte membrane does not contain any of the ligand-gated ion channels activated specifically by nicotinic cholinergic ligands, GABA or L–glutamate, making it an ideal expression system for use with these membrane proteins. Endogenous receptors activated by muscarinic cholinergic ligands are present. Activation of these muscarinic receptors induces two distinct currents, a large chloride current that is mimicked by inositol trisphosphate, and a smaller potassium current that is mimicked by intracellular injection of calcium. Both currents are readily separable pharmacologically from nicotinic receptors expressed following injection of exogenous messenger RNA.

The oocyte resting potential is around -50 to $-60\,mV$; the approximate equilibrium potentials of various monovalent ions are as follows: E_{K^+}, $-100\,mV$; E_{Cl^-}, $-25\,mV$; E_{Na^+}, $+80\,mV$.[27] Donnan equilibrium conditions are not fulfilled in the oocyte and active mechanisms are needed to preserve the intracellular ionic concentrations. Gradients of monovalent ions are maintained by an electrogenic Na^+–K^+ pump. Knowledge of the reversal potentials of major current carrying cations and anions is essential for the interpretation of the ionic basis of responses to neurotransmitters in oocytes injected with exogenous messenger RNA.

A large number of the neurotransmitters that are active in vertebrates are also effective at insect synapses. The conservation between phyla of the neurotransmitter roles of acetylcholine, GABA and L–glutamate is considerable, though differences in tissue specificity and pharmacology of the various ligand-gated ion channels that mediate many of their actions are detected. In vertebrates shared sequence homologies suggest many of the ligand-gated ion channels are members of a gene superfamily.[2] Those invertebrate ligand-gated ion channels cloned and sequenced to date show considerable homology to the vertebrate members of this family (Fig. 1 for example).

EXPRESSION OF TISSUE-SPECIFIC MESSENGER RNA FROM INSECT NERVE AND MUSCLE

The properties of in-situ insect nicotinic acetylcholine,[29,30] GABA[31–33] and L–glutamate receptors of insects have been characterised, mostly by means of

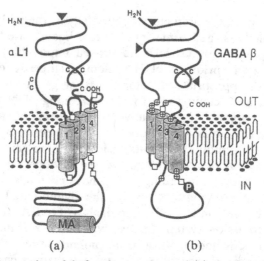

FIG. 1. Comparisons of models for the topology of (a) the locust αL1 subunit and (b) the GABA$_A$ receptor β-subunit in the membrane. Four membrane spanning helices in each subunit are shown as cylinders (1–4). The structure in the extracellular domain is drawn in an arbitrary manner; the presumed β-loop formed by the disulphide bond predicted at cysteines 128 and 142 (*Torpedo* AChR α-subunit numbering) is shown. Potential extracellular sites for N-glycosylation are indicated by triangles and a possible site for cAMP-dependent serine phosphorylation present in the GABA$_A$ receptor β-subunit is denoted by an encircled P. Those charged residues that are located close to the ends of the membrane spanning domains are shown as circles with positive charge marked, or as open squares for negative charges. Note the large excess of positive charge which will be at the mouths of the channel when four or five of the GABA$_A$ subunits are assembled to form the receptor. Note also that there will be a small excess of negative charge at the channel mouth in the corresponding structure formed from locust AChR α-subunits. Cysteine residues (C) are also shown corresponding to the vicinal pair in αL1 and one near the C-terminus (which may be involved in dimer formation[35]). MA, the predicted amphipathic helix peculiar to AChRs, is shown as forming part of the long cytoplasmic loop between M3 and M4 in all neuronal AChRs and is presumed to be involved in folding compactly the tertiary structure of the loop.
(From Ref. 2, reproduced by permission of Academic Press, 1989.)

physiological and biochemical methods. Recently functional expression of receptors for these transmitters has been detected in *Xenopus* oocytes following injection of insect RNA.

Acetylcholine receptors

Nicotinic acetylcholine receptors in insects, unlike those of vertebrates, are confined to the nervous system, where they are present in very high density.[29] These receptors gate non-selective cation channels allowing the flux of sodium, potassium and possibly calcium ions.[34] There are many cases where synaptic and extrasynaptic nicotinic acetylcholine receptors are blocked by α–bungarotoxin (see Ref. 30 for review). [^{125}I]α–Bungarotoxin has been used to affinity purify a 65 000 kDa polypeptide from locust[35] and cockroach[36] nervous and extrasynaptic nicotinic acetylcholine receptors are blocked by α–bungarotoxin (see Ref. 30 for review). [125I]α–Bungarotoxin has been used to affinity purify a 65 000 kDa polypeptide from locust[35] and cockroach[36] nervous tissue, though sedimentation and radiation inactivation studies indicate that the in-situ receptor is an oligomer of ~250 000 kDa.[35] Breer and colleagues have reconstituted a functional acetylcholine-gated cation channel in artificial lipid membranes using the affinity purified 65 000 kDa polypeptide.[37]

Injection of *Xenopus* oocytes with poly $(A)^+$ RNA isolated from the nervous tissue of young locusts (*Locusta migratoria*) has been reported to result in the expression of [125I]α–bungarotoxin binding sites and nicotine-induced $^{22}Na^+$ uptake.[38,39] We have examined the electrophysiological response to oocytes injected with total messenger RNA from the CNS of adult male cockroaches and have observed a nicotine-induced depolarisation (Fig. 2). Control oocytes, either injected with the equivalent volume of distilled water, or uninjected, show no such response to nicotine.

GABA receptors

γ-Aminobutyric acid (GABA) in insects, as in vertebrates, is a major inhibitory neurotransmitter, although in insects it has an inhibitory role on muscle as well as in the nervous system. GABA receptors in vertebrate central nervous tissues have been divided into two classes: $GABA_A$ receptors gate chloride channels and are modulated by benzodiazepines, barbiturates and steroids, while $GABA_B$ receptors act via G proteins to increase K^+ conductance, or Ca^{2+} conductance.[40] Most GABA receptors described to date in

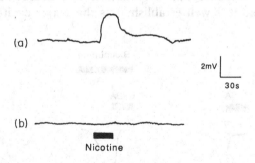

Fig. 2. Nicotine-induced depolarization recorded from *Xenopus* oocytes injected with RNA isolated from cockroach *Periplaneta americana* nervous system. (a) Response to nicotine (1.0×10^{-5} M) recorded from oocytes injected with RNA; (b) control showing no response to the same concentration to the same concentration of nicotine for an oocyte injected with water.

insects appear to resemble $GABA_A$ receptors in that they control chloride channels and may be modulated by benzodiazepines and barbiturates in at least some species,[31,33] although there is also a recent report of $GABA_B$-like activity in cockroach CNS.[41] There are, however, some pharmacological differences between insect GABA receptors and vertebrate $GABA_A$-type receptors, of which the first to emerge was an apparent insensitivity of at least some insect GABA responses to the $GABA_A$ antagonist bicuculline.[31-33] Also the GABA receptor linked benzodiazepine binding site is quite different from the corresponding site on vertebrate $GABA_A$ receptors.[42] In its high sensitivity to Ro 5-4864 and insensitivity to clonazepam, the insect site resembles more closely the peripheral benzodiazepine receptor of vertebrates.[43] Robinson and colleagues[44] have utilised the capacity of flunitrazepam to photoaffinity label its receptor and they have revealed two polypeptides of 45 kDa and 59 kDa. No reconstitution data are available to date for the insect GABA/benzodiazepine receptor.

As yet insect GABA receptor cDNA-derived messenger RNA has not been expressed in oocytes, though two laboratories have described the cloning and sequencing of putative *Drosophila melanogaster* GABA receptor subunits.[45,46] However, injection of messenger RNA from cockroach CNS results in chloride-mediated responses to GABA that are inhibited by picrotoxin and insensitive to bicuculline[47] (see Fig. 3). A study has also been performed of GABA responses to locust muscle messenger RNA injected into oocytes. Again this results in chloride-mediated responses which are insensitive to bicuculline but can be blocked by picrotoxin.[48] Thus messenger RNA from both insect nerve and muscle preparations is capable of expressing functional GABA receptors when injected into *Xenopus* oocytes.

L–Glutamate receptors

There is increasing evidence that L–glutamate is a neurotransmitter in insect nervous systems, and it is well established as the major excitatory neurotrans-

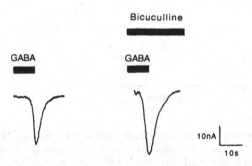

FIG. 3. GABA-induced current recorded from *Xenopus* oocytes injected with RNA isolated from cockroach, *Periplaneta americana*, nervous system. The response to GABA (5×10^{-4} M) in the absence and presence of bicuculline (1×10^{-4} M) are shown.

mitter at the locust neuromuscular junction.[49,50] L–Glutamate causes a de-polarising response, carried predominantly by Na^+ and K^+ ions, at *Schistocerca* neuromuscular junction synaptic receptors, and can also regulate a chloride dependent hyperpolarization at extrasynaptic sites.[51,52] The hyper-polarising response can be blocked by picrotoxin and is mimicked by ibotenate, whereas the depolarizing response can be mimicked by quisqualate, suggesting that two distinct L–glutamate receptors may mediate the responses. Studies on neurones from several species, notably the cockroach *Periplaneta* and the locust *Schistocerca,* reveal both hyperpolarising and depolarising actions of L–glutamate.[53–57] Recently Wafford *et al.*[58] using functionally distinct *Periplaneta* neurones demonstrated the existence of at least three distinct neuronal L–glutamate receptor subtypes, some of which are cell specific. Expression of insect (locust) muscle messenger RNA in oocytes reveals at least two classes of L–glutamate receptors: quisqualate may activate responses to Na^+, K^+ and Ca^{2+} ions, whereas ibotenate apparently activates a Cl^- conductance. L–Glutamate itself appears to increase the conductance of Na^+, K^+ and Cl^- ions.[48] L–Glutamate responses in oocytes injected with lobster muscle RNA have also been reported.[59] A response to L–glutamate of *Xenopus* oocytes injected with locust embryonic messenger RNA is shown in Fig. 4.

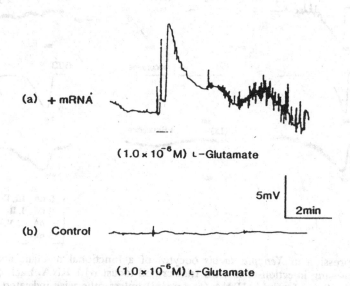

(a) + mRNA

$(1.0 \times 10^{-6}M)$ L–Glutamate

5mV

2min

(b) Control

$(1.0 \times 10^{-6}M)$ L–Glutamate

FIG. 4. Messenger RNA isolated from locust *Schistocerca gregaria* embryonic tissue results in expression of functional L–glutamate receptors following injection of 50 ng RNA (1 ng/nl solution in distilled water) into an oocyte of *Xenopus laevis*. (a) Response to L–glutamate (1.0×10^{-6} M) of oocyte injected with locust embryonic RNA. (b) Control oocyte showing no response to the same concentration of L–glutamate following injection of 50 nl of distilled water. From Ref. 46, reproduced by permission of Raven Press, New York (in press).

EXPRESSION OF cDNA–DERIVED RNA FROM A CLONED INSECT NICOTINIC ACETYLCHOLINE RECEPTOR α–SUBUNIT

The *Xenopus* oocyte expression system has been used extensively to assay the functional capacity of putative nicotinic acetylcholine receptors of vertebrates.

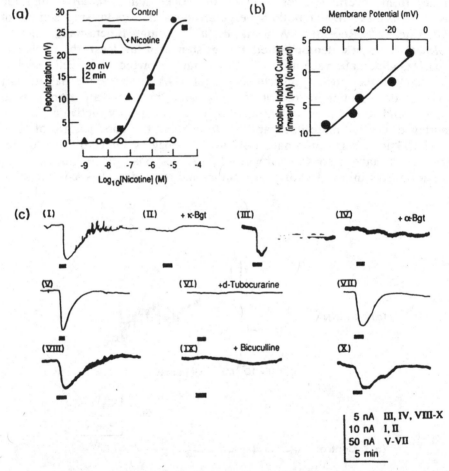

Fig. 5. Expression in *Xenopus laevis* oocytes of a functional nicotinic acetylcholine receptor following injection of in-vitro transcribed locust αL1 RNA. Each oocyte was injected with 50 ng of poly(A)$^+$RNA (at 1 ng/nl) unless otherwise indicated. All drugs tested were dissolved in Barth's medium. In all traces shown, the test concentration of bath-applied nicotine was 1.0×10^{-5} M unless otherwise stated. (a) Dose dependence of the nicotine-induced depolarisation of RNA-injected oocytes ($E_m = -60$ mV). Solid symbols represent data from three separate oocytes. Open circles depict data from control oocytes injected with 50 nl distilled water. Insert: control oocyte ($E_m = -45$ mV) injected with 50 nl distilled water fails to respond to 10^{-6} M nicotine whereas the

Using probes derived from vertebrate nicotinic acetylcholine receptor sequence two putative locust (*Schistocerca gregaria*) nicotinic receptor subunit clones have been isolated.[14] The functional expression of one of these, an α-like subunit, has been investigated in detail.

Functional expression of a cloned locust nicotinic acetylcholine receptor α–subunit

Marshall and colleagues[15] described the complete sequence of a locust (*Schistocerca gregaria*) nervous system nicotinic acetylcholine receptor subunit (αL1). The mature polypeptide comprises 534 amino acids which corresponds to $60 \cdot 641$ kDa (unglycosylated). As the locust subunit sequence contains two sites for the possible addition of N-linked sugars, its size is compatible with the 65 kDa subunit species purified by α–bungarotoxin affinity chromatography from *Locusta migratoria*[35] and *Periplaneta americana*.[30] The αL1 subunit contains in the proposed extracellular domain two adjacent cysteine residues which are characteristic of α(ligand binding) subunits. Hydropathy plots indicate four putative transmembrane domains M1–M4 (shown schematically in Fig. 1).

RNA derived from αL1 cDNA when injected alone into *Xenopus* oocytes resulted in the expression of functional nicotinic acetylcholine receptors gating cation channels[14,15] (see Fig. 3). This indicated that either homo-oligomeric receptors were being formed or that a hetero-oligomeric receptor was produced by combination with endogenous polypeptides. Recently, in studies on vertebrate CNS, very similar findings have been reported by Couturier *et al*.[60] for another nervous system nicotinic acetylcholine receptor subunit (α7) which forms a homo-oligomeric channel blocked by α–bungarotoxin.

RNA-injected oocyte is depolarised by 10^{-6} M nicotine. (b) Membrane potential dependence of the amplitude of nicotine-induced inward currents recorded from an RNA-injected oocyte. Nicotine (1.0×10^{-5} M) was applied for 2 min to an oocyte clamped at different membrane potentials and the peak current response was measured. (c) Blockade of the expressed insect receptor by a range of antagonists. Each antagonist was tested on a separate oocyte whose response to nicotine alone is shown first ($E_m = -60$ mV). The period of application of nicotine is denoted by the solid bar. Downward deflections indicate inward current. Nicotine response (I) is blocked (II) by a 30-min exposure to 10^{-7} M κ–Bgt. Nicotine response (III) is blocked (IV) by a 30-min exposure to 1.0×10^{-7} M α–Bgt. As with the predominant in-vivo nicotinic receptor (see Refs 61–63), the expressed receptor is blocked by α–Bgt and κ–Bgt (both at 10^{-7} M). Nicotine response (V) is blocked (VI) by a 15-min exposure to 1.0×10^{-6} M d–tubocurarine, but recovers (VII) after a 30-min wash in Barth's medium. Nicotine response (VIII) is blocked (IX) by a 15-min exposure to $5 \cdot 0 \times 10^{-6}$ M bicuculline. Rebathing in Barth's medium for 30 min (X) restores the response. (From Ref. 15, reproduced by permission of *EMBO J.*, **9**, 1990.)

TABLE 1
Pharmacology of Expressed and in-situ Insect Nicotinic Acetylcholine Receptors

| Ligand | Vertebrate nAChRs | | Insect nAChRs | | | |
	Muscle	CNS	Cloned locust	Locust cell body	Cockroach (cell body)	(synape)
Nicotine	A	A	A	A	A	A
Acetylcholine	A	A	A	A	A	A
α–Bungarotoxin	B	NE (except α-7)	B	B	B	B
κ–Bungarotoxin	NE	B ($\alpha 3\beta 2$; $\alpha 4\beta 2$)	B	B	B	B
d–Tubocurarine	B	B (weak)	B	B	B	B
Mecamylamine	B (weak)	B	B	B	B	—
Bicuculline		(GABA$_A$R blocker)	B	B	B	B
Strychnine		(Glycine R blocker)	B	B	—	B

A, agonist; B, blocker; NE, no effect; —, not tested; R, receptor.

Pharmacology of an expressed insect α–subunit, and an in-situ α–bungarotoxin-sensitive nicotinic receptors

The pharmacology of this expressed receptor is of interest. α–Bungarotoxin blocked the expressed receptor (Fig. 5) at concentrations similar to those required to block in-situ nicotinic receptors.[61-63] Mecamylamine, d–tubocurarine and other nicotinic receptor antagonists that block in-situ receptors of locust and cockroach were effective on the expressed receptor (Table 1). Bicuculline also blocked the expressed nicotinic receptor; bicuculline has been shown to block nicotinic responses in neurones of locust[64] and cockroach (Buckingham, S. D., pers. comm., 1990).

Other sequences that encode putative insect ACh receptor genes have been identified in *Drosophila*, again using probes derived from vertebrate nicotinic receptor subunits. These include the α-like subunit (ALS) which (when expressed as a fusion protein) does not bind α–bungarotoxin,[12] whereas ARD protein appears to encode a component of an α–bungarotoxin sensitive receptor.[11,65] However, at present these genes have not been functionally expressed. Recently other putative nicotinic receptor subunits have been described in *Drosophila*, so that to date the following sequences are known: ALS, Dα2 = SAD (α-like); ARD, SBD (non-α).[11-13,17,18] At very high concentrations (>10^{-3} M) nicotine-induced currents are detected, though the pharmacology is difficult to reconcile with that of any in-situ receptors. Co-expression studies with the various putative *Drosophila* receptor subunits will be of considerable interest.

EXPRESSION OF LIGAND-GATED ION CHANNELS OF *CAENORHABDITIS ELEGANS*

The free-living soil nematode *Caenorhabditis elegans* is the only animal for which the complete cell lineage and wiring diagram of the nervous system are known. As the physical gene map nears completion, and the sequence determination of its entire genome is already underway,[66] this invertebrate is well suited for genetic approaches to the study of ligand-gated ion channels. Mutants resistant to the potent nicotinic receptor agonist, levamisole, have been isolated. Resistance is due to mutations in 1 of 10 genes,[19] three of which (*unc-29, unc-38, Lev-1*) are found to encode structural subunit homologues of putative nicotinic acetylcholine receptor genes. To date, the ability to exploit the only known, viable, ligand-gated ion channel mutants has been limited by an inability to carry out electrophysiological experiments *in situ,* due to the small size of the organism. In an attempt to circumvent the problem, we have investigated *Caenorhabditis elegans* acetylcholine and GABA receptors, initially from normal (N2) worms, by injecting isolated RNA into oocytes of *Xenopus laevis.*

Acetylcholine receptors

RNA isolated from mixed stages of *Caenorhabditis elegans* when injected into *Xenopus* oocyte cytoplasm resulted in functional expression of nicotinic acetylcholine receptors. At concentrations above a threshold of around 1.0×10^{-6} M, levamisole acts on an expressed receptor that gates a cation channel[67] (see Fig. 6). Levamisole is much more effective than nicotine on this expressed receptor. This is also the case for in-situ acetylcholine receptors of the nematode *Ascaris suum.*[68] Such levamisole-induced depolarisations are reversibly reduced in amplitude by *d*–tubocurarine $(1.0 \times 10^{-5}$ M). In-situ cholinergic receptors of *Ascaris* are also blocked by *d*–tubocurarine.[69]

GABA receptors

GABA at concentrations above a threshold of approximately 1.0×10^{-6} M acts on an expressed receptor that gates a channel with a reversal potential close to that of the oocyte E_{Cl^-}, resulting in a depolarising response at resting potential (Fig. 6). This expressed GABA-gated chloride response is bicuculline insensitive; this is also the case for the in-situ receptors of *Ascaris.*[70]

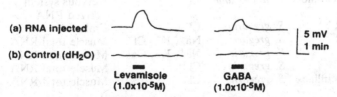

FIG. 6. Actions of levamisole and GABA on an (a) oocyte of *Xenopus laevis* injected with RNA from *C. elegans,* and (b) a control oocyte injected with distilled water. (From Ref. 67, reproduced by permission of *J. Physiol.,* **438,** 1991.)

L–Glutamate receptors

Recently Arena and colleagues[71] have shown that *Caenorhabditis elegans* poly (A)$^+$ RNA injected into *Xenopus* oocytes results in expression of L–glutamate-sensitive chloride channel. Respones to L–glutamate ($2 \cdot 0 \times 10^{-4}$ M) were not mimicked by kainate, N–methyl–D–aspartate or quisqualate, and were not blocked by MK–801. It will be of interest to examine further the pharmacology of this response to L–glutamate.

CONCLUSIONS

Invertebrates offer appropriate experimental material for several kinds of investigation on ligand-gated ion channels. The nervous systems of such organisms provide the richest source of certain members of this gene superfamily. Insects such as locusts and cockroaches with a large number of identified cells and pathways offer the prospect of ascribing function to the rich diversity of receptor subtypes that appear to exist for each neurotransmitter. Already tissue specific and cell-specific receptor subtypes have been demonstrated in these organisms *in situ*.

The powerful techniques of molecular cloning and genetics which have already contributed greatly to our understanding of vertebrate ligand-gated ion channels can readily be applied to invertebrate organisms such as *Drosophila melanogaster* and *Caenorhabditis elegans* with their well characterised genomes. Both organisms will facilitate chromosomal mapping and manipulation

TABLE 2

Insect and Nematode Ligand-Gated Ion Channels that have been Characterised *in situ* and also Expressed in *Xenopus* Oocytes Using RNA from Various Sources

Receptor/channel	Species	Ions	RNA source for functional expression in Xenopus *oocytes*	Ref.
nAChR (α–Bgt-sensitive)	*S. gregaria*	Na$^+$, K$^+$, Ca^{2+}	cDNA derived RNA	14, 15
nAChR	*P. americana*	Cationic	Nervous system derived RNA	This study
GABAR (Bicuculline-insensitive)	*P. americana*	Cl$^-$	Nervous system derived RNA	47
L–Glutamate R	*S. gregaria*	Cl$^-$	Embryo	46
L–Glutamate R	*S. gregaria*	Na$^+$, K$^+$, Cl$^-$	Muscle total RNA	48
Quisqualate R	*S. gregaria*	Na$^+$, K$^+$, Ca^{2+}	Muscle total RNA	48
Ibotenate R	*S. gregaria*	Cl$^-$	Muscle total RNA	48
GABAR (Bicuculline-insensitive)	*S. gregaria*	Cl$^-$	Muscle total RNA	48
nAChR	*C. elegans*	Cationic	Total RNA	67
GABAR	*C. elegans*	Anionic	Total RNA	67

of genes encoding ligand-gated channel subunits. Future studies on *Caenorhabditis elegans* offer the prospect of investigating further viable mutants of ligand-gated ion channels, and locating subunit genes on the almost complete physical gene map of this organism.

Studies with the *Xenopus* oocytes to date have permitted functional expression of invertebrate tissue-specific ligand-gated ion channels (Table 2). For example tissue-specific GABA-gated chloride channels have been expressed, using RNA prepared respectively from insect nervous tissue and muscle. Also, functional expression of a cloned invertebrate subunit of a ligand-gated ion channel (α–subunit of locust *Schistocerca gregaria* nicotinic acetylcholine receptor) has been achieved. As in-situ nicotinic receptors carry binding sites for nitromethylene insecticides it will be of interest to pursue further the expression of molecular targets of such pest control agents. Functional expression of an anthelmintic (levamisole) target site has also been demonstrated using *Xenopus* oocytes. Of equal significance perhaps is that expression studies on *Caenorhabditis elegans* offer the first prospect of direct experimental investigation of ligand-gated channels in an organism in which viable channel mutants can be obtained. Such studies will permit the correlation of a change in the primary sequence of a ligand-gated channel subunit with its physiological and behavioural consequences.

REFERENCES

1. Unwin, N., The structure of ion channels in membranes of excitable cells. *Neuron,* **3** (1989) 665–76.
2. Barnard, E. A., Darlison, M. G., Marshall, J. & Sattelle, D. B., Structural characteristics of cation and anion channels directly operated by agonists. In *Ion Transport,* ed. D. J. Keeling & C. D. Benham. Academic Press, San Diego, 1989, pp. 242–6.
3. Sattelle, D. B., Synaptic and extrasynaptic neuronal nicotinic receptors of insects. In *Nicotinic Acetylcholine Receptors in the Nervous System,* NATO ASI Series, Vol. H25, ed. F. Clementi, C. Gotti & E. Sher. Springer-Verlag, Berlin, 1988, pp. 241–56.
4. Claudio, T., Molecular genetics of acetylcholine receptor channels. In *Molecular Neurobiology,* ed. D. M. Glover and B. D. Hames. Oxford University Press, Oxford, UK, 1989, pp. 63–142.
5. Lindstrom, J., Schoepfer, R. & Whiting, P., Molecular studies of the neuronal nicotinic acetylcholine receptor family. *Mol. Neurobiol.,* **1** (1987) 218–337.
6. Seeburg, P. H., Widen, W., Verdooni, T. A., Pritchett, D. B., Werner, P., Herb, A., Luddens, H., Sprengel, R. & Sachmann, B., The GABAA receptor family: molecular and functional diversity. In *The Brain Cold Spring Harbor Symposia in Quantitative Biology,* Vol. LV, Cold Spring Harbor Press, Plainview, New York, 1991, pp. 29–40.
7. Betz, H., Ligand-gated ion channels in the brain: the amino acid receptor superfamily. *Neuron,* **5** (1990) 383–92.
8. Heinemann, S., Bettler, B., Boulter, J., Deneris, E., Gasic, C., Hartley, M., Hollman, M., Hughes, T. E., O'Shea-Greenfield, A. & Rogers, S., The glutamate receptor gene family. In *Excitatory Amino Acids Fidia Research Foundation Symposium Series,* Vol. 5, eds. B. S. Meldrum, F. Moroni, R. P. Simon & J. H. Woods. Raven Press, New York, 1991, pp. 109–13.

9. Barnard, E. A. & Bilbe, G., Functional expression in the *Xenopus* oocyte of messenger RNAs for receptors and ion channels. In *Neurochemistry: A Practical Approach,* ed. A. J. Turner & H. S. Bachelard. IRL Press, Oxford and Washington, DC, 1987, pp. 243–68.

10. Ballivet, M., Nef, P., Couturier, S., Rungger, D., Bader, C. R., Bertrand, D. & Cooper, E., Electrophysiology of a chick neuronal acetylcholine receptor expressed in *Xenopus* oocytes after cDNA injection. *Neuron,* **1** (1988) 847–52.

11. Hermans-Borgmeyer, I., Zopf, D., Ryseck, R. P., Hovermann, B., Betz, H. & Gundelfinger, E. D., Primary structure of a developmentally regulated nicotinic acetylcholine receptor protein from *Drosophila. EMBO J.,* **5** (1986) 1503–8.

12. Bossy, B., Ballivet, M. & Spierer, P., Conservation of neuronal nicotinic acetylcholine receptors from *Drosophila* to vertebrate central nervous systems. *EMBO J.,* **7** (1988) 611–18.

13. Gundelfinger, E. D., Hermans-Borgmeyer, I., Schloss, P., Sawruk, E., Udri, C., Vingron, M., Betz, H. & Schmitt, B., Ligand-gated ion channels of *Drosophila.* In *Nicotinic Acetylcholine Receptors in the Nervous System NATO ASI Series,* Vol. H25, ed. F. Clementi, C. Gotti & E. Sher. Springer-Verlag, Berlin, 1988, pp. 69–81.

14. Marshall, J., David, J. A., Darlison, M. G., Barnard, E. A. & Sattelle, D. B., Pharmacology, cloning and expression of insect nicotinic acetylcholine receptors. In *Nicotinic Acetylcholine Receptors in the Nervous System,* NATO ASI Series, Vol. H25, ed. F. Clementi, C. Gotti & E. Sher. Springer-Verlag, Berlin, 1988, pp. 257–81.

15. Marshall, J., Buckingham, S. D., Shingai, R., Lunt, G. G., Goosey, M. W., Darlison, M. G., Sattelle, D. B. & Barnard, E. A., Sequence and functional expression of a single α-subunit of an insect nicotinic acetylcholine receptor. *EMBO J.,* **9** (1990) 4391–8.

16. Sawruk, E., Hermans-Borgmeyer, I., Betz, H. & Gundelfinger, E., Characterization of an invertebrate nicotinic acetylcholine receptor gene: the *ard* gene of *Drosophila melanogaster. FEBS Letts.,* **235** (1988) 40–46.

17. Sawruk, E., Schloss, P., Betz, H. & Schmitt, B., Heterogeneity of *Drosophila* nicotinic acetylcholine receptors: SAD, a novel developmentally regulated α-subunit. *EMBO J.,* **9** (1990) 2671–7.

18. Baumann, A., Jonas, P. & Gundelfinger, E. D., Sequence of Dα2, a novel α-like subunit of *Drosophila* nicotinic acetylcholine receptors. *Nucleic Acids Research,* **18** (1990) 3640P.

19. Lewis, J., Wu, C. H., Berg, H. & Levine, J. H., The genetics of levamisole resistance in the nematode *Caenorhabditis elegans. Genetics,* **95** (1980) 905–28.

20. Gurdon, J. B., Lane, C. D., Woodland, H. R. & Marbaix, G., Use of frogs eggs and oocytes for the study of messenger RNA and its translation in living cells. *Nature,* **233** (1971) 177–80.

21. Smart, T. G., Houamed, K. M., Van Renterghem, C. & Constanti, A., mRNA directed synthesis and insertion of functional amino acid receptors in *Xenopus laevis* oocytes. *Biochem. Soc. Trans.,* **15** (1987) 117–22.

22. Blount, P. & Merlie, J. P., Molecular basis of the two non-equivalent ligand binding sites of the muscle nicotinic acetylcholine receptor. *Neuron,* **3** (1989) 3490–3507.

23. Blount, P., Smith, M. M. & Merlie, J. P., Assembly intermediate of the mouse muscle nicotinic acetylcholine receptor in stably transfected fibroblasts. *J. Cell Biol.,* **111** (1990) 2601–11.

24. Yu, X.-M. & Hall, Z. W., Extracellular domains mediating ε subunit interaction of muscle acetylcholine receptor. *Nature,* **352** (1991) 64–7.

25. Barnard, E. A., Miledi, R. & Sumikawa, K., Translation of exogenous messenger RNA coding for the nicotinic acetylcholine receptors produces functional receptors in *Xenopus* oocytes. *Proc. R. Soc. (Lond.) B,* **215** (1982) 241–6.

26. Gundersen, C. B., Miledi, R. & Parker, I., Voltage-operated channels by foreign messenger RNA in *Xenopus* oocytes. *Proc. R. Soc.* (*Lond.*) *B*, **220** (1983) 131–40.
27. Dascal, N., The use of *Xenopus* oocytes for the study of ion channels. *CRC Crit. Rev. Biochem.*, **22** (1987) 317–85.
28. Snutch, T. P., The use of *Xenopus* oocytes to probe synaptic communication. *TINS*, **11** (1988) 250–56.
29. Sattelle, D. B.,Acetylcholine receptors of insects. *Adv. Insect Physiol.*, **15** (1980) 215–315.
30. Breer, H. & Sattelle, D. B., Molecular properties and functions of insect acetylcholine receptors. *J. Insect Physiol.*, **33** (1987) 771–90.
31. Lummis, S. C. R., GABA receptors in insects. *Comp. Biochem. Physiol.*, **95C** (1990) 1–8.
32. Rauh, J. J., Lummis, S. C. R. & Sattelle, D. B., Pharmacological and biochemical properties of insect GABA receptors. *TIPS*, **11** (1990) 325–9.
33. Sattelle, D. B., GABA receptors of insects. *Adv. Insect Physiol.*, **22** (1990) 1–113.
34. David, J. A. & Sattelle, D. B., Ionic basis of membrane potential and of acetylcholine-induced currents in the cell body of the cockroach fast coxal depressor motor neurone. *J. Exp. Biol.*, **151** (1990) 21–39.
35. Breer, H., Kleene, R. & Hinz, G., Molecular forms and subunit structure of the acetylcholine receptor in the central nervous system of insects. *J. Neurosci.*, **5** (1985) 3386–92.
36. Sattelle, D. B. & Breer, H., Purification by affinity chromatography of a nicotinic acetylcholine receptor from the CNS of the cockroach *Periplaneta americana*. *Comp. Biochem. Physiol.*, **82O** (1985) 349–52.
37. Hanke, W. & Breer, H., Channel properties of an insect neuronal acetyl-choline receptor protein reconstituted in planar lipid bilayers. *Nature*, **321** (1986) 171–4.
38. Breer, H. & Benke, D., Synthesis of acetylcholine receptors in *Xenopus* oocytes induced by poly A-mRNA from locust nervous tissue. *Naturwissenchaften*, **72** (1985) 213–14.
39. Breer, H. & Benke, D., Messenger RNA from insect nervous tissue induces expression of neuronal acetylcholine receptors in *Xenopus* oocytes. *Mol. Brain Res.*, **1** (1986) 111–17.
40. Bowery, N., Classification of GABA receptors. In *The GABA Receptors*, ed. S. J. Enna. Humana Press, Clifton, New Jersey, 1983, pp. 173–213.
41. Hue, B., A novel synaptic GABA receptor in the cockroach CNS. *Pestic. Sci.* (1992) (in press).
42. Lummis, S. C. R. & Sattelle, D. B., Binding sites for [³H]GABA, [³H]flunitrazepam and [³⁵S]TBPS in insect CNS. *Neurochem. Int.*, **9** (1986) 287–93.
43. Le Fur, G., Peripheral benzodiazepine binding sites. In *GABA and Benzodiazepine Receptors*, ed. R. F. Squires. CRC Press, Boca Raton, Florida, 1988, pp. 15–34.
44. Robinson, T. M., MacAllan, D., Lunt, G. G. & Battersby, M., γ-Aminobutyric acid receptor complex of insect CNS: characterization of benzodiazepine binding sites. *J. Neurochem.*, **47** (1986) 1955–62.
45. Ffrench-Constant, R. H., Mortlock, D. P., Shaffer, C. D., MacIntyre, R. J. & Roush, R. T., Molecular cloning and transformation of cyclodiene resistance in *Drosophila*: an invertebrate γ-aminobutyric acid subtype A receptor locus. *Proc. Natl Acad. Sci. USA*, **88** (1991) 7209–13.
46. Sattelle, D. B., Marshall, K., Lummis, S. C. R., Leech, C. A., Miller, K. W. P., Anthony, N. M., Bai, D., Wafford, K. A., Harrison, J. B., Chapaitis, L. A., Watson, M. K., Benner, E. A., Vassallo, J. G., Wong, J. F. & Rauh, J. J., GABA and L-glutamate receptors of insect nervous tissue. In *Transmitter Amino Acid Receptors: Structures Transduction and Models for Drug Development*, ed. E. Costa. Raven Press, New York, 1991, 273–91.

218   D. B. Sattelle et al.

47. Lummis, S. C. R., Insect GABA receptors: characterisation and expression in *Xenopus* oocytes following injection of cockroach CNS mRNA. *Mol. Neuropharm.* (1992) (in press).
48. Fraser, S. P., Djamgoz, M. B. A., Usherwood, P. N. R., O'Brien, J., Darlison, M. G. & Barnard, E. A., Amino acid receptors from insect muscle: electrophysiological characterization in *Xenopus* oocytes following expression by injection of mRNA. *Mol. Brain Res.*, **8** (1990) 331–41.
49. Gration, K. A. F., Activation of ion channels in locust muscle by amino acids. In *Neuropharamacology of insects Ciba Foundation Symposium 88*, ed. M. O'Connor & J. Whelan. Pitman, London, 1982, pp. 240–59.
50. Usherwood, P. N. R., Neuromuscular transmitter receptors of insect muscle. In *Receptors for Neurotransmitters, Hormones and Pheronomes in Insects*, ed. D. B. Sattelle, L. M. Hall & J. G. Hildebrand. Elsevier, Amsterdam, 1980, pp. 141–53.
51. Cull-Candy, S. G., Two types of extrajunctional L-glutamate receptors in locust muscle. *J. Physiol (Lond.)*, **255** (1976) 449–64.
52. Anwyl, R., Permeability of the post-synaptic membrane of an excitatory glutamate synapse to sodium and potassium. *J. Physiol.*, **273** (1977) 367–88.
53. Walker, R. J., Neurotransmitter receptors in invertebrates. In *Receptors for Neurotransmitters, Hormones and Pheromones in Insects*, ed. D. B. Sattelle, L. M. Hall & J. G. Hildebrand. Elsevier/North Holland Biomedical Press, Amsterdam, 1980, pp. 41–57.
54. Giles, D. & Usherwood, P. N. R., The effects of putative amino acid neurotransmitters on somata isolated from neurons of the locust central nervous system. *Comp. Biochem. Physiol.*, **80C** (1985) 231–6.
55. Wafford, K. A. & Sattelle, D. B., Actions of putative amino-acid neurotransmitters on an identified insect motoneurone. *Neurosci. Lett.*, **63** (1986) 135–40.
56. Wafford, K. A. & Sattelle, D. B., L-Glutamate receptors on the cell body membrane of an identified insect motor neurone. *J. exp. Biol.*, **144** (1988) 449–62.
57. Horseman, B. G., Seymour, C., Bermudez, I. & Beadle, D. J., The effects of L-glutamate on cultured insect neurons. *Neurosci. Letts.*, **85** (1988) 65–70.
58. Wafford, K. A., Bai, D., Sepulveda, M.-I. & Sattelle, D. B., L-Glutamate receptors in the insect central nervous system. In *Excitatory Amino Acids Fidia Research Foundation Symposium Series*, Vol. 5, ed. B. S. Meldrum, F. Moroni, R. P. Simon & J. H. Woods. Raven Press, New York, 1991, pp. 275–9.
59. Kawai, N., Saito, M. & Ohsako, S., Differential expression of glutamate receptors in *Xenopus* oocytes injected with messenger RNA from lobster muscle. *Neurosci. Lett.*, **93** (1989) 203–7.
60. Couturier, S., Bertrand, D., Matter, J.-M., Hernandez, M.-C., Bertrand, S., Millar, N., Valera, S., Barkas, T. & Ballivet, M., A neuronal nicotinic acetylcholine receptor subunit (α7) is developmentally regulated and forms a homo-oligomeric channel blocked by α-BTX. *Neuron*, **5** (1990) 847–56.
61. Sattelle, D. B., Harrow, I. D., Hue, B., Pelhate, M., Gepner, J. I. & Hall, L. M., α-Bungarotoxin blocks excitatory synaptic transmission between cercal sensory neurones and giant interneurone 2 of the cockroach. *Periplaneta americana. J. exp. Biol.*, **107** (1983) 473–89.
62. Pinnock, R. D., Lummis, S. C. R., Chiappinelli, V. A. & Sattelle, D. B., K-Bungarotoxin blocks an α-bungarotoxin-sensitive nicotinic receptor in the insect central nervous system. *Brain Res.*, **458** (1988) 45–52.
63. Chiappinelli, V. A., Hue, B., Mony, L. & Sattelle, D. B., k-Bungarotoxin blocks nicotinic transmission at an identified invertebrate central synapse. *J. exp. Biol.*, **141** (1989) 61–71.
64. Benson, J. A., Transmitter receptors on insect neuronal somata. GABAergic and cholinergic pharmacology. In *Neurotox 1988: Molecular Basis of Drug and Pesticide Action*, ed. G. G. Lunt. Elsevier, Amsterdam, New York, 1988, pp. 193–205.

65. Schloss, P., Hermans-Borgmeyer, I., Betz, H. & Gundelfinger, E., Neuronal acetylcholine receptors in *Drosophila*: the ARD proteins is a component of a high affinity α-bungarotoxin binding complex. *EMBO J.*, **7** (1988) 2889–94.
66. Coulson, A., Sulston, J., Brenner, S. & Karn, J., Towards a physical map of the nematode *Caenorhabditis elegans*. *Proc. Natl Acad. Sci. USA*, **83** (1986) 7821–5.
67. Fleming, J. T., Riina, H. A. & Sattelle, D. B., Acetylcholine and GABA receptors of *Caenorhabditis elegans* expressed in *Xenopus* oocytes. *J. Physiol.*, **438** (1991) 371P.
68. Harrow, I. D. & Gration, K. F., Mode of action of the anthelmintics, morantel, pyrantel and levamisole on the muscle cell membrane of the nematode *Ascaris suum*. *Pestic. Sci.*, **16** (1985) 622–4.
69. Del Castillo, J., De Mello, W. C. & Morales, T. A., The physiological role of acetylcholine in the neuromuscular system of *Ascaris lumbricoides*. *Arch. Int. Physiol.*, **71** (1963) 741–57.
70. Holden-Dye, L., Hewitt, G. M., Wann, K. T., Krogsgaard-Larsen, P. & Walker, R. J., Studies involving avermectin and the 4-aminobutyric acid (GABA) receptor of *Ascaris suum* muscle. *Pestic. Sci.*, **24** (1988) 231–45.
71. Arena, J. P., Cully, D. F., Liu, K. K. S. & Paress, P. S., Properties of a glutamate-sensitive membrane current in *Xenopus* oocytes injected with *Caenorhabditis elegans* mRNA. *Pestic. Sci.* (1991) (in press).

SECTION 4

Molecular Structure

16

NMR Approaches to Protein Structure and Function

GORDON C. K. ROBERTS

Department of Biochemistry & Biological NMR Centre, University of Leicester, Leicester, LE1 9HN UK

INTRODUCTION

The revolution in NMR techniques over the last few years has enormously increased the power of NMR spectroscopy for the study of protein structure and interactions in solution. Three-dimensional structures of small proteins can be determined, and information on structure and dynamics, and on interactions with small molecules, obtained for considerably larger proteins. The barrier represented by the complexity of ^{1}H NMR spectra of proteins is gradually being eroded by the development of multidimensional NMR experiments and the use of isotope labelling with ^{2}H, ^{13}C and ^{15}N. The latter, together with the provision of the fairly substantial quantities of protein (tens of milligrams) required for NMR spectroscopy, has been greatly facilitated by the developments in molecular genetics and the ability to construct over-expression systems. In this paper I shall briefly outline some of these applications of NMR, illustrating them primarily with examples from our own recent work. A number of reviews are available which give further details of the methodology and a broader view of the wide range of biological applications.[1-6]

Most applications of NMR to the study of proteins involve the analysis of two-dimensional NMR spectra. The vast majority of these have a common structure: the diagonal corresponds to the conventional 'one-dimensional' spectrum, while the off-diagonal peaks, or cross-peaks, contain information about *connections* between resonances on the diagonal. The nature of these connections depend on the kind of two-dimensional experiment being carried out; one can observe *through-bond connections,* between the resonances of protons in the same amino-acid residue, *through-space connections,* between the resonances of protons (of the same or different residues) which are close together in space, and *exchange connections,* between the resonances of the same proton in two different environments (e.g. of ligand bound to a protein or free in solution).

223

STRUCTURE DETERMINATION

The sequence of steps involved in the determination of the three-dimensional structure of a small protein by NMR is shown in Fig. 1. The assignment of resonances to individual amino-acid residues is an essential first step in any NMR study of a protein, and for all but the smallest proteins this is still the

STEPS IN STRUCTURE DETERMINATION BY NMR

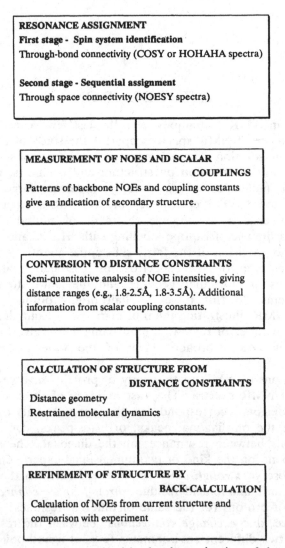

RESONANCE ASSIGNMENT
First stage - Spin system identification
Through-bond connectivity (COSY or HOHAHA spectra)

Second stage - Sequential assignment
Through space connectivity (NOESY spectra)

MEASUREMENT OF NOES AND SCALAR
                                    COUPLINGS
Patterns of backbone NOEs and coupling constants give an indication of secondary structure.

CONVERSION TO DISTANCE CONSTRAINTS
Semi-quantitative analysis of NOE intensities, giving distance ranges (e.g., 1.8-2.5Å, 1.8-3.5Å). Additional information from scalar coupling constants.

CALCULATION OF STRUCTURE FROM
                            DISTANCE CONSTRAINTS
Distance geometry
Restrained molecular dynamics

REFINEMENT OF STRUCTURE BY
                        BACK-CALCULATION
Calculation of NOEs from current structure and comparison with experiment

FIG. 1. The sequence of steps involved in the determination of the three-dimensional structure of a protein by NMR; for details see text.

rate-limiting step. The structural information is then obtained principally from measurements of the nuclear Overhauser effect (NOE) which provide constraints on internuclear distances.[7-9] This will be illustrated by our recent work[10] on a small (M_r 6988) IgG-binding domain of Protein G from *Streptococcus*.

The basic strategy for resonance assignment in the spectra of small proteins[2] involves two steps. First, the type of amino-acid residue from which a particular resonance arises is identified, by analysis of patterns of cross-peaks due to scalar coupling (a *through-bond* connection) in two-dimensional COSY or HOHAHA spectra. In COSY spectra, connections are observed between the resonances of protons separated by two or three chemical bonds, while in HOHAHA spectra connections can often be observed along the whole length of an amino-acid sidechain. Figure 2(a) shows a section of a HOHAHA spectrum of the protein G domain, illustrating the cross-peaks connecting the amide NH proton resonance to those of the sidechain protons. Thus, at the bottom of Fig. 2(a), one can see cross-peaks for a $C_\alpha H$ and a $C_\beta H_3$ characteristic of an alanine residue and, just above, the $C_\alpha H$, $C_\beta H$, and two $C_\gamma H_3$ cross-peaks of a valine residue. Secondly, sequence-specific assignments within each amino-acid type ('sequential assignments') are made by observation of NOEs (a *through-space* effect) between the NH proton of one residue and protons of the preceding residue in the sequence. Typically, NOEs are seen between protons within 5 Å of one another. The local conformation of the polypeptide chain determines which proton of residue i will be within <5 Å of the NH of residue $i + 1$, but whatever the conformation, an NOE is expected to at least one of the NH, $C_\alpha H$ or $C_\beta H$ protons.[2] Figure 2(b) shows part of a NOESY spectrum of the protein G domain in which a series of NH_i—NH_{i+1} NOEs allow one to trace connections along a section of the protein backbone. Since the type of amino-acid residue to which each of these NH protons belong has been identified in the first stage of assignment, one can position this section within the sequence of the protein—in this case, residues 33–40. For larger proteins, in which the degree of resonance overlap is greater, isotopic labelling is necessary (see below), but the assignment strategy is basically the same.

Since the nature of the NOEs between adjacent residues depends on local conformation, the sequential NOEs used to make resonance assignments can also provide information on secondary structure.[2] For example, a sequence of strong NH_i—NH_{i+1} NOEs, such as that seen in Fig. 1(b), and $C_\alpha H_i$—NH_{i+3} NOEs, accompanied by weak $C_\alpha H_i$—NH_{i+1} NOEs, is characteristic of an α-helix, while strong $C_\alpha H_i$—NH_{i+1} NOEs are characteristic of an extended conformation such as seen in a β-sheet. One can thus obtain a picture of the secondary structure of the protein (Fig. 3). It is interesting that this IgG-binding domain of Protein G has a quite different structure to that of the corresponding domain of Protein A, although both bind to the same site on the F_c portion of IgG.

Using longer-range NOEs—between residues which are not adjacent in the

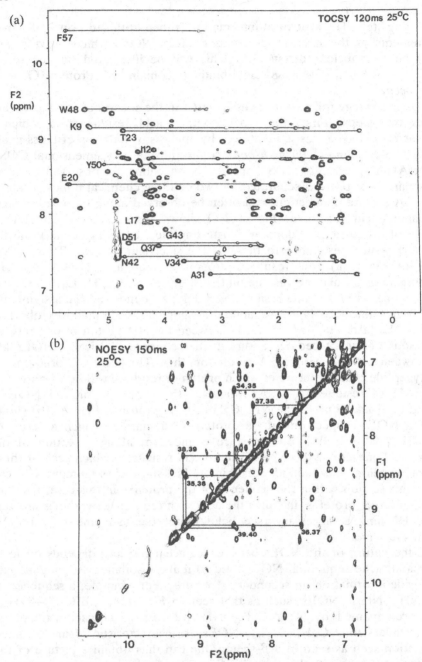

Fig. 2. Sections of two-dimensional ¹H NMR spectra of the IgG-binding domain of Protein G in 90% H₂O/10% ²H₂O, pH 4·2 (from Ref. 10). (a) Part of the HOHAHA spectrum (120 ms spin-lock time) showing direct and relayed connectivities between NH and aliphatic protons. The spin systems for selected residues are indicated. (b) Part of the NOESY spectrum (150 ms mixing time) showing a series of NH—NH NOEs between successive residues along the chain.

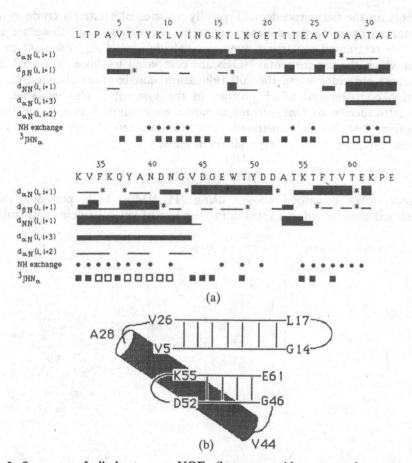

(a)

(b)

FIG. 3. Summary of all short-range NOEs (between residues up to three apart in the sequence) observed for the IgG-binding domain of Protein G (from Ref. 10). The relative strengths of the NOEs are indicated by the thickness of the bars. Potential NOEs obscured by overlaps are indicated by asterisks. The filled circles indicate amide protons which did not exchange with ^2H_2O over 12 h at 25°C, and are thus taken to be strongly hydrogen-bonded. The squares indicate scalar coupling constant data: filled squares denote $^3J_{N\alpha} > 8$ Hz, open squares denote $^3J_{N\alpha} < 6$ Hz.

sequence—a model of the three-dimensional structure can then be constructed. Approximate information on internuclear distances is obtained by semi-quantitative analysis of the experimental NOEs, typically by taking strong, medium and weak NOEs as corresponding to maximum internuclear distances of 2·5, 3·5 and 5 Å. Additional information on dihedral angles, particularly about the C_α—N and C_α—C_β bonds, which can be converted into distance constraints, is obtained from measurements of scalar coupling constants. A variety of computational procedures is used to convert these distance constraints to a structural model (Refs 2, 5, 6, 12), and there is as yet no general consensus on the best procedure. Typically a series of relatively crude models are obtained by using the distance geometry algorithm,[11] and these are then refined by restrained molecular dynamics calculations.[12] A refinement procedure in which the experimental NOEs are compared to those calculated from the current structure using the full relaxation matrix (thus allowing for the relaxation contributions of *all* protons in the system) is also valuable. The recent introduction of *time-average* distance constraints,[13] as opposed to the assumption that all the constraints must be satisfied simultaneously, is an important advance in obtaining realistic models.

ISOTOPE LABELLING

The procedures described above, using ^{1}H NMR, have provided three-dimensional structures for a substantial number of small proteins and isolated

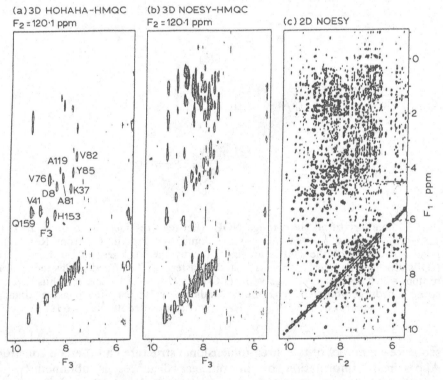

FIG. 4. A comparison between a two-dimensional NOESY spectrum (c) and identical regions from 'slices' through three-dimensional ^{15}N-^{1}H spectra (a), (b) of uniformly ^{15}N-labelled dihydrofolate reductase (from Ref. 17). Spectrum (a) shows a slice through a HOHAHA/HMQC spectrum, showing through-bond cross-peaks between NH, C_αH and some C_βH, and (b) [directly comparable to (c)] a slice through a NOESY/HMQC spectrum, showing through-space cross-peaks from NH to other protons, both taken at an ^{15}N chemical shift of 120·1 ppm. The greater simplicity of the slices from the three dimensional spectra is evident, even with the lower digital resolution which leads to some spreading of the cross-peaks along the vertical axis.

domains of larger proteins. The problems of overlap in the spectrum limit this approach to proteins of less than about 100–120 residues. Selective substitution with ^{2}H can simplify the ^{1}H spectrum and provide direct assignment to residue type.[1,14] The most powerful approach is, however, isotope labelling with ^{13}C and ^{15}N, in conjunction with three-dimensional (and now four-dimensional) NMR spectroscopy (e.g. Refs 15, 16); this will be crucial in extending the usefulness of NMR to larger proteins. In effect, the two-dimensional ^{1}H spectra are 'spread out' in a third dimension by the ^{13}C or ^{15}N chemical shift. Individual 'slices' at a particular ^{13}C or ^{15}N chemical shift are thus much less crowded and easier to analyse, as illustrated in Fig. 4 by two-dimensional, and sections of three-dimensional, NOESY spectra of dihydrofolate reductase, a protein of 162 residues.[17] The combination of isotope labelling and multi-dimensional NMR has been used for the determination of three-dimensional structures of proteins of up to 150 residues,[18] and extension up to 250 residues should be possible.

In addition to general labelling, specific labelling with stable isotopes provides a method for extending the power of the NMR technique to relatively large proteins. For example, in chloramphenicol acetyltransferase, a trimeric enzyme of M_r 75 000 which is the effector of chloramphenicol resistance in bacteria, we have used enzyme labelled with 2-^{13}C-histidine in conjunction with two-dimensional ^{1}H-^{13}C heteronuclear multiple quantum coherence spectra to resolve the resonances of six of the seven histidine residues.[19] Specific labelling of *ligand* molecules with ^{13}C or ^{15}N also provides a valuable method for studying the conformation and environment of the bound ligand (see below).

STUDIES OF LIGAND BINDING

A wide range of NMR experiments can provide information on different aspects of the binding of small molecules to proteins, including the conformation of the bound ligand, individual protein–ligand contacts, and ligand-induced conformational changes.[1] Several of these can be illustrated by experiments on dihydrofolate reductase (dhfr). This is a small (M_r 18–25 000) dehydrogenase which catalyses the reduction of dihydrofolate (and, less effectively, folate) to tetrahydrofolate, using NADPH as coenzyme. It is pharmacologically important as the 'target' for the important 'anti-folate' group of drugs such as methotrexate, trimethoprim and pyrimethamine. As a result, there have been intensive studies of this enzyme over the last decade, including detailed structural studies by both crystallography and NMR.[20–22]

The contribution of individual interactions: comparison of related complexes

The crystal structure[20] shows that the binding of methotrexate involves a number of important ion-pairs and a network of hydrogen bonds between enzyme and ligand which must contribute to binding and specificity. The contributions of these individual interactions, such as the His28—γ-carboxylate and Arg57—α-carboxylate ion-pairs, can be studied by using appropriate methotrexate analogues.[23,24] The γ-amide of methotrexate binds nine times less tightly to the enzyme than methotrexate itself. Of the assigned resonances from the protein, only those of His28 and Leu27 differ between the γ-amide and methotrexate complexes, indicating that these complexes are essentially isostructural, and that the difference in binding constant is a reasonable measure of the contribution of the His28—γ-carboxylate ion pair to the overall interaction energy. The α-amide of methotrexate binds 100-fold less tightly than methotrexate, but the differences in the NMR spectrum indicate that in this case the whole of the p-aminobenzoyl-glutamate moiety binds differently in the analogue, and the His28–γ-carboxylate ion-pair is also broken, so that the difference in binding energy cannot be simply interpreted.

Similar structural comparisons are important in interpreting the effects of site-directed mutagenesis. Comparison of the structures of mutant and wild-type dhfr has revealed considerable variation in the structural consequences of amino-acid substitutions. Studies of the Asp26→Asn (D26N) mutant, show that any structural changes in the enzyme–methotrexate complex are very small and local (and may result from the displacement of a bound water molecule).[25] Close structural similarity to the wild-type enzyme has also been observed for the D26E mutant,[26] although here the differences, while still local, are somewhat larger. In dhfr W21L, small chemical shift differences are seen for resonances of seven residues which are not close to the site of substitution—some as much as 10 Å away.[26] Finally, the T63Q substitution involves a residue which makes a hydrogen bond to the 2'-phosphate of the coenzyme. NMR analysis shows that the conformational effects are again slight, but, surprisingly, they are concentrated in the active site and *not* around the adenine binding site.[27] This is thus a striking example of the effects of a mutation being transmitted a substantial distance (c. 25 Å) through a protein. Current work is aimed at the use of NOEs to describe the changes in solution structure resulting from amino-acid substitution more precisely—perhaps as precisely as ±0·2–0·3 Å.

Intermolecular NOEs

Intermolecular NOEs can provide very valuable information on the proximity between nuclei of the ligand and specific nuclei of the protein, and hence on the mode of binding of the ligand. We have used this approach successfully in a wide range of proteins, including phospholipase A_2,[28] dihydrofolate reductase,[29] the *E. coli trp* repressor,[30] and chloramphenicol acetyltransferase.[31] In the last of these examples, the large size of the protein (M_r 75 000) required the use of an isotopically labelled ligand, di(^{13}C-acetyl)-chloramphenicol, and this is a generally useful procedure. The presence of the ^{13}C allows one to 'edit' the NOESY spectrum so that only the NOEs involving the protons attached to the ^{13}C are observed. This combination of isotope labelling and NOEs provides a direct and simple identification of the 'contact' residues of the protein. In the case of chloramphenicol acetyltransferase, the individual residues giving NOEs to the acetyl groups of the bound ligand were identified by site-directed mutagenesis.

The conformation of the bound ligand

Most small molecules, of course, exist in solution as a mixture of a large number of different conformations, only one of which, in general, will exist in the complex formed when the small molecule binds to a protein. (One tends to think of the ligand having 'a' conformation when bound although, as described below, substantial fluctuations in this conformation can and do occur.) This conformational selection process is obviously an important part of molecular recognition. A variety of NMR parameters can be used to glean information on the conformation of small molecules bound to proteins.[1] Of these, the most reliable and generally useful is again the NOE, in this case between nuclei of the ligand. These measurements can be made particularly readily by using the transferred NOE experiment,[32,33] in which the NOE effects between nuclei in a bound ligand are transferred to the more easily detected nuclei in the free ligand by virtue of the exchange of ligand molecules between the bound and free states. For example, the *syn* and *anti* conformations about the glycosidic bond of nucleotides can be distinguished by the observation of NOEs between the ribose C1'-H and the protons of the aromatic ring. In the case of the nicotinamide ring of the nicotinamide coenzymes, an NOE to the C2-H is expected for the *anti* conformation and to the C6-H for the *syn* conformation. Using the transferred NOE experiment, a C1'-H—C2-H NOE but *not* a C1'-H—C6-H NOE was observed for the nicotinamide ring of NADP$^+$ bound to dhfr, demonstrating that the conformation about the glycosidic bond is *anti* in the complex.[34] In the case of the thio analogue of NADP$^+$, both these NOEs were observed,[34] indicating that this particular ligand exists in a mixture of conformations in the bound state.

In favourable cases, one can study the binding of distinct conformational isomers of the ligand. We have recently been studying the binding to dhfr of a number of analogues of the anti-malarial pyrimethamine (2,4-diamino,5-[4'-chlorophenyl], 6-ethyl pyrimidine) synthesized by Stevens and his colleagues at Aston.[35,36] Rotation about the phenyl–pyrimidine bond of pyrimethamine is, as expected, slow, and four non-equivalent aromatic proton resonances are seen for pyrimethamine bound to the enzyme. In this case, of course, the two rotational isomers about the phenyl–pyrimidine bond are intrinsically identical. This is, however, no longer the case in unsymmetrically substituted compounds such as 2,4-diamino,5-[3'-nitro, 4'-fluorophenyl], 6-ethyl pyrimidine (FNP). We have found that this compound binds to the enzyme in two conformational states, A and B, the ratio A:B in the binary complex being 60:40. Comparison of the chemical shifts of the protons of the phenyl ring in pyrimethamine and its analogues bound to dhfr reveals a very similar pattern in each case, as shown in Fig. 5. This indicates that the phenyl ring protons of FNP are experiencing essentially the same environment in each of the two forms A and B as that experienced by the corresponding protons in pyrimethamine. This can only be true if forms A and B represent the two conformational isomers resulting from rotation by *ca.* 180° about the pyrimidine–phenyl bond of FNP, with the 2,4-diaminopyrimidine ring being bound in the same way in each case.

Measurements of NOEs between an *ortho*-proton of pyrimethamine (which,

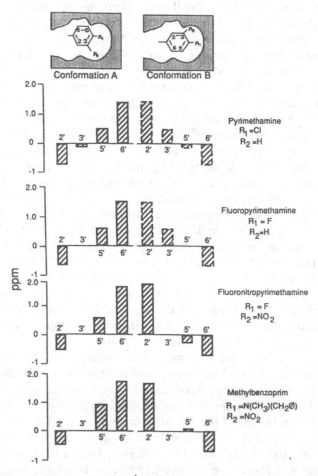

FIG. 5. Schematic representation of the ^1H chemical shift differences between bound and free species for (a) pyrimethamine, (b) fluoropyrimethamine, (c) fluoronitropyrimethamine (FNP) and (d) methylbenzoprim. Note that for the symmetrically substituted compounds (a) and (b) the two orientations of the ring are equivalent; the chemical shifts have been indicated as dashed bars for conformation **B** to facilitate comparison with the data for the other compounds. From Ref. 36.

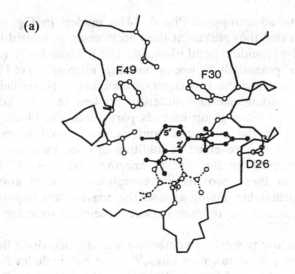

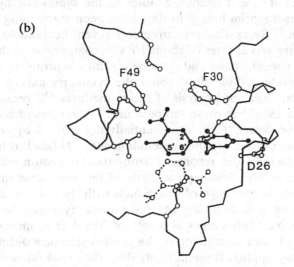

FIG. 6. Model of the binding site in the complex of dhfr with 2,4-diamino, 5-[3′-nitro, 4′-fluorophenyl], 6-ethyl pyrimidine in conformations (a) **A** and (b) **B**. The position of the nicotinamide ring of the coenzyme is indicated by dashed bonds. The model takes as its starting point the crystallographic data[20] for the *L. casei* dhfr–methotrexate–NADPH complex. From Ref. 36.

from its chemical shift, corresponds to the 2′-proton of FNP in conformation A) and the protons of Phe30 allow us to distinguish the two isomers of the unsymmetrically substituted compounds. We can thus construct models of the two complexes of FNP, and these are depicted in Fig. 6. These models account, in a semi-quantitative way, for the large difference in chemical shift between the protons on either side of the phenyl ring (cf. Fig. 5), as resulting from the magnetic anisotropy of Phe30. (The models in Fig. 6 are slightly over-simplified in that they represent the two isomers as related by a rotation about the phenyl–pyrimidine bond of *exactly* 180°). From Fig. 6 it can be seen that the polar, and potentially hydrogen-bonding, nitro group of FNP is in very different environments in the two isomeric complexes. Nevertheless, the two complexes have binding energies differing by less than $0.3 \, \text{kcal mol}^{-1}$. In conformation A, the nitro group extends partly into the binding site of the nicotinamide ring of the coenzyme and as one would expect the slight preference for form A is reversed by addition of coenzyme, an A:B ratio of $0.3:0.7$ being observed for the ternary enzyme–FNP–NADP$^+$ complex.[35,36] The coexistence of these two distinct complexes between unsymmetrically substituted pyrimethamine analogues and the enzyme has important implications for understanding the relation between chemical structure and binding affinity.

NMR is particularly powerful in detecting and characterising fluctuations in conformation over a wide range of rates,[1,4,5] and this includes fluctuations in the conformations of bound ligands. Studies of the symmetrically substituted benzyl ring of trimethoprim bound to dhfr have been made using [*m*-methoxy ^{13}C]-trimethoprim.[37] Relaxation measurements reveal fluctuations of $\pm 25-35°$ about the symmetry axis at rates of about $10^9 \, \text{s}^{-1}$, comparable to those seen for aromatic rings of phenylalanine and tyrosine residues in proteins. Again, as for the amino-acid residues, 180° 'flips' about the symmetry axis of the ring are much rarer events; lineshape analysis of the *m*-methoxy ^{13}C resonances shows that they occur at $250 \, \text{s}^{-1}$ at room temperature.[37] The rates of both the rapid oscillations and the slower 'flips' are essentially determined by fluctuations in the structure of the protein around the binding site. ^{15}N-labelled trimethoprim can be used to identify the resonance from the N_1 proton which forms a hydrogen bond to Asp26; lineshape analysis of this resonance then yields an estimate of the rate of exchange of this proton with the solvent, and hence of the rate of making and breaking of this crucial hydrogen bond between inhibitor and enzyme. This occurs at a rate of $34 \, \text{s}^{-1}$ at room temperature,[37] and clearly results from a fluctuation in the protein structure distinct from that involved in the ring flipping. It seems likely that when *both* fluctuations happen to occur simultaneously then the trimethoprim will dissociate from the enzyme; the measured dissociation rate constant is $1.7 \, \text{s}^{-1}$ at 298 K. The rates of all these fluctuations in the structure of the enzyme-trimethoprim complex are affected by coenzyme binding,[37] and also by amino-acid substitutions;[26] in fact, it appears that structural fluctuations are significantly more sensitive to the effects of amino-acid substitutions than is the time-average structure.

Conformational equilibria

Changes in chemical shifts and NOEs of resonances from the protein produced by ligand binding can be interpreted in terms of ligand-induced conformational changes. For example, the binding of the coenzyme NADPH to dhfr can be shown to lead to a movement of helix C.[38] In phospholipase A$_2$, changes in intramolecular NOEs indicate that there are slight local changes in conformation when the inhibitor binds—notably a movement of Tyr69, which moves to 'cap' the active site.[28]

In some cases, where two or more conformations have similar energies and interconvert slowly, a conformational equilibrium can be observed directly. Both the enzyme–folate and the enzyme–folate–NADP$^+$ complexes exist in solution as a mixture of two or three conformations which interconvert sufficiently slowly for separate NMR signals to be observed, and whose proportions are pH-dependent.[29,39] Two-dimensional NOE experiments permitted the identification of one crucial structural difference between these conformations: the orientation of the pteridine ring.[29] The first step is to identify the resonance position of the pteridine 7-proton in each of the three conformations of the complex; this is done by means of a two-dimensional exchange experiment, relying on magnetisation transfer by chemical exchange of folate molecules between the free state and the three bound states. Having identified these resonance positions, one can then seek NOEs between the 7-proton and protons of the protein in each of the three conformations. In the conformations denoted I and IIa, NOEs are observed between the pteridine 7-proton of bound folate and the two methyl groups of Leu27. Similar NOEs are observed for the 7-proton of methotrexate in the enzyme–NADP$^+$–methotrexate complex (although in the latter case, but not the former, NOEs are also observed to the methyl protons of Leu19). In conformations I and IIa of the enzyme–folate–NADP$^+$ complex, the pteridine ring of folate therefore has an orientation in the binding site similar (but not identical) to that seen for methotrexate in the crystal structure. In conformation IIb, by contrast, *no* NOEs are observed between the pteridine 7-proton and these methyl groups; given the structural constraints of the pteridine ring binding site, this can only be explained if the ring has turned over by approximately 180° about an axis along the C2—NH$_2$ bond.[29] The stereochemistry of reduction indicates that only the latter is a catalytically productive mode of binding for the substrate. By contrast, the complex of folate and NADP$^+$ with the mutant dhfr D26N exists in only a single conformation over the pH range 5·5–7·5; this strongly suggests that Asp26 is the group responsible for the pH-dependence of the conformational equilibrium of this complex.[25] The single conformation seen for this mutant corresponds to the low-pH conformer of the wild-type complex, as one would expect from a comparison of the aspartate and asparagine sidechains. The very low k_{cat} for dhfr D26N (*ca.* 1% of wild-type) thus arises not only because Asp26 is, most probably, a proton donor in the reaction,[22] but also because the mutant is 'locked' in a catalytically non-functional conformation.

APPROACHES TO MEMBRANE PROTEINS

The methods outlined above, those of so-called 'high-resolution' NMR spectroscopy, are applicable to proteins in solution, where the tumbling of the molecule leads to (relatively) sharp resonance lines. Proteins embedded in membranes will naturally tumble orders of magnitude more slowly, and the same methods are not directly applicable. There are two possible approaches to the study of membrane proteins: the use of high-resolution methods to study individual isolated domains of the protein in solution, and the use of solid-state NMR methods to study the intact protein *in situ*; these will be discussed only very briefly here, principally to provide references to some of the recent literature.

Studies of isolated domains

The use of NMR to study individual domains of large and complex proteins, including those of the blood-clotting system, complement proteins, and DNA-binding proteins has become well-established over the last few years;[40] a simple example was discussed above.[10] A similar approach can readily be applied to isolated domains of membrane proteins, using the methods outlined above to determine the structure of the domains in solution. For example, the structure of the 'complement control protein' or 'short consensus repeat', a small domain which appears in the extracellular portions of the ELAM-1 cell adhesion receptor and the interleukin-2 receptor, as well as in complement proteins, has recently been determined by NMR.[41] The developments in NMR are such that the ligand-binding domains of neurotransmitter receptors will soon be within reach of this technique. Peptides corresponding to trans-membrane segments of membrane proteins have also been studied by NMR, in organic or mixed organic–aqueous solvents to mimic the non-polar environment of the membrane—for example, transmembrane helices of the sodium channel[42] and of bacteriorhodopsin.[43] Small membrane proteins or peptides can also be studied in detergent micelles.[44]

Solid-state NMR studies of membrane proteins

For relatively very immobile molecules such as those in membranes, well-resolved NMR spectra can only be obtained for 'magnetically dilute' nuclei, such as ^{13}C introduced at specific sites, by the special techniques of solid-state NMR, including high-speed magic angle spinning, cross polarisation and high-power decoupling (for reviews, see Refs 45, 46). Recently, a very powerful new solid-state NMR method, the rotational resonance technique, has been introduced,[47,48] which allows one to measure the distance (over about 5 Å) between, for example, two ^{13}C nuclei in a membrane protein. In common with other solid-state NMR methods, this has the disadvantage of requiring *specific* isotope labelling and also substantial quantities of protein. It is therefore unlikely that these methods yet provide a route to the determination of the three-dimensional structure of a membrane protein. However, it is clear that the rotational resonance technique in particular has considerable potential in, for example, identifying contacts between ligand and protein.

ACKNOWLEDGEMENTS

Many people, whose names appear in the reference list, have carried out the work described here; my thanks to all of them, and particularly to Lu-Yun Lian, John Arnold, Jeremy Derrick, Bill Primrose, Vasudevan Ramesh, Janette Thomas and Ji-chun Yang in Leicester, Jim Feeny and Berry Birdsall (NIMR, Mill Hill) and Julie Andrews (Manchester). Work at Leicester has been supported by SERC, MRC, the Wellcome Trust, the Leverhulme Trust, MITI (Japan), du Pont and Fisons Pharmaceuticals.

REFERENCES

1. Jardetzky, O. & Roberts, G. C. K., *NMR in Molecular Biology*. Academic Press, New York, 1981.
2. Wuthrich, K., *NMR of Proteins and Nucleic Acids*. Wiley, New York, 1986.
3. Bax, A., Two-dimensional NMR and protein structures. *Ann. Rev. Biochem.*, **58** (1989) 223.
4. Oppenheimer, N. J. & James, T. L., *Methods in Enzymology*, Vol. 176, Academic Press, New York, 1989.
5. Oppenheimer, N. J. & James, T. L., *Methods in Enzymology*, Vol. 177, Academic Press, New York, 1989.
6. Wagner, G., *Prog. Nucl. Magn. Reson. Spectrosc.*, **22** (1990) 101.
7. Wuthrich, K., *Science*, **243** (1989) 45.
8. Clore, G. M. & Gronenborn, A. M., Determination of three-dimensional structures of proteins in solution by nuclear magnetic resonance spectroscopy. *Protein Eng.*, **1** (1987) 275.
9. Kaptein, R., Boelens, R., Scheek, R. M. & Gunsteren, W. F. van, Protein structures from NMR. *Biochemistry*, **27** (1988) 5389.
10. Lian, L.-Y., Yang, J.-C., Derrick, J.P., Sutcliffe, M. J., Roberts, G. C. K., Murphy, J. P., Goward, C. R. & Atkinson, T., Sequential ^1H NMR assignments and secondary structure of an IgG–binding domain from Protein G. (In press).
11. Crippen, G. M. & Havel, T. M., *Distance Geometry and Molecular Conformation*. John Wiley, Chichester, 1988.
12. Scheek, R. M., Gunsteren, W. F. van & Kaptein, R., *Methods in Enzymology*, **177** (1989) 204.
13. Torda, A., Scheek, R. M. & Gunsteren, W. F. van, *J. Mol. Biol.*, **214** (1990) 223.
14. Birdsall, B., Arnold, J. R. P., Jimenez Barbero, J., Frenkiel, T. A., Bauer, C. J., Tendler, S. J. B., Carr, M. D., Thomas, J. A., Roberts, G. C. K. & Feeney, J., The ^1H NMR assignments of the aromatic resonances in complexes of *Lactobacillus casei* dihydrofolate reductase and the origins of their chemical shifts. *Eur. J. Biochem.*, **191** (1990) 659.
15. Zuiderweg, E. R. P. & Fesik, S. W., Heteronuclear three-dimensional NMR spectroscopy of the inflammatory protein C5a. *Biochemistry*, **28** (1989) 2387.
16. Kay, L. E., Clore, G. M., Bax, A. & Gronenborn, A. M., *Science*, **249** (1990) 411.
17. Carr, M. D., Birdsall, B., Frenkiel, T. A., Bauer, C. J., Jimenez-Barbero, J., McCormick, J. E., Feeney, J. & Roberts, G. C. K., Dihydrofolate reductase: sequential assignments and secondary structure in solution. *Biochemistry* **30** (1991) 5335.

18. Clore, G. M., Wingfield, P. W. & Gronenborn, A. M., High-resolution three-dimensional structure of Interleukin 1β in solution by three- and four-dimensional nuclear magnetic resonance spectroscopy. *Biochemistry,* **30** (1991) 2315.

19. Derrick, J. P., Lian, L.-Y., Roberts, G. C. K. & Shaw, W. V., Identification of the C_2-^{1}H histidine NMR resonances in chloramphenicol acetyltransferase by a [13]C-^{1}H heteronuclear multiple quantum coherence method. *FEBS Lett.,* **280** (1991) 125.

20. Bolin, J. T., Filman, D. J., Matthews, D. A., Hamlin, R. C. & Kraut, J., Crystal structures of *Escherichia coli* and *Lactobacillus casei* dihydrofolate reductase refined at 1·7 Å resolution. *J. Biol. Chem.,* **257** (1982) 13650.

21. Roberts, G. C. K., NMR and mutagenesis studies of the specificity and activity of dihydrofolate reductase. In *Protein Engineering,* ed. M. Ikehara. Scientific Societies Press, Tokyo, 1990, p. 45.

22. Roberts, G. C. K., NMR and mutagenesis studies of dihydrofolate reductase. In *Chemistry and Biology of Pteridines 1989; Pteridines and Folic Acid Derivatives,* ed. H. C. Curtius, S. Ghisla & N. Blau. de Gruyter, Berlin, 1990, pp. 681–93.

23. Antonjuk, D. J., Birdsall, B., Cheung, H. T. A., Clore, G. M., Feeney, J., Gronenborn, A. M., Roberts, G. C. K. & Tran, T. Q., A ^{1}H NMR study of the role of the glutamate moiety in the binding of methotrexate to dihydrofolate reductase. *Br. J. Pharmacol.,* **81** (1984) 309.

24. Hammond, S. J., Birdsall, B., Feeney, J., Searle, M. S., Roberts, G. C. K. & Cheung, H. T. A., Structural comparisons of complexes of methotrexate analogues with *lactobacillus casei* dihydrofolate reductase by two-dimensional ^{1}H NMR at 500 MHz. *Biochemistry,* **26** (1987) 8585.

25. Jimenez, M. A., Arnold, J. R. P., Andrews, J., Thomas, J. A., Roberts, G. C. K., Birdsall, B. & Feeney, J., Dihydrofolate reductase; control of the mode of substrate binding by asparate 26. *Protein Eng.,* **2** (1989) 627.

26. Birdsall, B., Andrews, J., Ostler, G., Tendler, S. J. B., Feeney, J., Roberts, G. C. K., Davies, R. W. & Cheung, H. T. A., NMR studies of differences in the conformations and dynamics of ligand complexes formed with mutant dihydrofolate reductases. *Biochemistry,* **28** (1989) 1353.

27. Thomas, J. A., PhD thesis, University of Leicester, 1991.

28. Bennion, C., *et al.,* Design and syntheses of some substrate analogue inhibitors of phospholipase A_2 and investigations by NMR and molecular modelling into the binding interactions of the enzyme-inhibitor complex. *J. Med. Chem.* (1992) (in press).

29. Birdsall, B., Feeney, J., Tendler, S. J. B., Hammond, S. J. & Roberts, G. C. K., Dihydrofolate reductase: multiple conformations and alternative modes of substrate binding. *Biochemistry,* **28** (1989) 2297.

30. Hyde, E. I., Ramesh, V., Frederick, R. & Roberts, G. C. K., NMR studies of the activation of the *Escherichia coli trp* repressor. *Eur. J. BioChem.* **20** (1991) 569.

31. Lian, L.-Y., Derrick, J. P., Roberts, G. C. K. & Shaw, W. V., Analysis of the binding of 1,3 diacetyschloramphenicor to chloramphenicor acetyltransferase by heteronuclear multiple quantum coherence nmr and site directed mutagenesis. (In preparation).

32. Albrand, J. P., Birdsall, B., Feeney, J., Roberts, G. C. K. & Burgen, A. S. V., The use of transferred nuclear Overhauser effects in the study of the conformation of small molecules bound to proteins. *Int. J. Biol. Macromol.,* **1** (1979) 37.

33. Clore, G. M. & Gronenborn, A. M., *J. Magn. Reson.,* **48** (1982) 402.

34. Feeney, J., Birdsall, B., Roberts, G. C. K. & Burgen, A. S. V., Use of transferred nuclear Overhauser effect measurements to compare the binding of coenzyme analogues to dihydrofolate reductase. *Biochemistry,* **22** (1983) 628.

35. Tendler, S. J. B., Griffin, R. J., Birdsall, B., Stevens, M. F. G., Roberts, G. C. K. & Feeney, J., Direct ^{19}F NMR observation of the conformational selection of optically active rotamers of the antifolate compound fluoronitropyrimethamine bound to the enzyme dihydrofolate reductase. *FEBS Lett.,* **240** (1988) 201.

36. Birdsall, B., Tendler, S. J. B., Arnold, J. R. P., Feeney, J., Griffin, R. J., Carr, M. D., Thomas, J. A., Roberts, G. C. K. & Stevens, M. F. G., NMR studies of multiple conformations in complexes of *L. casei* dihydrofolate reductase with analogues of pyrimethamine. *Biochemistry,* **29** (1990) 9660.
37. Searle, M. S., Forster, M. J., Birdsall, B., Roberts, G. C. K., Feeney, J., Cheung, H. T. A., Kompis, I. & Geddes, A. J., Dynamics of trimethoprim bound to dihydrofolate reductase. *Proc. Natl Acad. Sci. USA,* **85** (1988) 3787.
38. Hammond, S. J., Birdsall, B., Searle, M. S., Roberts, G. C. K. & Feeney, J., Dihydrofolate reductase: ^1H resonance assignments and coenzyme-induced conformational changes. *J. Mol. Biol.,* **188** (1986) 81.
39. Birdsall, B., DeGraw, J., Feeney, J., Hammond, S. J., Searle, M. S., Roberts, G. C. K., Colwell, W. T. & Crase, J., ^{15}N and ^1H NMR evidence for multiple conformations of dihydrofolate reductase with its substrate, folate. *FEBS Lett.,* **217** (1987) 106.
40. Baron, M., Norman, D. G. & Campbell, I. D., *Trends Biochem. Sci.,* **16** (1991) 13.
41. Barlow, P. N., Baron, M., Norman, D. G., Day, A. J., Willis, A. C., Sim, R. B. & Campbell, I. D., Secondary structure of a complement control protein module by two-dimensional ^1H NMR. *Biochemistry,* **30** (1991) 997.
42. Mulvey, D., King, G. F., Cooke, R. M., Doak, D. F., Harvey, T. M. & Campbell, I. D., High resolution ^1H NMR study of the solution structure of the S4 segment of the sodium channel protein. *FEBS Lett.,* **257** (1989) 113.
43. Barsukov, I. L., Abdulaeva, G. V., Arseniev, A. S. & Bystrov, V. F., *Eur. J. Biochem.,* **192** (1990) 321.
44. Schiksnis, R. A., Bogusky, M. J., Tsang, P. & Opella, S. J., Structure and dynamics of the Pf1 filamentous bacteriophage coat protein in micelles. *Biochemistry,* **26** (1987) 1373.
45. Smith, S. O. & Griffin, R. G., *Ann. Rev. Phys. Chem.,* **39** (1988) 511.
46. Opella, S. J. & Stewart, P. L., *Methods in Enzymology,* **176** (1989) 242.
47. Raleigh, D. P., Levitt, M. H. & Griffin, R. G., *Chem. Phys. Lett.,* **146** (1988) 71.
48. Raleigh, D. P., Creuzet, F., Das Gupta, S. K., Levitt, M. H. & Griffin, R. G., *J. Amer. Chem. Soc.,* **111** (1989) 4502.

17

From Molecular Biology to Molecular Modelling: The Nicotinic Acetylcholine Receptor

DAVID J. OSGUTHORPE,[a] GEORGE G. LUNT[b] & VICTOR B. COCKCROFT[a]

[a]*Molecular Graphics Unit*, [b]*Department of Biochemistry, University of Bath, Claverton Down, Bath, BA2 7AY, UK*

INTRODUCTION

Currently, a molecular biologist is able to sequence DNA at the rate equivalent to 300 amino acids per day. This is leading to an explosion in the number of protein sequences available. Moreover, cross-hybridisation studies mean that once a protein has been sequenced it is relatively easy to find related sequences within the same organism or the same gene in other organisms. This way, for many proteins whole families of sequences are available from a wide spectrum of evolutionarily different organisms.

One well-studied system for which extensive sequences have become available is the nicotinic acetylcholine (nACh) receptor. The nACh receptor allows cations to flow across the membrane into the cytoplasm of a cell after binding the native ligand acetylcholine and other agonists. As such it is crucial to the function of the nervous system of many animals, including insects, and thus provides a target for insecticidal compounds. The sequence data indicates that it is a member of a large, homologous and functionally similar set of proteins, the ligand-gated ion channel (LGIC) superfamily. Due to the large body of experimental data available for one particular form of this receptor, the nAChR from *Torpedo* electric organ, it is the prototype for understanding the relation between structure and function within the LGIC class of receptors. Using the superfamily concept it should be possible to extrapolate ideas from these studies to all other LGIC receptors. The superfamily presently includes the nACh receptor,[1] the GABA$_A$-type receptors for γ-aminobutyric acid (GABA)[2], and the glycine receptors.[3] In the latter two cases, although the receptors function similarly the channels, at least in vertebrates, are selective for anions.

The major unsolved problem with these receptors is that although there is a wealth of sequence data available, the current structural data are limited to a resolution of between 8 and 15 Å,[4,5] which is not enough to determine the position of the backbone atoms of the protein. This makes it impossible to

241

rationalise in structural terms the activity of the various ligands or to rationally design ligands which are specific to a particular receptor.

Furthermore, although using X-ray crystallography it is now relatively easy to determine the structure of a globular protein, given that crystals of the protein can be grown, only a few structures of transmembrane proteins have been solved. The only high-resolution structure for a transmembrane protein that has so far been determined is that for the photosynthetic reaction centre.[6] Additionally, structures for bacteriorhodopsin,[7] a light-harvesting complex[8] and a porin[9] have been determined to reasonable resolution ($\approx$3–6 Å) from electron microscopy. These are the only transmembrane structures known to date.

Our objective has been to provide a structural rationalisation of the biochemical, pharmaceutical and biophysical information available for the nACh receptor by developing explicit three-dimensional models of the receptor at atomic level detail. The overall shape of the receptor has been defined by the structural studies based on electron microscopy[10] which show the receptor is composed of five subunits arranged symmetrically to give a barrel-shape with a cylindrical extracellular funnel and a centrally located ion channel. The sequence information shows that the subunits are homologous to each other and are contained within a radius of 19% identity. Furthermore, the extracellular region (the first $\approx$250 amino acids) and the transmembrane region are significantly homologous, not less than 25% identity. Using the sequence data from molecular biology, the modelling was started from a multiple amino acid sequence alignment of the LGIC subunits, of which over 70 are now known (from cloning studies) for different species, for different agonist ligands and for receptors of differing ion selectivity.

Two regions of the extracellular portion of the *Torpedo* nACh receptor have been suggested by biochemical studies to be involved in agonist binding. The first is the region around the paired Cysteines 192–193 of the α-subunit which are labelled by various affinity ligands, including (4-*N*-maleimide)benzyl-trimethylammonium (MBTA) and bromoacetylcholine,[11] after selective reduction of the disulphide bridge between them. In addition, a peptide extending from position 185 to 196 has been shown to bind α-bungarotoxin.[12] The second region is from position 125 to 147 of the α-subunit. This peptide contains a 15-residue stretch of sequence termed the 'Cys-loop', so called because a disulphide bridge links Cysteine residues at positions 128 and 142.[11] A synthetic peptide to this region was shown to interact with acetylcholine and α-bungarotoxin.[13] More recently, Madhok *et al.*[14] have reported that for the brain nACh receptor high-affinity binding of nicotine is specifically inhibited by antibodies raised to a peptide covering positions 3–12 of the Cys-loop of neuronal α-subunits.

Using the thesis that all LGIC receptors have evolved from a single proto-receptor, we have concentrated on the Cys-loop region, which is the most conserved stretch of amino acid sequence in the diverse set of sequences of LGIC subunits known to date; 4 of its 15 residue positions are invariant. In contrast, only 14 residues are invariant in the entire N-terminal extracellular region of LGIC subunits, which is >200 residues long. Additionally, at position 11 of the Cys-loop an invariant aspartic acid residue occurs, which is one of only two invariant acidic residue positions present in the extracellular region of LGIC subunits. Therefore, this region must be important to the function of this receptor, either by being necessary for linking the binding of the agonist to the opening of the ion channel or by being part of the binding site itself. The presence of the invariant aspartate suggests the latter, as this would be a good candidate for the proposed anionic binding site, which would interact with the charged nitrogen that occurs in all LGIC agonists and most competitive antagonists. Some earlier proposals have been made[15,16] about the structure of this region in the α-subunit of the *Torpedo* nACh receptor but these did not include the mutual pairing of Cysteines 128–142. With the establishment of the concept of a LGIC superfamily we have been able to extend our analysis outwards from the nACh receptor and take into account the extensive biochemical, pharmacological and biophysical information for all known members of the receptor superfamily.[17]

THE Cys-LOOP MODEL

To create a specific model of the Cys-loop we used sequence conservation, spatial conservation and secondary structure prediction techniques to develop a model of its structure. From the sequence alignment a two-residue periodicity in the conservation of hydrophobic positions suggested a β-strand structure, with an exposed hydrophilic face and buried hydrophobic face. The disulphide bridge residues 1 and 15 of the loop thus require a turn to occur around residues 8–9, where a conserved Pro residue occurs at position 9. By examining similar sequence turns occurring in known protein structures, in particular the requirement from the sequences for either phenylalanine or tyrosine at the preceding residue position, the starting conformation for this turn was chosen to be a type of VIa β-turn.[18] This defined the conformation of the backbone of the Cys-loop, the side-chain angles were taken from standard values for the β-sheet structure.[19]

Energy minimisation from this starting structure provided us with the basic structural model of the Cys-loop structure, see Fig. 1. Additional tests were performed on this structure to determine the structural stability of this loop, i.e. we wished to know if this loop could adopt many conformations or was limited to a few. This was done by performing a short (5 ps) molecular dynamics simulation at high temperature (600 K) to allow the molecule to attempt to find any nearby accessible (low energy) conformations. The Cys-loop remained close to the β-sheet structure as initially defined. As a further set of tests, new loops were created by mutating the residues to Ala, first for all residues except Cys and Pro, then for all residues except Cys and

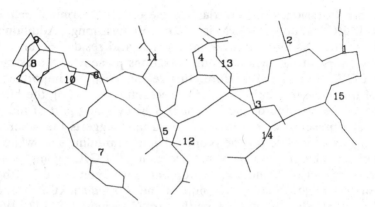

Fig. 1. The energy minimised model of the Cys-loop structure of the α2-subunit of chick brain nACh receptor. Numbering refers to the position of residues within the Cys-loop.

similar high-temperature dynamics simulations performed. Only in the latter case was there significant conformational mobility, as one would expect.

We then went back to the data from molecular biology to investigate how this structural model fitted in with the Cys-loop sequences for the other members of the LGIC superfamily. These models were constructed by mutating the residues of the initial nAChR Cys-loop structure according to the sequence of the new receptor and this was followed with energy minimisation.

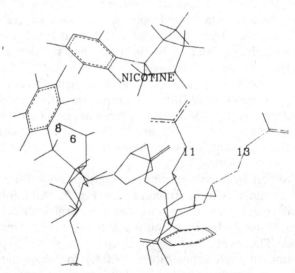

Fig. 2. Energy minimised model of (−)nicotine docked onto the Cys-loop of the α2-subunit of chick brain nACh receptor. The residue numbers identify residues important in binding and are relative to the first Cys residue of the Cys-loop being numbered 1, as in Fig. 1.

All of the LGIC Cys-loop sequences had the same hydrophobic/hydrophilic separation, with the invariant Pro at position 9 preceded in all cases by an amino acid containing a benzene ring, i.e. Phe or Tyr (position 8) and the invariant Asp at position 11. Interestingly, it was also possible to assign differences in sequence on the hydrophilic face to the different requirements for binding the ligands in the different receptors. Thus a hydrogen bonding residue (glutamine or histidine) occurs at position 13, whereas the residue at position 6 changes for each member of the LGIC superfamily, suggesting it is important in defining the selectivity of each LGIC member for its agonist. In the β-subunit of the GABA$_A$ receptor, it is an arginine residue, a lysine residue in the 48 kDa subunit of the glycine receptor and a threonine residue in the α-subunit of the nACh receptor. Figure 2 shows the residues listed above and how they relate to the binding of one ligand, nicotine. The different size, charge and hydrogen bonding capability of these residues can be associated with a specific interaction with each of the different ligands that bind to that receptor, thus providing a method of discriminating between the various ligands.

EXTENDED BINDING SITE MODEL

So far, the modelling has created an explicit, atomic-level detail model of this small section of the protein that we propose forms part of the binding site for agonists. However, there are other data available, mainly from biochemical studies, which can be used to extend this model. In particular, the region around Cys 192–193 has long been identified as part of the binding site, from, for example, studies using reagents specific for the binding site that alkylate these cysteines in the *Torpedo* nACh receptor. For these reagents the active part of the molecule is in a position equivalent to the bromomethyl group of bromoacetylcholine and is, therefore, probing for residues in a limited region of space around the binding site. We have modelled this region as being close to the agonist binding site and part of a binding cleft. This region is important for toxin binding, such as bungarotoxin, but the large size of these toxins means they only need to bind close to the active site to occlude it.

Additional information on the binding site cleft can be provided by pharmacophore mapping using the extensive range of agonists and antagonists known for the LGIC family. By superimposing the differing agonists and antagonists for all LGIC members that are known to act at the native agonist binding site, a spatial map of the binding site cleft can be constructed. This was used in the positioning of other components of the agonist binding site around the Cys-loop hydrophilic surface. The superimposition was accomplished using a three-point pharmacophore model for agonist and competitive antagonist binding, which includes the charged nitrogen of all agonists, and the electronegative and electropositive atoms of the polarised π-electron system found close to the charged nitrogen, which is also found in all agonists and some antagonists. Interestingly, the agonists can be seen to be small molecules compared to the antagonists, which occupy a much larger volume of space. This has an implication for the mechanism of ion channel opening, as will be discussed later.

TRANSMEMBRANE HELICES MODELLING

From hydrophobicity profiles calculated from the sequence data, four regions of the nAChR (M1–M4) have been identified as possible transmembrane helices,[20] and one region (MA) has been identified as a highly charged amphiphilic helix.[21] Helices are the most likely secondary structure to be found in low-dielectric medium as the helical structure allows the backbone hydrogen bonding groups to interact favourably and yet it only requires short-range contacts. This has been demonstrated experimentally by helix-coil transition studies on various amino acid polymers in differing solvents, which show that in low-dielectric solvents the helical conformation dominates. Hence the general assumption that the most likely secondary structure in a membrane will be helical.

In order to model the transmembrane region we chose to start by defining which of M1–M4 and MA is the helix which is believed to line the ion channel. (Remember the ion channel in the LGIC receptors is thought to be created by the combination of subunits, with each subunit contributing essentially one helix to the ion channel.) Models of the amphiphilic helix show that all the charged residue side-chains are aligned down one side of the helix surface and because of this it has been suggested that it is the helix which lines the ion channel.[21,22] Recent work by Numa[23,24] demonstrated that partial ion channel activity still existed even after the deletion (by molecular biological techniques) of the MA helix region. Having removed MA as a contender for the ion-channel lining helix, attention now has focussed on the M2 helix. This contains a number of Ser and Thr residues which can be lined up along one side of the helix surface. Thus, the hydroxyl side-chains of these residues could replace, or interact with, the solvent shell surrounding the ion as it is passing through the channel. Also, novel peptides have been synthesised based on the M2 sequence which have been shown to act as ion channels.[25]

We have used two approaches in modelling the transmembrane helices. The first was to use a known four-helical bundle structure, myohemerythrin. Although this is a globular protein, its structure could represent the packing of the M1–M4 helices in the membrane (but not its sequence of course). In particular, its helices are of a similar length to that required for membrane-spanning helices and, more importantly, the helix packing (an anti-parallel bundle) corresponds to that required for putting the helices M1–M3 in the membrane. This is because the loops connecting helices M1–M2 and M2–M3 are very short, which means that the helix pairs M1–M2 and M2–M3 must be lying anti-parallel in the membrane. Thus we could generate a model for the transmembrane helices by superimposing the four helices M1–M4 onto the known helices of myohemerythrin. Information from molecular biology was used at the superimposition step to fit the pattern seen in the LGIC sequences onto the patterns found in an alignment of myohemerythrin sequences. Considerations included subtype conservation, variability, hydrophobicity and size, as well as preference of amino acids to occur at the helix termini.[26] In addition, biochemical data assigning residue positions of M2 as sites lining the ion channel were used.[27]

The other approach to constructing a model of the transmembrane helices was to perform a molecular dynamics simulation of the M1–M3 helices and observe if and how these helices packed together during the course of the simulation. The initial structure for this simulation was generated by assigning the helical conformation (phi −60, psi −40) to residue positions 209–236, 244–265 and 275–299 (*Torpedo* α-subunit sequence numbering) with the intervening linker regions being set to the extended conformation (phi −120, psi −120). A high-temperature simulation (800 K) was chosen to speed up the crossing of energy barriers and the surrounding lipid molecules were not included, although the dielectric constant of 1 used in these simulations better reflects the low-dielectric nature of the transmembrane region than simulations where the excluded solvent is water. As we wished to maximise the effect of long-range interactions the non-bond and electrostatic interactions were not truncated by a cut-off function. This increases the amount of computer time required to perform the simulation significantly and was made possible at Bath by the availability of a TITAN superminicomputer in the Molecular Graphics Unit. A 20 ps trajectory was calculated and the results analysed using the FOCUS program.[28]

The analysis showed that over the course of this simulation the three helical segments did indeed fold into an antiparallel bundle. A detailed analysis of the final structural model from the simulation showed several features that were consistent with experimental data. In particular, the partly hydrophilic surface of the M2 segment was exposed and could contribute to the wall of the ion channel, as suggested by experiment.[29-32] Interestingly, rather than being internally packed there was a sufficient gap in the final structure for the M4 helix to be packed to give a four-helix bundle arrangement. However, it was apparent that there was some distortion of the helical regions due to the breaking of hydrogen bonds. This may reflect the high temperature used in the dynamics simulation and suggests the final structure could have been improved by performing a simulated annealing experiment by reducing the temperature gradually in the final stages of the simulation.

Five copies of the former model were docked together with, in each case, M2 forming the central ion channel lining helix to model the complete receptor ion channel (see Fig. 3). M1 is the most tightly packed helix, making both intra-subunit contacts (M2 and M4), and inter-subunit contacts (M2 and M3 of next subunit). This leads to the M4 helix lying on the outside of the helix bundles and making no contacts with the helices of adjacent subunits. Thus,

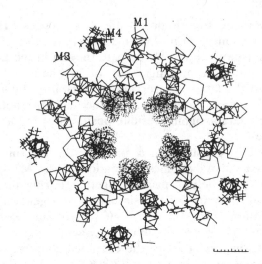

Fɪɢ. 3. View looking from the cytoplasmic side of the receptor through the ion channel of the model of the transmembrane region. The helices M1–M4 of one subunit are labelled, the helices in the other subunits being in the same relative positions, M2 always being the central helix. Only Cα atoms of the backbone are shown plus all atoms of selected side chains. The total length of the scale bar is 10 Å.

from the model there seem to be few structural constraints on this helix and the sequence alignments show that this is the least conserved helix. Further, Numa has replaced this helix with a hydrophobic helix from the insulin receptor and retained activity,[24] which again indicates there are few structural constraints on this helix. In this model, the M1 and M3 helix of adjacent subunits pack onto each other in the whole receptor such that the intracellular end of the M1 helix (see DiPaola *et al.*[32]) and the extracellular end of the M3 helix may also contribute significantly to the surface region of the pore of the channel in the open conformation. The cysteine residue at position 222 (*Torpedo* numbering) is placed such that it would face towards the lipid bilayer at the interfaces between adjacent subunits in the whole oligomer. Thus, reagents which label this residue (see Marquez *et al.*[33]) may do so by approach from the lipid phase rather than from the aqueous pore of the channel. The invariant proline residue at position 221 faces towards the centre of the bundle and is not anticipated to be an intersubunit contact residue.

EXTRACELLULAR DOMAIN MODELLING

At this point, we have a model for a small portion of the extracellular domain, the agonist binding site, and a model for the transmembrane domain. We would like to construct a model of the whole extracellular domain which could be incorporated with the transmembrane domain to produce a model which could be used to understand the overall function of the receptor, i.e. how binding the agonist leads to the opening of the ion channel. However, using standard techniques no homology has been found between the extracellular domain and any protein of known structure. Thus, as currently no reliable solution to the *ab-initio* protein folding problem is known we need to adopt novel techniques. The first assumption we used in modelling the extracellular domain is that, as it appears to be entirely external to the membrane from electron micrographs, it is highly likely that its structure is determined by the same rules as for normal globular proteins, for which we have a large body of structural data.

In order to solve this problem, among others, we have recently developed a novel method for scanning a database of known structures using a probe sequence based on aligned sequences of unknown structure.[34] This method requires having available a set of sequences for the unknown structure which are around 30–50% homologous, so that a multiple sequence alignment algorithm can produce a good alignment, yet not so similar that too few amino acids are substituted. This alignment of sequences is used to produce a predicted surface accessibility at each amino acid position in the sequence (based on a scale of accessibilities defined for each amino acid). These accessibilities are then used to scan the computed accessibilities of known structures in the database to find segments of known structures which correspond in accessibility to the unknown structure.

We chose a subset of the LGIC sequences which fulfilled the above requirements and applied this method to them. This resulted in the protein, of known structure, pyrophosphatase (PPase) being identified as having a high degree of similarity to the extracellular domain of the nACh receptor. A pair-wise alignment of these two sequences was prompted by this result and this gave an 18% identity conservation between them. This low level of sequence homology has in other modelling work[35,36] been used to generate structures by homology modelling which have been shown to be correct overall by subsequent X-ray studies.

Using this alignment of nAChR and PPase, we were able to generate a model for the entire extracellular domain of the receptor, by amino acid mutation of the PPase structure. Examination of the resulting structure shows features which are in agreement with the biochemical data. For example, the Main Immunogenic Region (MIR) occurs on the surface of this protein as does the Cys-loop. A problem that does occur in this structure is that the region around Cys 192–193 is not close to the Cys-loop, which is not consistent with the extended model we have constructed previously. However, this has led to a suggestion of the mode of action of the receptor, as discussed next.

FUNCTIONAL MODEL OF THE RECEPTOR

The function of these receptors is to bind a ligand, e.g. acetylcholine in the case of nAChR, which causes an ion channel in the receptor to open. None of the modelling described so far has suggested how this might occur. A very important piece of evidence regarding function is that the ion channel in the nACh receptor can open and close in the absence of a ligand. This indicates that all the structural requirements for ion channel opening must be possessed by the receptor itself, and that opening does not directly depend on the ligand. Indeed the ligand appears to act simply as an equilibrium shifter, by moving the position of the open/close equilibrium further to open than closed. Two alternative functional hypotheses can be envisaged. In the first case, each individual subunit is quite flexible and ligand binding in the extracellular domain causes a conformational change to occur mainly in this subunit (possibly with some partial allosteric-type effect on neighbouring subunits) which allows the ion channel to open. The binding site for the ligand occurs on the subunit itself and does not involve other subunits. The other possibility is that each subunit is quite rigid and it is motion of the subunits relative to each other which causes the ion channel to open. In this case, the suggestion would be that the binding site occurs at the interface between subunits, as by definition the individual subunits are rigid, i.e. they do not change conformation greatly on ligand binding, which would make it difficult to transfer the information of a ligand binding at a site distant from the interface region between subunits to the other subunits. These are the extreme cases, as of course it is possible that the actual mechanism involves a mixture of these two cases, i.e. there is conformational change in the subunit where binding occurs but with significant relative motion between subunits as well.

Using the PPase model, it is possible to construct a model of the nACh receptor in which the ligand binding site does not contain the Cys-loop and Cys 192–193 in close proximity by creating a binding site between subunits, with the Cys-loop coming from the peptide chain of one subunit and Cys 192–193 from another subunit. As none of the currently available biochemical, electrophysiological and pharmaceutical data can absolutely eliminate this as a possible model for the nACh receptor, experiments need to be devised to probe the possibility of binding sites existing between subunits. It also provides a rationalisation for the observation that antagonists of these receptors appear to be large, whereas agonists are small. A large molecule binding to a site between subunits could prevent the relative motion of the subunits, hence preventing the ion channel from opening, which would create an antagonist. The agonists themselves could change the equilibrium of the open/closed states by acting as a 'molecular lubricant', and thus allowing the relative motion to occur more freely, but they must of necessity be small. An implication of these ideas is that there are five possible binding sites per nACh receptor, yet the model does not require that all five of them need to be occupied to open the channel, which would explain the experimental observation that only two molecules of acetylcholine are required to open the channel.

CONCLUSIONS

We have shown how modelling can be used to tie together as much information as possible from the current glut of information that is available from molecular biology. The main benefit of the construction of explicit atomic models, besides giving insights into receptor structure and function, is that it allows specific experiments to be designed and prioritised to test the model and to comprehend and verify current experimental data. This modelling is a first step in the understanding of the structure–activity relationships of receptor ligands, whilst taking into consideration the receptor sites themselves.

ACKNOWLEDGEMENTS

We thank the SERC and Shell Research (UK) Ltd for financial support.

REFERENCES

1. Noda, M., Takahashi, H., Tanabe, T., Toyosato, M., Furutani, Y., Hirose, T., Asai, M., Inayama, S., Miyata, T. & Numa, S., Primary structure of α-subunit precursor of *Torpedo californica* acetylcholine receptor deduced from cDNA sequence, *Nature*, **299** (1982) 793.
2. Schofield, P. R., Darlison, M. G., Fujita, N., Burt, D., Stephenson, F. A., Rodriguez, H., Rhee, L. M., Ramachandran, J., Reale, V., Glencorse, T. A., Seeburg, P. H. & Barnard, E. A., Sequence and functional expression of the GABA_A receptor shows a ligand gated receptor superfamily, *Nature*, **328** (1987) 221.
3. Grenningloh, G., Rienitz, A., Schmitt, B., Methfessel, C., Beyrether, K., Gundelfinger, E. D. & Betz, H., The strychnine binding subunit of the glycine receptor shows homology with the nicotinic acetylcholine receptor, *Nature*, **328** (1987) 215.
4. Unwin, N., The structure of ion channels in membranes of excitable cells, *Neuron*, **3** (1989) 655.
5. Mitra, M., McCarthy, M. P., & Stroud, R. M., Three-dimensional structure of the nicotinic acetylcholine receptor and location of the major associated 43-kD cytoskeletal protein, determined at 22 Å by low dose electron microscopy and X-ray diffraction to 12·5 Å, *J. Cell. Biol.*, **109** (1989) 755.
6. Deisenhofer, J., Epp, O., Mikki, K., Huber, R. & Michel, H., Structure of the protein subunits in the photosynthetic reaction centre of *Rhodopseudomonas viridis* at 3 Å resolution, *Nature*, **318** (1985) 618.
7. Henderson, R., Baldwin, J. M., Ceska, T. A., Zemlin, F., Beckmann, E. & Downing, K. H., Model for the structure of bacteriorhodopsin based on high-resolution electron cryomicroscopy, *J. Mol. Biol.*, **213** (1990). (In press).
8. Kuhlbrandt, W. & Wang, Da Neng, Three-dimensional structure of plant light-harvesting complex determined by electron crystallography, *Nature*, **350** (1991) 130.
9. Jap, B. K., Walian, P. J. & Gehring, K., Structural architecture of an outer membrane channel as determined by electron crystallography, *Nature*, **350** (1991) 167.
10. Toyoshima, C. & Unwin, N., Ion channel of acetylcholine receptor reconstructed from images of postsynaptic membranes, *Nature*, **336** (1988) 247.
11. Kao, P. N. & Karlin, A., Acetylcholine receptor binding site contains a disulphide cross-link between adjacent half-cystinyl residues, *J. Biol. Chem.*, **261** (1986) 8085.

12. Neumann, D., Barchan, D., Fridkin, M. & Fuchs, S., Analysis of ligand binding to the synthetic dodecapeptide 185–196 of the acetylcholine receptor α subunit, *Proc. Natl. Acad. Sci. USA,* **83** (1986) 9250.

13. McCormick, D. J. & Atassi, M. Z., Localisation and synthesis of the acetylcholine-binding site in the α-chain of the *Torpedo californica* acetylcholine receptor', *Biochem. J.,* **224** (1984) 995.

14. Madhok, T. C., Chao, C. C., Matta, S., Hong, A. & Sharp, B. M., Monospecific antibodies against a synthetic peptide predicted from the alpha-3 nicotinicreceptor cDNA inhibit binding of [3H]nicotine to rat brain nicotinic cholinergic receptor, *Biochem. Biophys. Res. Com.,* **165** (1989) 151.

15. Smart, L., Meyers, H., Hilgenfeld, R., Saenger, W. & Maelicke, A., A structural model for the ligand-binding sites at the nicotinic acetylcholine receptor, *FEBS Lett.,* **178** (1984) 64.

16. Luyten, W. H. M. L., A Model for the Acetylcholine Binding Site of the Nicotinic Acetylcholine Receptor, *J. Neuroscience Res.,* **16** (1986) 51.

17. Cockcroft, V. B., Osguthorpe, D. J., Barnard, E. & Lunt, G. G., Modelling of Agonist binding to the Ligand Gated Ion-Channel Superfamily of Receptors, *Proteins,* **8** (1990) 386.

18. Chou, P. Y. & Fasman, G. D., β-turns in proteins, *J. Mol. Biol.,* **115** (1977) 135.

19. Sutcliffe, M. J., Hayes, F. R. F. & Blundell, T. L., Knowledge based modelling of homologous proteins, part II: rules for the conformation of substituted sidechains, *Protein Eng.,* **1** (1987) 385.

20. Claudio, T., Ballivet, M., Patrick, J. & Heinemann, S., Nucleotide and deduced amino acid sequence of *Torpedo californica* acetylcholine receptor γ subunit, *Proc. Natl. Acad. Sci. USA,* **80** (1983) 1111.

21. Finer-Moore, J. & Stroud, R. M., Amphipathic analysis and possible formation of the ion channel in an acetylcholine receptor, *Proc. Natl. Acad. Sci.,* **81** (1984) 155.

22. Stroud, R. M. & Finer-Moore, J., Acetylcholine receptor structure, function and evolution, *Ann. Rev. Cell Biol.,* **1** (1985) 317.

23. Mishina, M., Tobimatsu, T., Imoto, K., Tanaka, K., Fulita, Y., Fukuda, Y., Hirose, T., Inayama, S., Takahashi, T., Kuno, M. & Numa, S., Location of functional regions of acetylcholine receptor α-subunit by site-directed mutagensis, *Nature,* **313** (1985) 364.

24. Tobimatsu, T., Fujita, Y., Fukuda, K., Tanaka, K., Mori, Y., Konno, T., Mishina, M. & Numa, S., Effects of substitution of putative transmembrane segments on nicotinic acetylcholine receptor function, *FEBS. Lett.,* **222** (1987) 56.

25. Lear, J. D., Wasserman, Z. R. & DeGrado, W. F., Synthetic peptide models for protein ion channels, *Science,* **240** (1988) 1177.

26. Richardson J. S. & Richardson, D. C., Amino acid preferences for specific locations at the ends of alpha helices, *Science,* **240** (1988) 1648.

27. Guy, H. R. & Hucho, F., The ion channel of the nicotinic acetylcholine receptor, *TINS,* **10** (1987) 318.

28. Dauber-Osguthorpe, P., Sessions, R. B. & Osguthorpe, D. J., FOCUS, (Finally One Can Understand Simulations); Molecular dynamics analysis program. Molecular Graphics Unit, University of Bath, Bath, UK (1988).

29. Charnet, P., Labarca, C., Leonard, R. J., Vogelaar, N. J., Czyzyk, L., Gouin, A., Davidson, N. & Lester, H., An open-channel blocker interacts with adjacent turns of α-helices in the nicotinic acetylcholine receptor, *Neuron,* **2** (1990) 87.

30. Giraudat, J., Dennis, M., Heidmann, T., Haumont, P.-Y., Lederer, F. & Changeux, J.-P., Structure of the high-affinity binding site for noncompetitive blockers of the acetylcholine receptor: [³H]chlorpromazine labels homologous residues in the β and δ chains, *Biochemistry,* **26** (1987) 2410.

31. Hucho, F., Oberthur, W. & Lottspeich, G., The ion-channel of the nicotinic acetylcholine receptor is formed by the homologous helices MII of the receptor subunits, *FEBS Lett.,* **205** (1986) 137.

32. DiPaola, M., Kao, P. N. & Karlin, A., Mapping of the α-subunit site photolabeled by the noncompetitive inhibitor [³H]quinacrine azide in the active state of the nicotinic acetylcholine receptor, *J. Biol. Chem.*, **265** (1990) 11017.

33. Marquez, J., Iriate, A. & Martinez-Carrion, M., Covalent modification of a critical sulphydryl group in the acetylcholine receptor: cysteine-222 of the α-subunit, *Biochemistry*, **28** (1989) 7433.

34. Cockcroft, V. B., A modelling study of ligand-gated ion channel receptors, PhD thesis, University of Bath, 1992.

35. Pearl, L. H. & Taylor, W. R., A structural model for the retroviral proteases, *Nature*, **329** (1987) 351.

36. Blundell, T. & Pearl, L., Retroviral proteases: A second front against AIDS, *Nature*, **337** (1989) 596.

SECTION 5

Resistance Mechanisms

18

Mechanisms of Resistance to Pyrethroids and DDT in a Japanese Strain of the Housefly

Young-Joon Ahn

Department of Agrobiology, College of Agriculture, Seoul National University, Suweon 441-744, Korea

&

Eisuke Funaki,* Naoki Motoyama

Faculty of Horticulture, Chiba University, Matsudo, Chiba-ken 271, Japan

INTRODUCTION

Synthetic pyrethroids were introduced into pig farms in Japan in the late 1970s for the control of housefly populations, most of which had already developed resistance to organophosphorus and carbamate insecticides. Despite their extensive and intensive use, only a slight or no decrease in susceptibility to pyrethroids has been found until recently, owing probably to the open-type structure of Japanese pig farms. The structure allows flies to take refuge easily during pyrethroid sprays, which helps to preserve susceptible genes in a fly population.

The first report on the development of pyrethroid resistance in Japan appeared in 1984.[1] A colony in a pig farm in Mashiko city where pyrethroid sprays failed to control houseflies was found to be resistant to several pyrethroids and DDT. The field colony was further selected with resmethrin in the laboratory for 10 generations. The resultant strain, referred to as Mashiko-res10, showed an extremely high level of resistance to the compound.[2]

The present paper reports mechanisms by which the Japanese housefly strain exerts resistance to pyrethroids and DDT.

CROSS-RESISTANCE

LD$_{50}$ values of various pyrethroids and DDT applied topically to susceptible CSMA(S) and Mashiko-res10(R) strains are presented in Table 1. It is obvious

* Present address: Agrochemical Research Department, Ube Research Laboratory, Ube Industries Ltd, Tokiwadai, Ube, Yamaguchi-ken 755, Japan.

TABLE 1
Cross Resistance to Pyrethroids and DDT

Insecticide	LD_{50} ($\mu g/fly$)		R/S
	S (susceptible)	R (resistant)	
Remethrin	0·029	85	2 931
Permethrin	0·037	21	568
Phenothrin	0·046	>93	>2 022
Tetramethrin	1·0	>165	>165
Ethofenprox	0·072	26	333
Allethrin	0·51	46	90
Pyrethrin	0·70	32	46
Fenvalerate	0·042	90	2143
Fluvalinate	0·063	>83	>1 221
Cyclopromethrin	0·071	>162	>2 282
Tralomethrin	0·008	>148	>18 500
DDT	0·21	>200	>952

that the Mashiko-res10 strain is resistant to pyrethroids regardless of the presence or absence of an α-cyano group as well as to DDT, suggesting the involvement of a common mechanism for both pyrethroids and DDT resistance. This was the first incidence of a high level of pyrethroid resistance ever reported in Japan.

GENETICS

The mode of inheritance of resmethrin resistance was studied by crossing the CSMA and Mashiko-res10 strains. The LD-P line of the F_1 progeny was compared to those of the susceptible and resistant strains. The results suggested that pyrethroid resistance in the Mashiko-res10 strain was genetically incompletely dominant. Later studies revealed the incomplete dominance consisted of at least two factors, a dominant factor on chromosome V and a recessive factor on chromosome III.

Linkage group analysis was carried out using a susceptible strain referred to as *aabys* which had visible mutant markers on all five chromosomes. Thirty-two phenotypes separated in F_2 were selected by phenotype for several generations to obtain 32 genotypes. The flies were then subjected to a toxicity test with resmethrin for a linkage group analysis. The results clearly demonstrated that major factors for pyrethroid resistance in the Mashiko-res10 strain were located on chromosome III and V, which were found to be recessive and dominant, respectively (Fig. 1).

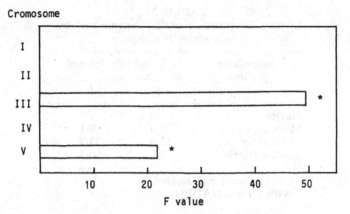

FIG. 1. Linkage group analysis of dominant and recessive factors for resmethrin resistance in R strain. ★ significant at 1% level.

METABOLISM

In-vivo dynamics of ^{14}C-fenvalerate

The in-vivo dynamics of ^{14}C-fenvalerate applied topically at two different doses were investigated. The lower dose, 0.012 μg/fly, was equivalent to LD_{10} of the CSMA strain, while the higher dose, 0.066 μg/fly, was equivalent to LD_{90} of the same strain. At various times following the application of ^{14}C-fenvalerate, the radioactivity in the external wash, internal extract, and water-soluble and chloroform-soluble excreta were determined. Assuming a one-compartment model, first-order rate constants for cuticular penetration (k_{pen}), and for excretion of water-soluble metabolite (k_{exc}) were computed (Table 2).

At either dose, there was no significant difference in the penetration of fenvalerate between the strains. In contrast, there was a remarkable difference

TABLE 2

First Order Rate Constant for Cuticular Penetration (k_{pen}) and Excretion of Water-Soluble Metabolites (k_{exc}) of Fenvalerate

Rate constant	Fenvalerate applied (μg/fly)	k (h^{-1}) S	k (h^{-1}) R	R/S
k_{pen}	0.012^a	0.18	0.20	1.1
	0.066^b	0.15	0.14	0.9
k_{exc}	0.012	0.16	0.60	3.8
	0.066	0.005	0.28	56

^{a} Equivalent to LD_{10} of S strain.
^{b} Equivalent to LD_{90} of S strain.

TABLE 3
In-vitro Degradation of ^{14}C-Fenvalerate[a] by
Subcellular Fractions

Subcellular fraction	% Metabolite formed	
	S	R
Nuclei	4·0	7·3
Mitochondria	5·0	9·1
Microsome[b]	17·0	33·7
Soluble fraction	6·6	10·3

[a] Chlorophenyl ring labeled.
[b] With 1·1 mM NADPH.

in the excretion rate constant. The Mashiko-res10 strain showed a much higher rate for excretion of water-soluble metabolites than the CSMA strain. The difference became greater at the higher dose. The results implied that an increased metabolism could be a factor for pyrethroid resistance in this strain.

In-vitro degradation of ^{14}C-fenvalerate

The in-vitro metabolism of ^{14}C-fenvalerate by subcellular fractions was studied in the resistant and susceptible strains. ^{14}C-Fenvalerate labeled at two different positions, i.e. the chlorophenyl-ring and the phenoxyphenyl-ring, was used as the substrate. Metabolites were separated by silica gel TLC with two solvent systems following the method of Ohkawa *et al.*[3]

The results of the in-vitro studies are presented in Table 3, and are in support of the conclusion obtained from the in-vivo studies mentioned above. The Mashiko-res10 strain showed an increased rate of metabolite formation as compared to the CSMA strain in all the subcellular fractions. Microsomes fortified with NADPH, in particular, were the most active and this was the fraction which gave the largest interstrain difference.

The effect of NADPH, CO, and piperonyl butoxide (PBO) on metabolite

TABLE 4
Effect of Cofactor and Inhibitors on In-vitro Degradation of ^{14}C-Fenvalerate
by Microsomes of R Strain

Cofactor	Inhibitor	% Metabolite formed[a]				
		I	II	III	IV	Total
None	None	1·2	—	—	20·7	21·9
1·1 mM NADPH	None	8·2	4·5	6·7	22·2	42·6
1·1 mM NADPH	CO	3·0	1·3	3·8	10·6	21·9
1·1 mM NADPH	0·3 mM PBO	1·9	1·3	1·6	8·3	11·7

[a] I, unknown; II, 4HO-PBacid; III, 4HO-fenvalerate; IV, PBacid.

formation by microsomes from the resistant strain was studied using phenoxyphenyl-ring labeled ^{14}C-fenvalerate as the substrate (Table 4). The addition of NADPH almost doubled the total metabolite formation. In contrast, CO and PBO inhibited the in-vitro degradation significantly, suggesting that a large portion of the reaction was mediated by the cytochrome P_{450}-dependent monooxygenase system. The most abundant metabolite found was PBacid produced by oxidative cleavage of the carboxyl ester bond.

Substrate specificity of cytochrome P_{450} species

Yachiyo is an organophosphorus resistant strain known to have high cytochrome P_{450}-dependent monooxygenase activity for diazinon degradation.[4] It is of interest to know if the enzyme system also shows high activity for pyrethroid degradation. Table 5 shows an interstrain comparison in degradation activity of microsomes from three strains, i.e. Mashiko-res10(Py-R), Yachiyo(OP-R), and CSMA(S). Chlorophenyl-ring labeled ^{14}C-fenvalerate was used as the substrate in this study. The Mashiko-res10 strain showed the highest activity in total metabolite formation. The major metabolite was Cl-Vacid formed by the oxidative cleavage of the carboxyl ester bond of fenvalerate.

An interesting finding was that the Yachiyo strain was not as active as the Mashiko-res10 strain with regard to fenvalerate degradation. In particular, the formation of Cl-Vacid in the Yachiyo strain was far less than that in the Mashiko strain, indicating a difference in substrate specificity of cytochrome P_{450} isozymes between the pyrethroid and organophosphorus resistant strains. In support of this conjecture, CO-binding difference spectra of cytochrome P_{450} preparations from the three strains gave a slight difference in λ_{max} values. The λ_{max} for the Yachiyo strain was 451 nm, while those for the Mashiko-res10 and CSMA strains were 450 and 451·5 nm, respectively.

A proposed pathway of fenvalerate metabolism by housefly microsomes is illustrated in Fig. 2. The first phase takes place through two routes, i.e. oxidative cleavage of the carboxyl ester bond resulting in the production of PBacid and Cl-Vacid, and oxidative hydroxylation of phenoxyphenyl-ring

TABLE 5
Interstrain Comparison of ^{14}C-Fenvalerate Degradation by Microsomes in the Presence of NADPH

Strain	% Metabolite formed[a]				
	A	B	C	D	Total
Mashiko(R)	5·4	2·6	6·9	17·7	32·6
Yachiyo(OP-R)	3·7	1·9	6·7	5·9	18·2
CSMA(S)	1·8	0·4	4·6	4·4	11·2

[a] A, unknown; B, 4HO-Cl-Vacid; C, 4HO-fenvalerate; D, Cl-Vacid.

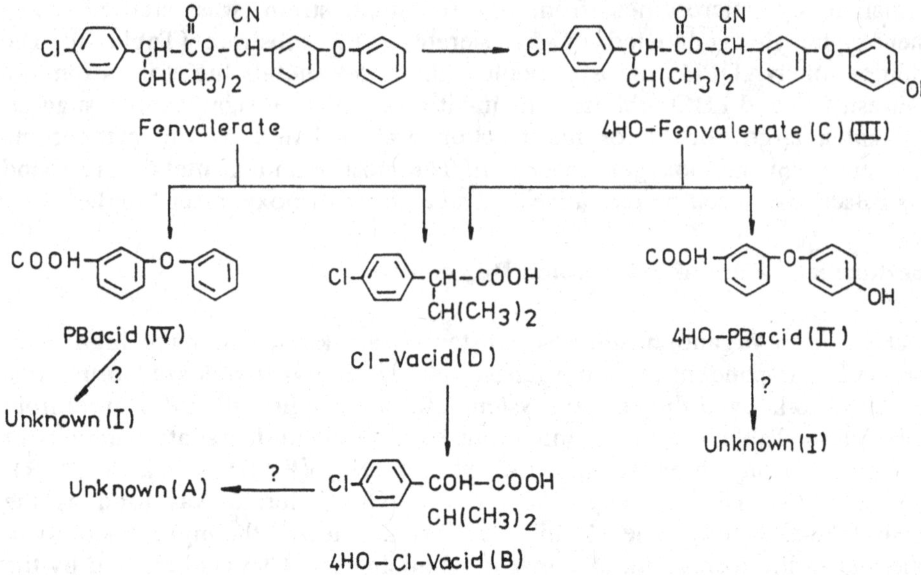

FIG. 2. Proposed metabolic pathways of fenvalerate by housefly microsomes.

resulting in the production of 4HO-fenvalerate. The Mashiko-res10 strain has a cytochrome P_{450} species which shows high activity especially for the former reaction.

Significance of metabolic factor

The importance of an increased metabolism for pyrethroid resistance can be estimated by use of a specific inhibitor. The effect of PBO, a well known inhibitor of cytochrome P_{450}, on the toxicity of resmethrin to the CSMA and Mashiko-res10 strains was studied. A remarkable decrease in the LD_{50} was observed in both strains (Table 6). However, the extent of this synergistic

TABLE 6
Effect of Piperonyl Butoxide (PBO) on Toxicity of Resmethrin

Strain	LD_{50} ($\mu g/fly$)		SE^b
	Without	*With PBO^a*	
S	0·015	0·002 0	8
R	85	0·35	243

^{a} 2·5 μg of PBO applied topically 1 h prior to resmethrin.
^{b} LD_{50} without PBO/LD_{50} with PBO.

TABLE 7
Synergistic Effect of Piperonyl Butoxide (PBO)
Applied Simultaneously with Pyrethroids at 1:4
Ratio to R Strain

Insecticide	LD_{50} ($\mu g/fly$)	SE^a
Resmethrin	0·20	425
Permethrin	0·24	48
Ethofenprox	0·30	87
Fenvalerate	0·79	114
Fluvalinate	0·41	>202
Tralomethrin	0·17	>871

^{a} LD_{50} without PBO/LD_{50} with PBO.

effect (SE), judged by the ratio of the LD_{50} obtained without PBO, and with PBO, was drastically different between the strains. The SE value in the Mashiko-res10 strain was 243, while that in the CSMA strain was only 8, indicating that the increase in the oxidative metabolism of fenvalerate plays a very important role as a mechanism of pyrethroid resistance.

Table 7 shows LD_{50} of various pyrethroids which were applied as a mixture with PBO at a 1 to 4 ratio to the Mashiko-res10 strain. The addition of PBO produced again a significant synergism, decreasing the LD_{50} values of all the pyrethroids tested. The level of LD_{50} reached low enough to restore the efficacy of pyrethroids in controlling the resistant flies.

KDR FACTOR

It is well established that reduced nerve sensitivity may be the most common and important mechanism for pyrethroid resistance.[5-9] Since nerve sensitivity is influenced by the gene *kdr* in the housefly, it is generally referred to as the *kdr* factor. The factor or *kdr*-type factor confers cross-resistance to pyrethroids and DDT,[6,8] both of which have a common site of action. It is now known that there is another alleic factor, referred to as *super-kdr*, in houseflies, which has the same characteristics as *kdr* but shows stronger resistance to pyrethroids than *kdr*.[10] Genetic analysis demonstrated that the gene *kdr* is located on the third chromosome in houseflies.[9]

Since the first appearance of pyrethroid-resistance houseflies in Japan, we have been interested to determine whether or not the Japanese colony has the *kdr*-type mechanism. The nerve sensitivity of the five strains to several pyrethroids and DDT was measured by the method described previously.[8,9,11] The mean time between the chemical application and the onset time of the effect for susceptible and resistant strains is a suitable criterion for the measurement of nerve sensitivity and the genetic analysis of resistance to both pyrethroids and DDT.[8,9,11]

TABLE 8
Nerve Sensitivity of Housefly Strains to Pyrethroids Applied at
1×10^{-6} M to the Exposed Thoratic Ganglia

Chemical	Type	Latency (min) ± SD		
		CSMA	Mashiko-res 10	R/S[a]
Permethrin	I	8·8 ± 1·7	46·3 ± 5·0	5·3
Resmethrin	I	3·7 ± 1·5	>34·7 ± 8·3	>9·4
Tetramethrin	I	1·8 ± 0·6	>60·6	>33·3
Cypermethrin	II	3·6 ± 1·3	>40·4 ± 6·6	>11·2
Deltamethrin	II	8·8 ± 2·7	>60·0	>6·8
Cyclopromethrin	II	5·0 ± 2·2	>48·4 ± 4·0	>9·7
Fenvalerate	II	8·1 ± 3·2	>60·655	>7·4

[a] R/S: Latency of the Mashiko-res10 strain/latency of the
CSMA strain.

Comparison in CNS sensitivity

Nerve sensitivity of the CSMA and Mashiko-res10 strains to several pyrethroids and DDT was investigated at 1×10^{-6} M. The resistant strain showed
reduced nerve sensitivity to DDT and all the pyrethroids tested, regardless of
their structures (Table 8), indicating that the *kdr* factor may be present in the
Japanese colony of the housefly. However, it has been generally acknowledged
that the factor *kdr* may not be present in Japanese colonies of the housefly,
although nerve insensitivity to DDT was once reported with the DDT-resistant
strain in 1958.[11]

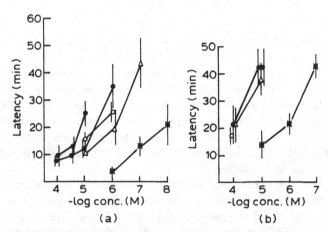

FIG. 3. Dosage–nerve sensitivity curves of CSMA(∗), Mashiko(△), Mashiko-res
10(●). *kdr*(○) and super-*kdr*(■) strains of houseflies determined by resmethrin (A)
and DDT (B) application to the exposed thoracic ganglia. Vertical bars represent
standard deviation of 7–11 individuals. Latency: the time at which the nerve response
appeared after insecticide treatment. Room temperature: 25 ± 1°C.

Dosage–nerve sensitivity curves at various concentrations of resmethrin and DDT are shown in Fig. 3. The data showed that there was a great interstrain difference in nerve response. Resmethrin produced repetitive discharge in *c.* 4 min in the susceptible strain at 1×10^{-6} M, whereas much higher concentrations were required to produce similar effects in the resistant strains. At 1×10^{-8} M, the chemical produced a nerve response in *c.* 21 min in the susceptible strain, whereas no response appeared in the resistant strains except the Mashiko strain, even at 1×10^{-7} M. The nerve insensitivity ranking for resmethrin was in order of *super-kdr* > Mashiko-res10 > *kdr* > Mashiko > CSMA strains. The two Japanese colonies of the housefly were also far less sensitive to DDT than the CSMA strain. However, in contrast to their sensitivity to resmethrin, their sensitivity to DDT was similar to that of the *super-kdr* strain.

Recently, it has been suggested that DDT and pyrethroids bind to the same receptor, but at allosteric sites.[12] In the present study, the Mashiko-res10, which showed a cross-resistance to pyrethroids as well as to DDT, was found to have a nervous system insensitive to all these insecticides. Both qualitative and quantitative differences in nerve response to insecticides were observed among the strains. The ganglia of all resistant strains showed a similar degree of response to DDT as judged by the latent period (Fig. 3), while their response to pyrethroids varied depending upon the strains. The fact that the ganglia of the Mashiko-res10 strains were less sensitive to resmethrin than those of the original Mashiko strain indicates that the *kdr* factor may evolve towards *super-kdr,* which confers stronger resistance than the *kdr,* by selection with pyrethroids.

Effect of metabolic inhibitor

It is of great interest to determine whether or not enhanced detoxication by enzyme is involved in pyrethroid resistance in addition to the reduced nerve sensitivity by the *kdr* gene. Since the pretreatment of the Mashiko colony with the oxidase inhibitor PBO reduced the resistance level significantly, a possible involvement of local detoxication of resmethrin and DDT in the CNS was investigated in the Mashiko-res10 strain using the oxidase inhibitor PBO and the DDT dehydrochlorinase inhibitor DMC, respectively (Fig. 4 and Table 9). There was no great difference in frequency of motor discharge and latent period between the Mashiko-res10 flies treated with resmethrin only and with resmethrin plus PBO or with DDT only and with DDT plus DMC, indicating no detoxication by the enzyme in the CNS.

Role of *kdr* factor

A simple rapid detection method of *kdr* in the field may be of great importance for practical use of pyrethroids, because of the recessive nature of *kdr.*[13] To separate houseflies with or without *kdr,* both bioassay and electrophysiological methods were used.

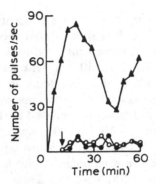

FIG. 4. Time response curves of the impulses in the motor nerve of CSMA and Mashiko-res10 strains of houseflies to resmethrin (1×10^{-5} M) and PBO (2×10^{-5} M) applied to the exposed thoracic ganglia. ▲: CSMA, resmethrin only; ●: Mashiko-res 10, resmethrin only; ○: Mashiko-res 10, PBO and resmethrin. Seven individuals were tested. The arrow indicates the latent time. Room temperature: $25 \pm 1°C$.

The bioassay method described by Sawicki (pers. comm., 1990) was used.[14] A mixture of 1 µg DDT, 1 µg PBO (oxidase inhibitor), and DMC (DDT dehydrochlorinase inhibitor) was applied. Flies which survived were regarded as those having the *kdr* factor. The most effective electrophysiological determination involved measuring latent time and number of bursts per second. This technique was used when 10^{-5} M permethrin was applied directly to the exposed thoracic ganglia. The houseflies were counted as *kdr*-type if the motor nerve response appeared over 8 min or the number of bursts per second was <40 (Fig. 5). A comparison of *kdr* frequency determined by bioassay and electrophysiological methods shows that results obtained by either method are roughly in the same range (Table 10). These data prove the bioassay method used to estimate *kdr* frequency in various housefly strains is roughly in accordance with *kdr* frequency determined by the electrophysiological method, suggesting that the bioassay method is a simple alternative for *kdr* detection.

TABLE 9

Effect of Pretreatment with DMC on Nerve Sensitivity of Housefly Strains to DDT Applied to the Exposed Thoracic Ganglia

Strain	Latency (min) ± SD		
	DDT (1×10^{-5} M)	*DDT* (5×10^{-6} M)	*+DMC* (1×10^{-5} M)
CSMA	13·4 ± 4·9	—[a]	—[a]
Mashiko	37·8 ± 6·2	—[a]	30·4 ± 4·8
Mashiko-res10	42·3 ± 6·9	43·0 ± 6·5	39·8 ± 8·9

[a] Not determined.

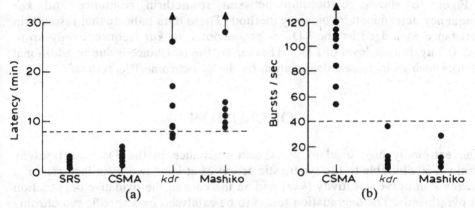

FIG. 5. Nerve sensitivity of two susceptible SRS and CSMA and two *kdr*-type strains of houseflies after saline solutions containing 10^{-5} M permethrin were applied to the exposed thoracic ganglia.

TABLE 10

Comparison of *kdr* Frequency Determined by Bioassay and Electrophysiological Methods

Strain	kdr Frequency (%)		
	Latency	Burst	Bioassay
CSMA	9	5	0
Rutgers	12	19	11
Yachiyo	55	95	85
Mashiko-res10	84	95	100
kdr	83	100	92

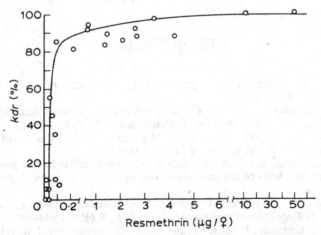

FIG. 6. Relationship between resmethrin resistance and *kdr* frequency.

Figure 6 shows relationship between resmethrin resistance and *kdr* frequency determined by bioassay method. These data indicate that resmethrin resistance as judged by the LD_{50} is proportional to *kdr* frequency only up to 0·2–0·3 μg/female level of LD_{50}. Thereafter, the resistance is due to additional factors such as increased degradation by the cytochrome P_{450} system.

CONCLUSION

The extremely high level of pyrethroid resistance in the Japanese housefly strain is attributable to the synergistic action of at least two main factors, i.e. a decrease in nerve sensitivity (*kdr*) and an increase in the oxidative degradation of pyrethroids. The degradation seems to be catalysed by a specific cytochrome P_{450} species present only in the pyrethroid resistant strain but not in the organophosphate resistant strain.

The *kdr* factor generally results in cross-resistance to all the pyrethroids and DDT. Both qualitative and quantitative differences exist in nerve responses to pyrethroids among resistant strains equipped with the *kdr* factor. The target site insensitivity may be associated with changes in the nerve membrane phospholipids,[15] a reduced number of binding receptors,[12] reduced Ca-ATPase susceptibility,[16] or lowered susceptibility of sodium channels for pyrethroids and DDT.[17] For a better understanding of the nature of the *kdr* factor, further investigations are needed on the exact mode of nerve insensitivity at molecular and nerve membrane levels.

The occurrence of the *kdr* factor poses a problem for the future use and development of pyrethroids. However, resistance level brought about by this *kdr* factor alone appears at most in the range of 0·2–0·3 μg/female in terms of LD_{50} of resmethrin (Fig. 6). Suppression of additional metabolic factors by specific inhibitors may restore the efficacy of pyrethroids to some extent.

REFERENCES

1. Motoyama, N., Pyrethroid resistance in a Japanese colony of the housefly. *J. Pestic. Sci.*, **9** (1984) 523–6.
2. Funaki, E. & Motoyama, N., Cross resistance to various insecticides of the housefly selected with a pyrethroid. *J. Pestic. Sci.*, **11** (1986) 219–22.
3. Ohkawa, H., Kaneko, H., Tsuji, H. & Miyamoto, J., Metabolism of fenvalerate (Sumicidin) in rats. *J. Pestic. Sci.*, **4** (1979) 143–55.
4. Oi, M., Dauterman, W. C. & Motoyama, N., Biochemical factors responsible for an extremely high level of diazinon resistance in a housefly strain. *J. Pestic. Sci.*, **15** (1990) 217–24.
5. Farnham, A. W., Genetics of resistance of houseflies (*Musca domestica*) to pyrethroids 1. Knock-down resistance. *Pestic. Sci.*, **8** (1977) 631–6.
6. Miller, T. A., Kennedy, J. M. & Collins, C., CNS insensitivity to pyrethroids in the resistant *kdr* strain of houseflies. *Pestic. Biochem. Physiol.*, **12** (1979) 224–30.

7. DeVries, D. H. & Georghiou, G. P., Decreased nerve sensitivity and decreased cuticular penetration as mechanism of resistance to pyrethroids in a (1*RS*)-*trans*-permethrin-selected strain of the housefly. *Pestic Biochem. Physiol.*, **15** (1981) 234–41.
8. Ahn, Y. J., Shono, T., Hido, O. & Fukami, J., Mechanism of resistance to pyrethroids and DDT in the housefly, *Musca domestica* L. *Pestic. Biochem. Physiol.*, **31** (1986) 46–53.
9. Ahn, Y. J., Shono, T. & Fukami, J., Linkage group analysis of nerve insensitivity in a pyrethroid resistant strain of housefly. *Pestic. Biochem. Physiol.*, **26** (1986) 231–7.
10. Sawicki, R. M., Unusual response of DDT-resistant houseflies to carbinol analogues of DDT. *Nature (London)*, **275** (1978) 443–4.
11. Yamasaki, T. & Narahashi, T., Resistance of houseflies to insecticides and the susceptibility of nerve to insecticides. Studies on the mechanism of action of insecticides (XVII). *Botyu-Kagaku*, **23** (1958) 146–57.
12. Chang, C. P. & Plapp, F. W., DDT and pyrethroids: receptor binding and mode of action in the housefly. *Pestic. Biochem. Physiol.*, **20** (1983) 76–85.
13. Ahn, Y. J., Shono, T. & Fukami, J., Inheritance of pyrethroid resistance in a housefly strain from Denmark. *J. Pestic. Sci.*, **11** (1986) 591–6.
14. Sawicki, R. M. (Per. comm). The International Congress of Pesticide Chemistry, 1990).
15. Chiang, C. & Devonshire, A. L., Changes in membrane phospholipids, identified by Arrhenius plots of acetylcholinesterase and associated with pyrethroid resistance (*kdr*) in housefly. *Pestic. Sci.*, **13** (1982) 156–60.
16. Ghiasuddin, S. M., Kadous, A. A. & Matsumura, F., Reduced sensitivity of Ca-ATPase in the DDT-resistant strain of the German cockroach. *Comp. Biochem. Physiol.*, *(Ser. C)*, **68** (1981) 15–20.
17. Salgado, V. L., Irving, S. N. & Miller, T. A., Depolarization of motor nerve terminals by pyrethroids in susceptible and *kdr*-resistant houseflies. *Pestic Biochem. Physiol.*, **20** (1983) 100–114.

19

Isolation of Insect Genes Coding for Voltage-Sensitive Sodium Channels and Ligand-Gated Chloride Channels by PCR-Based Homology Probing

DOUGLAS C. KNIPPLE, KEVIN E. DOYLE, JOSEPH E. HENDERSON & DAVID M. SODERLUND

Department of Entomology, New York State Agricultural Experiment Station, Cornell University, Geneva, New York 14456, USA

INTRODUCTION

Two important insecticide target sites in the insect nervous system are the voltage-sensitive sodium channel and the GABA (γ-aminobutyric acid) receptor/chloride ionophore complex.[1] The former molecule is a critical component of nerve excitation, mediating the transient inward flux of sodium ions during the nerve action potential. The latter mediates synaptic inhibition in response to GABA, the major inhibitory neurotransmitter of vertebrate and invertebrate nervous systems. Target site insensitivity to insecticidal compounds acting at these sites has been described and genetically isolated in a number of insect species.[2] The isolation and characterization of the genes encoding these nerve membrane macromolecules in insects is, therefore, of critical importance for understanding the molecular basis of target site resistance mechanisms as well as the modes of action of insecticides in susceptible insects.

Recent molecular investigations in vertebrate species and *Drosophila* have shown that, within a given organism, voltage-sensitive sodium channels are encoded by distinct but structurally related genes comprising a gene family.[3] Similarly, GABA receptor/chloride channels of mammals, which have been shown to be heteromultimeric, are the products of a gene family that is itself a component of a larger superfamily encoding other ligand-gated ion channels.[4] Comparative sequence analysis both within and between species for both types of membrane macromolecules has identified small regions of conserved amino acid sequence, or 'signature motifs', which are presumed to be of critical functional importance. In this paper we describe the use of the polymerase chain reaction[5] (PCR) to isolate DNA sequences from insect species that are homologous to conserved sequence elements of identified vertebrate sodium channel and GABA receptor genes. We also discuss the advantages of PCR-based homology probing,[6,7] which employs degenerate oligonucleotide primers deduced from conserved sequence domains of characterized sodium channel and GABA receptor genes, over methods that employ low stringency hybridization with heterologous probes to detect DNA sequence homology.

ISOLATION OF PUTATIVE VOLTAGE-SENSITIVE SODIUM CHANNEL GENE SEQUENCES OF INSECTS

Knowledge of the primary structure of voltage-sensitive sodium channels has been advanced by the cloning and sequencing of the sodium channel gene expressed in the electroplax of *Electrophorus electricus*,[8] three distinct sodium channel genes expressed in rat brain,[9,10] and two additional genes expressed in rat skeletal[11] and cardiac[12] muscle. Comparative sequence analysis of the inferred amino acid sequences of these genes has identified four major repeated internal homology units, each of which contains a series of six amino acid sequences having appropriate properties for the formation of membrane-traversing helices. Other conserved structural elements that have been postulated to be involved in the voltage-sensitive gating of the channel, its inactivation, and the determination of its cation selectivity have been identified. Conservation of these structural elements and a high level of amino acid sequence similarity has also been shown for a gene encoding the dihydropyridine receptor/calcium channel of rabbit skeletal muscle.[13] Taken together, these findings strongly suggest that the genes encoding sodium and calcium channels are related members of a multigene family that arose from a common ancestral gene.

The first cloned insect genes with structural features similar to vertebrate sodium channel genes were isolated from *Drosophila melanogaster*. The *DSC1* gene was isolated by hybridization screening under low stringency conditions using vertebrate sodium channel cDNA probes.[14-16] Partial sequence analysis of this gene showed conservation of the major structural features identified for vertebrate sodium channel genes,[14] but to date no genetic or other functional criterion that establishes the physiological significance of the *DSC1* gene product has been met. More recently, the gene encoded by the *para* locus has been cloned by a molecular genetic approach, and sequence analysis has shown that it too possesses substantial structural similarities to characterized vertebrate sodium channel genes.[17] The deduced amino acid sequences of *para* and *DSC1* are about as similar to each other (approximately 50% identical residues or conservative substitutions) as either is to any of the characterized vertebrate sodium channel sequences.[17] In contrast to *DSC1*, the *para* gene has been clearly implicated in sodium channel function on the basis of genetic, pharmacological and electrophysiological analyses of mutant phenotypes.[18]

We initiated a program to investigate the structure and function of sodium channel genes of the house fly, because sodium channel structural variants have been implicated as the basis for the well characterized insecticide resistance mechanism known as *kdr*, which confers resistance to pyrethroids and DDT in that species.[2] Given the preponderance of evidence suggesting that the *para* gene product is the predominant sodium channel of the *Drosophila* nervous system, we developed an experimental design based on the use of PCR with degenerate oligonucleotide primers to amplify specifically *para*-homologous sequences present in the house fly genome.[19] As the first step in the design of selective target primers, we identified all regions of seven or more amino acid residues that are identical between the *para* and rat brain 1 sodium channels, but different from *DSC1*. Because the amplification of sequences by PCR is optimal when the target sequences on the template DNA are relatively closely spaced, all regions of less than 500 base pairs (bp) that are delimited by two such conserved target sequences were then identified. Among these, a region coding for an interval between membrane-spanning domains IS5 and IS6 was selected, because the codon ambiguity of its terminal amino acid sequences is relatively low (Fig. 1). All nucleotide sequence combinations capable of encoding each of these conserved motifs were then deduced and the least ambiguous stretches of sequence were selected for the design and synthesis of degenerate target primers. Because each degenerate primer comprised all of the nucleotide sequence combinations specifying the conserved 'signature

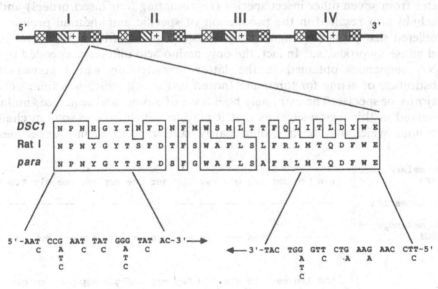

Fig. 1. Design of oligonucleotide primers for the selective amplification of *para*-homologous sequences from the house fly genome. *Top*. A schematic representation of sodium channel gene structure showing four internally repeated homology units, each of which contains a series of six postulated membrane-spanning sequence domains.[8] *Middle*. The amino acid sequences shown correspond to a region of the postulated pore-forming region (see text and Refs 21–23) between sequence domains IS5 and IS6 of the inferred sequences of the rat brain 1 cDNA,[9] the *para* sodium channel cDNA,[17] and *DSC1* genomic sequence.[14] *Bottom*. The nucleotide sequences of the degenerate primer pairs used to screen for *para*-homologous sequences in the house fly. Reproduced with permission of Wiley-Liss, Inc. from PCR-generated sodium channel gene probe for the house fly. *Arch. Insect Biochem. Physiol.*, **16** (1991) by D. C. Knipple, L. L. Payne and D. M. Soderlund.[19]

motif' from which it was deduced, one molecular species present in each primer pool was expected to have perfect sequence complementarity to one strand of the template DNA. During PCR, this specific primer would then anneal to the target sequence on denatured template DNA, thereby priming the synthesis of DNA by the thermostable *Taq* polymerase. The nascent strands produced would be available to serve as templates for synthesis in repeated cycles of primer annealing, synthesis and denaturation, resulting in the exponential amplification of the DNA between the target primers.

When PCR reactions using genomic DNAs from *Drosophila*, house fly, and calf thymus were performed with the above experimental procedure, amplification products of the predicted size were obtained in all cases, but none resulted from reactions lacking template DNA.[19] DNA sequence analysis revealed a surprisingly high level of nucleotide sequence identity with only 16 differences found between the house fly amplification product and the corresponding region of the *para* gene. These differences occurred exclusively in third base positions of codons, and produced no change in amino acid sequence (Fig. 2). Similar experiments performed with genomic DNA templates from seven other insect species (representing four insect orders) and an arachnid also resulted in the production of specific amplification products of predicted size possessing high levels of sequence similarity to the *Drosophila* and house fly products.[20] In fact, the only amino acid difference encoded by the DNA sequences obtained in the latter investigation was a conservative substitution of serine for threonine (noted in Fig. 2), which was found in the majority of species. The extremely high level of amino acid sequence similarity observed in this region suggests that it plays a critical role in sodium channel function, which has been evolutionarily conserved. Recently experimental

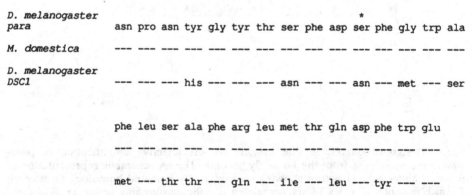

FIG. 2. Comparison of the inferred amino acid sequence of the PCR-amplified putative sodium channel gene fragment from the house fly[19] (*center*) with the corresponding sequences of the *para*[17] and *DSC1*[14] loci of *Drosophila*. The asterisk in the *para* sequence indicates the serine residue that was found to be replaced by threonine in the inferred sequences of PCR amplification products obtained from DNA of *Aedes aegypti*, *Lymantria dispar* (one of two variants found), *Plutella xylostella*, *Trichoplusia ni*, *Periplaneta americana*, *Leptinotarsa decemlineata*, and *Tetranychus urticae*.[20]

evidence has demonstrated that the homologous region of voltage-gated potassium channels (i.e. the 'H5' region between membrane-spanning domains S5 and S6) constitutes the pore-forming region.[21-23] These findings suggest that the interval between IS5 and IS6 of the voltage-sensitive sodium channel very likely involves the same function.

We have used the house fly amplification product as a probe in order to isolate clones from genomic and cDNA gene libraries prepared from wild type (i.e. pyrethroid-susceptible) and *kdr* house fly strains. Our ongoing characterizations of these clones will seek to identify structural differences between wild type and *kdr* sodium channels and to demonstrate the relationship between these differences and the functional properties of sodium channels relating to the mode of action of insecticides acting at that site.

AMPLIFICATION OF *Drosophila melanogaster* SEQUENCES HOMOLOGOUS TO MAMMALIAN GABA RECEPTOR GENES

Molecular cloning and characterization of mammalian GABA receptor genes has provided evidence for genes encoding four distinct but homologous subunits (α, β, γ, and δ).[4] These subunit genes, together with the gene for the strychnine-binding subunit of the glycine-gated chloride channel,[24] comprise a family of ligand-gated chloride channel genes possessing four regions postulated to encode membrane-spanning domains with similar properties. There is a much lower overall level of amino acid sequence conservation among identified vertebrate members of the ligand-gated chloride channel gene family than among members of the sodium channel gene family. As a consequence, a single octapeptide 'signature motif' present in transmembrane domain 2 (Fig. 3) is the only continuous stretch of six or more invariant amino acids encoded by all of the characterized members of the ligand-gated chloride channel gene family. The absence of multiple highly conserved sequence elements in these genes prevented the design of experiments using the standard PCR method employing two degenerate target primers to amplify homologous sequences present in the genomes of insect species. We therefore devised the following strategy based on a recently described 'single site' PCR procedure[26] to amplify sequences present in the *Drosophila* genome that are adjacent to the site encoding the octapeptide signature motif of the ligand-gated chloride channel gene family.

The initial step in this procedure involves the complete digestion of genomic DNA with a restriction endonuclease that cuts with a relatively high frequency and yields 5' protruding ends (Fig. 4, Step 1). The termini of the resultant restriction fragments are then modified by ligation of an adaptor comprised of two synthetic oligonucleotides, designated '5' anchor template' and '3' tailed linker' (Fig. 4, Step 2). This DNA is then employed as a template for PCR

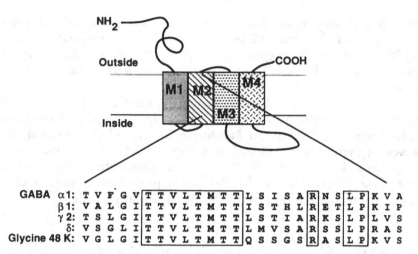

FIG. 3. Conceptual model of a ligand-gated chloride channel subunit, showing the four postulated transmembrane domains (M1–M4). Inferred amino acid sequences in M2 are shown for four rat brain GABA receptor subunits[25] and for the strychnine-binding subunit of the rat brain glycine receptor.[24] Boxes indicate conserved amino acids.

using a mixture of oligonucleotides encoding the signature motif as the forward primer, and an oligonucleotide that is identical to the first 20 bases of the anchor template as the reverse primer. During the initial PCR cycle (Fig. 4, Step 3), chain elongation is initiated by annealing of the specific target primer to any complementary sequences on the genomic DNA fragments, and is terminated at the anchor template sequence appended to the downstream restriction fragment terminus. The noncomplementary 3′ end of the tailed linker prevents chain elongation of re-annealing DNA strands and there is thus no sequence complementary to the anchor primer that is available to initiate chain elongation in the reverse direction. In the second PCR cycle (Fig. 4, Step 4), the specific products of the first cycle are now available to serve as templates for reverse priming by the anchor primer, and forward priming by the target sequence primer occurs as in the previous cycle. Subsequent PCR cycles (Fig. 4, Step 5) result in geometric amplification of fragments of defined lengths bounded by the target sequence primer and the anchor template sequence downstream from it. Since the distance between the target sequence and the downstream restriction site to which the anchor template is ligated is anticipated to vary among different members of a multigene family, several fragments of discrete lengths are predicted to result from the coordinate amplification of sequences derived from multiple transcription units.

Our initial experiments with this procedure employed a 512-fold degenerate target primer that incorporated codon bias for the two most abundantly used leucine codons in *Drosophila* (Henderson, J. E., Soderlund, D. M. & Knipple, D. C., manuscript in preparation). This primer was used along with the anchor primer to amplify putative ligand-gated chloride channel gene sequences from

1. Digest genomic DNA with restriction enzyme to produce 5' protruding ends:

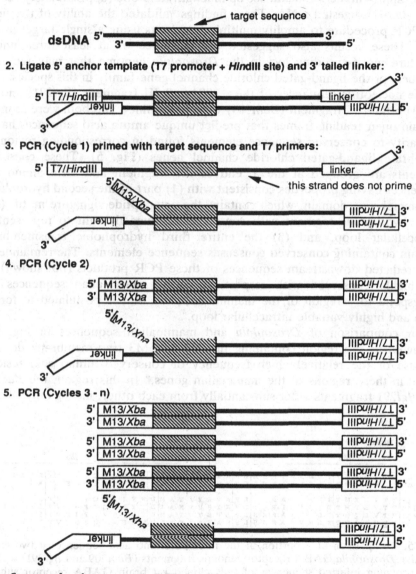

FIG. 4. Strategy for the geometric PCR amplification of genomic DNA sequences using a single chloride channel-specific primer based on the procedure of Roux and Dhanarajan;[26] see text for explanation.

genomic DNA that had been digested with either *Bgl*II or *Bam*HI prior to adaptor ligation. Electrophoretic analysis of the amplification products obtained from these templates revealed two DNA fragments from *Bgl*II-digested DNA (approximately 130 and 300 bp in length) and one (approximately 320 bp) from *Bam*HI-digested DNA. These findings validated the ability of the single site PCR procedure to amplify multiple fragments using a single target primer pool. These results also suggested the existence of at least three unique signature motif-encoding loci in the *Drosophila* genome that may represent members of the ligand-gated chloride channel gene family in this species.

The nucleotide sequences of the ~300 bp *Bgl*II fragment (*Bgl*291) and the ~320 bp *Bam*HI fragment (*Bam*309) were determined and both were found to contain open reading frames that predict unique amino acid sequences having similarity to conserved sequence elements found in corresponding regions of vertebrate ligand-gated chloride channel genes (Fig. 5). These conserved elements are grouped at the 5′ end of the fragment and form amino acid sequences having properties consistent with (1) part of the second hydrophobic transmembrane domain, which contains the octapeptide signature motif, (2) a short hydrophilic segment with conserved residues thought to represent an extracellular loop, and (3) the entire third hydrophobic transmembrane domain containing conserved consensus sequence elements. The remainder of the predicted downstream sequences of these PCR products (not shown) are not highly conserved with respect to known mammalian sequences and correspond to a region of the mammalian gene that is postulated to form a large and highly variable intracellular loop.[4]

Our comparison of *Drosophila* and mammalian sequences in Fig. 5 is restricted to the region containing the M2 and M3 transmembrane domains because of the relatively high frequency of conserved amino acid residues found in these regions of the mammalian genes.[4] In this region, the *Bam*309 and *Bgl*291 fragments differ substantially from each other, having only 18 of 46

```
        target primer
        - - - - - - - ->
              M2                                                    M3
         ┌─────────────┐┌─────────────────┐      ┌───────────────┐  ┌───────┐
Bam309 : T T V L T M T T L M S S T N A A L P K I S Y V K S I D V Y L G T C F V M V F A S L L G K Q T
Bgl291 : T T V L T M T F L G L E A R T D L P K V S Y P T A L D F F V F L S F G F I F A T I L Q F A V

GABAα1: T T V L T M T T L S I S A R N S L P K V A Y A T A M D W F I A V C Y A F V F S A L I E F A T
    β1: T T V L T M T T I S T H L R E T L P K I P Y V K A I D I Y L M G C F V F V F L A L L E Y A F
    γ2: T T V L T M T T L S T I A R K S L P L V S Y V T A M D L V F S V C F I F V F S A L V E Y G T
     δ: T T V L T M T T L M V S A R S S L P R A S A I K A L D V Y F W I C Y V F V F A A L V E Y A F
                                                                            *        *
```

Fig. 5. Comparison of a portion of the inferred amino acid sequence of two cloned putative *Drosophila* GABA receptor genomic fragments (*Bam*309 and *Bgl*291) with the corresponding inferred sequences of four cloned rat brain GABA receptor subunits representing different receptor subtypes.[25] Boxes show residues common to all six sequences; boldface symbols indicate residues conserved between either *Bam*309 or *Bgl*291 and at least one of the four mammalian sequences; asterisks indicate conserved mammalian sequences not found in either *Bam*309 or *Bgl*291. The positions of the target sequence for PCR amplification and the extent of inferred transmembrane domains M2 and M3 are indicated above the sequences.

amino acids in common. In contrast, somewhat higher levels of identity (ranging from 24 to 30 identical residues out of 46) are found in all pairwise comparisons of the four GABA receptor subunits shown in Fig. 5 and the corresponding region of the glycine receptor subunit[24] (glycine receptor sequences not shown). Moreover, the amino acid sequences encoded by the *Bam*309 and *Bgl*291 fragments in this region are less similar to each other than either sequence is to the corresponding regions encoded by genes representing the four classes of mammalian GABA receptor subunits (20–25 identical residues out of 46) or of the strychnine-binding subunit of the glycine-gated chloride channel[24] (21 and 25 identical residues). Whereas the *Bam*309 and *Bgl*291 sequences clearly appear to represent members of the ligand-gated chloride channel gene family, it is not possible to derive homology relationships from the limited sequence information currently available that would specifically implicate either fragment as part of a gene coding for a specific class of GABA receptor subunit.

To gain further information on the identity and function of the genes represented by the discrete amplification products obtained by this procedure, we initiated studies to determine if any of these fragments mapped to regions of the *Drosophila* genome that encode potentially relevant functions identified previously on the basis of genetic analysis. Biotinylated probes were therefore prepared from isolated PCR products and hybridized *in situ* to *Drosophila* polytene chromosome squashes. The results of preliminary mapping experiments suggest that the *Bam*309 amplification product may be of particular interest in the context of insecticide resistance phenomena, because it consistently labeled a site on the left arm of chromosome 3 at region 66F (not shown), a region to which a cyclodiene resistance-conferring gene hypothesized to involve an altered GABA receptor has been mapped.[27] The *Bgl*291 probe consistently labeled region 75A, a region to which a clone homologous to a vertebrate glycine receptor cDNA has been mapped,[28] and also labeled other chromosomal loci with varying degrees of reproducibility (not shown). In addition to the latter cytogenetic mapping results, evidence for the existence of a family of sequences in *Drosophila* related to *Bgl*291 was obtained from Southern blot analysis of genomic DNA in which several unique restriction fragments were labeled strongly under moderately high stringency hybridization and wash conditions by radiolabeled *Bgl*291 probe (not shown). These results suggest that additional members of this family of putative ligand-gated chloride channel genes remain to be isolated.

In order to isolate additional coding sequences upstream and downstream from the sequences present on *Bgl*291 and *Bam*309, we used a PCR-based procedure to amplify sequences present in several stage-specific cDNA libraries made in λgt10. Specific primers have been made corresponding to unique sequences on opposite strands of the two amplification products. PCR using the downstream-directed primer and the λgt10 sequencing primer resulted in single fragments corresponding to the 3′ ends of the gene containing the *Bgl*291 and *Bam*309 sequences, whereas amplifications using the upstream primer and the λgt10 sequencing primer produced an array of products of different sizes corresponding to the 5' ends of partial cDNAs of variable lengths that were present in the libraries. Characterization of the cDNAs isolated by this method and the other genomic amplification products obtained in this investigation are currently in progress and will be reported in detail elsewhere.

CONCLUSIONS

We have shown that the PCR-based homology probing methodology permits isolation of invertebrate DNA sequences that are homologous to characterized vertebrate genes encoding voltage-sensitive sodium channels and GABA-gated chloride channels. In contrast to approaches based on low stringency hybridizations of heterologous probes, this method is highly specific, rapid and straightforward. Furthermore, because the products obtained by PCR represent amplified sequences of the template DNA, they can be used directly as conspecific hybridization probes under normal (i.e. high stringency) conditions for Northern and Southern analyses, for in-situ hybridization to polytene chromosomes, and for the isolation of genomic or cDNA clones for full structural and functional characterization.

The two applications of PCR described here also serve to illustrate different strategies that can be employed in the analysis of multigene families that code for insecticide target sites. First, the isolation of *para*-homologous sequences from the house fly and other arthropod species shows how appropriate selection of target sequences can be used to isolate a specific member of a gene family for which functional criteria have been firmly established. In contrast, the isolation of sequences homologous to vertebrate GABA receptor genes shows how the PCR methodology can be used as a point of entry for an entire gene family. The 'single site' PCR approach actually can facilitate the latter strategy, because the randomness of the positions of downstream restriction sites relative to target sequences provides a discriminating criterion for identification of products derived from discrete transcription units.

Our straightforward isolation of *para*-homologous sequences from a variety of arthropod species illustrates the power of PCR-based homology probing to provide points of entry for the isolation of insecticide target site genes in species of economic or medical importance or species in which resistance due to target site insensitivity is either known or suspected. This approach, together with the development of heterologous expression assays for cloned genes, opens the way for studies of insecticide–target site interactions in species in which conventional physiological or biochemical approaches are not possible. In principle, the more elaborate single-site PCR strategy that we have used to isolate GABA receptor-homologous sequences in *Drosophila* should also permit the straightforward isolation of these conserved sequence elements from virtually any species, but this remains to be tested in practice.

Finally, as of yet the genes we have isolated can be postulated to encode insecticide target sites only on the basis of structural criteria. Rigorous demonstration of their physiological significance is still required and can only be provided by genetic evidence or by the functional expression of cloned genes. In this regard, the existence of target site insensitivity mutants provides a potentially valuable resource for efforts to show that particular genes encode pharmacologically important proteins.

ACKNOWLEDGMENTS

This work was supported in part by grants from the US Department of Agriculture (89-37263-4425), the US Army Research Office/DOD University Research Instrumentation Program (DAAL03-87-G-0030), and the Cornell Biotechnology Program, which is sponsored by the New York State Science and Technology Foundation, a consortium of industries, the US Army Research Office, and the National Science Foundation, and also by a gift from Rhone-Poulenc Ag Company, Research Triangle Park, NC. We thank L. Payne and P. Herrick for their technical assistance in these studies, and we thank K. Nelson and K. Poole for insect rearing.

REFERENCES

1. Soderlund, D. M., Bloomquist, J. R., Wong, F., Payne, L. L. & Knipple, D. C., Molecular neurobiology: implications for insecticide action and resistance. *Pestic. Sci.*, **26** (1989) 359–74.
2. Soderlund, D. M. & Bloomquist, J. R., Molecular mechanisms of insecticide resistance. In *Pesticide Resistance in Arthropods*, ed. R. T. Roush & B. E. Tabashnik. Chapman and Hall, New York, 1990, pp. 58–96.
3. Catterall, W. A., Structure and function of voltage-sensitive ion channels, *Science*, **242** (1988) 50–61.
4. Olsen, R. W. & Tobin, A. J., Molecular biology of GABA$_A$ receptors, *FASEB J.*, **4** (1990) 1469–80.
5. Saiki, R. K., Gelfand, D. H., Stoffel, S., Scharf, S. J., Higuchi, R., Horn, G. T., Mullis, K. B. & Erlich, H. A., Primer-directed enzymatic amplification of DNA with a thermostable DNA polymerase. *Science*, **239** (1988) 487–91.
6. Gould, S. J., Subramani, S. & Scheffler, I. E., Use of the DNA polymerase chain reaction for homology probing: isolation of partial cDNA or genomic clones encoding the iron–sulfur protein of succinate dehydrogenase from several species. *Proc. Natl. Acad. Sci. USA*, **86** (1989) 1934–8.
7. Kamb, A., Weir, M., Rudy, B., Varmus, H. & Kenyon, C., Identification of genes from pattern formation, tyrosine kinase, and potassium channel families by DNA amplification. *Proc. Natl. Acad. Sci. USA*, **86** (1989) 4372–6.
8. Noda, M., Shimizu, S., Tanabe, T., Takai, T., Kayano, T., Ikeda, T., Takahashi,

H., Nakayama, H., Kanaoka, Y., Minamino, N., Kangawa, K., Matsuo, H., Raftery, M. A., Hirose, T., Inayama, S., Hayashida, H., Miyata, T. & Numa, S., Primary structure of *Electrophorus electricus* sodium channel deduced from cDNA sequence. *Nature*, 312 (1984) 121–7.

9. Noda, M., Ikeda, T., Kayano, T., Suzuki, H., Takeshima, H., Kurasaki, M., Takahashi, H. & Numa, S., Existence of distinct sodium channel messenger RNAs in rat brain. *Nature*, 320 (1986) 188–92.

10. Auld, V., Goldin, A. L., Krafte, D. S., Marshall, J., Dunn, J. M., Catterall, W. A., Lester, H. A., Davidson, N. & Dunn, R. J., A rat brain Na^+ channel α subunit with novel gating properties. *Neuron*, 1 (1988) 449–61.

11. Trimmer, J. S., Cooperman, S. S., Tomiko, S. A., Zhou, J., Crean, S. M., Boyle, M. B., Kallen, R. G., Sheng, Z., Barchi, R. L., Sigworth, F. J., Goodman, R. H., Agnew, W. S. & Mandel, G., Primary structure and functional expression of a mammalian skeletal muscle sodium channel. *Neuron*, 3 (1989) 33–49.

12. Rogart, R. B., Cribbs, L. L., Muglia, L. K., Kephart, D. D. & Kaiser, M. W., Molecular cloning of a putative tetrodotoxin-resistant rat heart Na^+ channel isoform. *Proc. Natl Acad. Sci. USA*, 86 (1989) 8170–74.

13. Tanabe, T., Takeshima, H., Mikami, A., Flockerzi, V., Takahashi, H., Kangawa, K., Kojima, M., Matsuo, H., Hirose, T. & Numa, S., Primary structure of the receptor for calcium channel blockers from skeletal muscle. *Nature*, 328 (1987) 313–18.

14. Salkoff, L., Butler, A., Wei, A., Scavarda, N., Giffen, K., Ifune, C., Goodman, R. & Mandel, G., Genomic organization and deduced amino acid sequence of a putative sodium channel gene in *Drosophila*. *Science*, 237 (1987) 744–9.

15. Okamoto, H., Sakai, K., Goto, S., Takasu-Ishikawa, E. & Hotta, Y., Isolation of *Drosophila* genomic clones homologous to the eel sodium channel gene. *Proc. Japan Acad. Ser. B*, 63 (1987) 284–8.

16. Ramaswami, M. & Tanouye, M. A., Two sodium-channel genes in *Drosophila*: implications for channel diversity. *Proc. Natl Acad. Sci. USA*, 86 (1989) 2079–82.

17. Loughney, K., Kreber, R. & Ganetzky, B., Molecular analysis of the *para* locus, a sodium channel gene in *Drosophila*. *Cell*, 58 (1989) 1143–54.

18. Stern, M., Kreber, R. & Ganetzky, B., Dosage effects of a *Drosophila* sodium channel gene on behavior and axonal excitability. *Genetics*, 124 (1990) 133–43.

19. Knipple, D. C., Payne, L. L. & Soderlund, D. M., PCR-generated sodium channel gene probe for the housefly. *Arch. Insect Biochem. Physiol.*, 16 (1991) 45–53.

20. Doyle, K. E. & Knipple, D. C., PCR-based phylogenetic walking: isolation of *para*-homologous sodium channel gene sequences from seven insect species and an arachnid. *Insect Biochem.*, 21 (1991) 689–96.

21. Yool, A. J. & Schwartz, T. L., Alteration of ionic selectivity of a K^+ channel by mutation of the H5 region. *Nature*, 349 (1991) 700–704.

22. Yellen, G., Jurman, M. E., Abramson, T. & MacKinnon, R., Mutations affecting internal TEA blockade identify the probable pore-forming region of a K^+ channel. *Science*, 251 (1991) 939–42.

23. Hartmann, H. A., Kirsch, G. E., Drewe, J. A., Taglialatela, M., Joho, R. H. & Brown, A. M., Exchange of conduction pathways between two related K^+ channels. *Science*, 251 (1991) 942–4.

24. Greeningloh, G., Rieniz, A., Schmitt, B., Methfessel, C., Zensen, M., Bey-reuther, K., Gundelfinger, E. D. & Betz, H., The strychnine-binding subunit of the glycerine receptor shows homology with nicotinic acetylcholine receptors. *Nature*, 328 (1987) 215–20.

25. Shivers, B. D., Killisch, I., Sprengel, R., Sontheimer, H., Kohler, M., Schofield, P. R. & Seeburg, P. H., Two novel $GABA_A$ receptor subunits exist in distinct neuronal subpopulations. *Neuron*, 3 (1989) 327–37.

26. Roux, K. H. & Dhanarajan, P., A strategy for single site PCR amplification of dsDNA: priming digested cloned or genomic DNA from an anchor-modified restriction site and a short internal sequence. *BioTechniques,* **8** (1990) 48–57.
27. ffrench-Constant, R. H. & Roush, R. T., Gene mapping and cross-resistance in cyclodiene insecticide-resistant *Drosophila melanogaster* (Mg.), *Genet. Res., Camb.,* **57** (1991) 17–21.
28. Betz, H., 13th meeting of the International Society of Neurosciences, Sydney, Australia, 15 July 1991 (abstract). *J. Neurochem.,* **57** (1991) S1.

20

Cloning of an Invertebrate GABA Receptor from the Cyclodiene Resistance Locus in *Drosophila*

R. H. FFRENCH-CONSTANT

Department of Entomology, 237 Russell Laboratories, University of Wisconsin, Madison, Wisconsin 53706, USA

&

R. T. ROUSH

Department of Entomology, Comstock Hall, Cornell, University, Ithaca, New York 14853, USA

INTRODUCTION

For reasons which are unclear the invertebrate GABA receptor has until recently remained uncloned via heterologous probing with vertebrate clones or via the polymerase chain reaction (PCR) using primers based on vertebrate consensus sequences. Due to the lack of success with these approaches and in view of the fact that cyclodienes have long been thought to interact with the GABA$_A$ receptor/chloride ionophore complex, we chose to clone the gene responsible for conferring cyclodiene resistance as a putative invertebrate GABA receptor locus in the genetic model *Drosophila melanogaster*. The present paper reviews the cloning of this gene and the identification of a GABA receptor as the gene product. Its implications in the understanding of the molecular basis of insecticide action and resistance are discussed.

GABA RECEPTOR BIOLOGY IN VERTEBRATES

The GABA$_A$ receptor is the major receptor for the inhibitory neurotransmitter γ-aminobutyric acid (GABA) and is a member of a gene superfamily of ligand-gated ion channels. Much of what is known about the molecular biology of GABA$_A$ receptors comes from their recent cloning in a number of vertebrates.[1] The GABA$_A$ receptor is a hetero-oligomeric protein composed of several distinct polypeptide types (α, β, γ and δ) which show 20–40% identity to each other. Further, each polypeptide type is associated with a family of genes (e.g. $\beta_{1,2,3}$) which have 60–80% amino acid sequence identity.

The first successful cloning of the GABA$_A$ receptor resulted from purification of proteolytic fragments from purified bovine brain receptor protein and partial sequencing, followed by oligonucleotide probing of a bovine brain cDNA library.[2] This resulted in the isolation of the first α and β subunits. Subsequently, further α and β subunits[3,4] and clones for the novel γ and δ subunits have been isolated with degenerate oligonucleotide probes.[5,6]

RATIONALE OF CLONING INVERTEBRATE GABA RECEPTOR VIA CYCLODIENE RESISTANT MUTANT

Both mammalian and insect GABA-activated chloride channels are sensitive to cyclodienes based on the results of radioligand binding studies.[7] The binding sites for [^{3}H]dihydropicrotoxinin (DHP) in cockroach and vertebrate CNS are also blocked by a variety of cyclodienes.[7] Thus, cyclodienes and picrotoxinin (PTX) appear to bind at the same site on the GABA$_A$ receptor.

These findings are confirmed by the cross-resistance to PTX observed in cyclodiene-resistant insects. Further, differences in the dissociation constant (K_d) and the saturation level (B_{max}) of the PTX binding site of cyclodiene susceptible and resistant strains of the German cockroach *Blatella germanica* have been detected.[8] These results indicate that the binding site in the insensitive strain has only one tenth the affinity of that in the susceptible, suggesting that the cyclodiene/PTX binding site on the GABA$_A$ receptors of cyclodiene resistant insects is altered. Therefore, we reasoned that cloning of the gene altered in cyclodiene resistant *Drosophila* mutants would allow us to clone the insect GABA$_A$ receptor itself, as well as advancing our understanding of target site insensitivity to these compounds.

MUTANT ISOLATION AND CHARACTERIZATION

A mutant strain of *D. melanogaster* showing high levels (4270-fold) of resistance to dieldrin, a cyclodiene, was isolated by screening field-collected populations.[9] The mutant shows a semi-dominant phenotype following contact exposure to dieldrin. A dose of 30 μg of dieldrin applied to the inside of a 20 ml glass vial will discriminate completely between homozygous insensitive (Rdl^R/Rdl^R hereafter R/R, showing 15% mortality) and heterozygous individuals (R/S showing 100% mortality). In contrast, the R/S and homozygous sensitive (S/S) flies can best be distinguished at 0·5 μg (Fig. 1).

Following repeated backcrossing to a susceptible strain, the mutant still showed cross resistance to a range of cyclodienes, lindane and the GABA antagonist PTX (approximately 200 000-fold). In contrast the backcrossed strain showed no resistance to insecticides in other chemical classes including DDT, propoxur and malathion.[10]

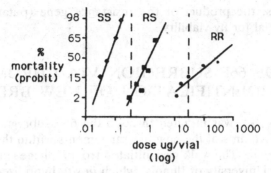

Fig. 1. Dose mortality regressions with dieldrin for homozygous susceptible *S/S* (*n* = 1705), homozygous insensitive (*R/R*) (*n* = 1832) and their F1 heterozygous progeny (*n* = 1372).

LOCALIZATION OF GENE ON POLYTENE CHROMOSOME MAP

One of the key advantages for the use of *Drosophila* as a tool in the cloning of insecticide resistance genes is the ability to localize any gene on the detailed polytene chromosome cytological map.[11] The cyclodiene resistance gene was localized on this map following recombinational and deficiency mapping.[10] Following approximate localization of the gene at map unit 25 on chromosome III by recombinational mapping,[9] a number of deficiencies from this region were screened to see if they uncovered the gene.

If a deficiency uncovers the resistance gene (i.e. in the absence of any susceptible gene product or sensitive receptors), full levels of resistance are displayed. Thus, *R/Df* flies for the deficiency *Df(3L)29A6* displayed full insensitivity (15% mortality at 30 µg dieldrin), whereas similar flies for the overlapping deficiency *Df(3L)AC1* were fully susceptible at this dose (100% mortality). The region of *Df(3L)29A6* not shared by *Df(3L)AC1* corresponds to one subregion, 66F, which must therefore contain the gene.

GENERATING NEW BREAKPOINTS UNCOVERING THE GENE

In order further to localize the gene within subregion 66F, new rearrangements uncovering the resistance gene were made with γ-irradiation. By irradiating males at 4000 rad and crossing them to *R/R* females, all the expected *R/S* progeny will die at a dose of 30 µg dieldrin, except for any flies heterozygous for resistance and a new rearrangement. One in approximately 5000 flies screened carried a new rearrangement uncovering the resistance gene. All new rearrangements obtained failed to complement each other, that is to say, when crossed between themselves no progeny carrying two rearranged chromosomes were viable. Thus, the product of the resistance gene (putatively a GABA receptor) is essential for fly viability.

CLONING OF 66F SUBREGION VIA CHROMOSOMAL WALK AND IDENTIFICATION OF NEW BREAKPOINTS

A chromosomal walk[12] was initiated through the 66F subregion, with the aim of finding the breakpoints of the new rearrangements within the subregion and thus locating the gene. The walk was initiated from a phage clone λ121 (a kind gift of G. Collier, University of Illinois) which *in situ* hybridized to 66F1,2, the beginning of the region. The walk was undertaken in a cosmid genomic library containing 35–45 kb inserts in a modified CosPer vector (a kind gift of J. Tamkun, University of California). This vector has flanking *Not1* restriction sites (a rare restriction enzyme in *Drosophila*) in the polylinker, which allow for easy insert isolation and end fragment identification, thus simplifying chromosomal walking. It also carries P-element repetitive ends and a wild-type copy of the white gene (w^+) conferring red eyes, for use as a marker in P-element mediated germline transformation (see below).

The positions of the new breakpoints were located by probing fragments from the walk to genomic Southern blots of wild-type and rearrangement strains. Deficiency breakpoints are apparent as a single novel band in the rearrangement strain; inversions or insertions are apparent as two novel bands. A single new deficiency breakpoint of *Df(3L)Rdl-2* was located within cosmid 2 of the walk. However, a cluster of three breakpoints, each showing two novel bands on a genomic Southern, were located in cosmid 6. Two of these rearrangements were cytologically visible. One, *In(3L)Rdl-11*, is a simple inversion with one break within 66F and the other at 63E-F. The second, *Rdl-20*, is a complex rearrangement with one breakpoint within 66F and at least three others outside the region. Since both of these rearrangements have only one break within 66F, these breakpoints must mark the location of the gene.

RESCUE OF THE SUSCEPTIBLE PHENOTYPE BY P-ELEMENT MEDIATED GERMLINE TRANSFORMATION

In order to test the hypothesis that cosmid 6 carried a complete copy of the susceptible gene, this cosmid was used in P-element mediated germline transformation. Since the cosmid library was made from a susceptible strain, insertion of a cloned copy of the S allele into flies heterozygous for resistance and *Df(3L)29A6* (*R/Df*) would be expected to rescue susceptibility to 30 μg dieldrin by effectively restoring an *R/S* genotype. These flies were generated via the cross scheme in Fig. 2. Following successful isolation of a w^+ (red eye colour) transformant bearing an insert on chromosome II, the insert was

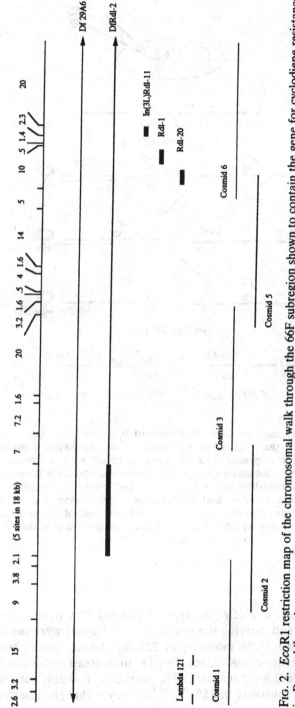

FIG. 2. *Eco*R1 restriction map of the chromosomal walk through the 66F subregion shown to contain the gene for cyclodiene resistance in *Drosophila melanogaster*. The extent of the overlapping cosmids in the walk are shown below the restriction map in kilobases. The solid boxes represent the locations of chromosomal breakpoints uncovering the resistant phenotype (see text) and sequences in the direction of the arrows extending from the boxes are deleted. Breakpoints without arrows correspond to inversions or insertions. The other breakpoints of these rearrangements lie outside the cloned region.

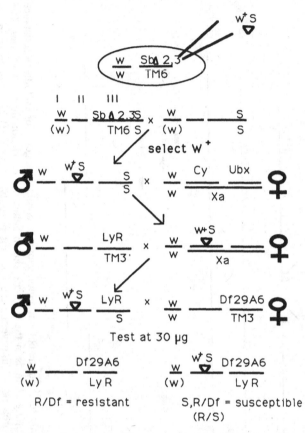

FIG. 3. Cross scheme used to prove that cosmid 6 contained a functional wild-type (susceptible) copy of the cyclodiene resistance gene. Embryos containing an endogenous source of transposase ($\Delta 2, 3$) were injected with a cosmid carrying a susceptible copy of the gene and a copy of the mini-white eye color gene as a selectable marker. Successful G_0 transformants with red eyes (w^+) were subsequently crossed into flies hemizygous for resistance and a deficiency uncovering the resistance gene $Df(3L)29A6$. These flies carrying the $w + S$ insert are susceptible to a dose of 30 μg dieldrin whereas their sibs without the insert are resistant to this dose (see text for data).

crossed into flies of the R/Df genotype. Resistant flies heterozygous for the deficiency (R/Df) and carrying the insert (w^+, red eyed) were susceptible to a dose of 30 μg dieldrin (93% mortality of 221 flies tested, corresponding to the expected R/S mortality of 100%, see Fig. 1). In contrast, their sibs lacking the insert (w, white eyed) were resistant (7% mortality, $n = 214$, corresponding to the expected R/Df mortality of 15%[10]). This proves that the inserted cosmid 6 must carry a functional copy of the susceptible gene.

ISOLATION AND IDENTIFICATION OF GABA cDNA

We used the 10 kb EcoR1 restriction fragment from cosmid 6, containing two of the three clustered rearrangement breakpoints, to screen a cDNA library made from RNA collected from 0–24-hour-old embryos (a kind gift of Nick Brown, University of Cambridge, UK). We isolated a number of cDNAs, one of which spanned all three of the breakpoints, as shown by probing blots of restriction digests of candidate cDNAs with gel purified restriction fragments from cosmid 6. Translation of partial sequence data from this cDNA corresponding to 150 amino acids shows it to have 64% amino acid similarity and 45% identity to the β_3 subunit of the rat GABA$_A$ receptor. The sequence we have determined so far corresponds to the extracellular domain of the protein and shows conservation of the cystine bridge found in all members of the gene superfamily.[1] We are proceeding to determine the full sequence of this cDNA.

CONCLUSIONS AND FUTURE DIRECTIONS

This study has shown that clustered chromosomal breakpoints of three rearrangements uncovering the resistance gene occupy exactly the same position as a locus coding for a GABA$_A$ receptor subunit, and that the susceptible phenotype of the gene can be rescued following germline transformation of DNA flanking these breakpoints. This is consistent with the hypothesis that cyclodiene resistance is due to an altered cyclodiene/PTX binding site on the GABA receptor.[8]

The cloning of cDNAs from this locus will allow for functional expression of mRNA derived from this and other subunit cDNAs in *Xenopus* oocytes and insect cell lines. This will facilitate not only advances in our understanding of invertebrate GABA receptor pharmacology but will also aid in the rational design of drugs and pesticides active at this site.[13] This is especially significant, as the insect GABA receptor continues to form an attractive target site for insecticides due to pronounced differences in its pharmacology with that of vertebrates.[14] Further, production of protein and thus antibodies to insect GABA receptors will enable a study of GABA receptor distribution in the insect nervous system, which although well studied in vertebrates, remains poorly understood in invertebrates.

Precise localization of the mutation conferring resistance and therefore presumably the identification of the cyclodiene/PTX binding site will be determined from analysis of resistant mutants. Finally, given that cyclodiene resistance is perhaps the most widespread example of insecticide resistance, occurring in more than 280 species of arthropods and vertebrates,[10,15,16] clones of this gene will also allow us to compare resistance-associated mutations in a wide range of different organisms.

REFERENCES

1. Olsen, R. W. & Tobin, A. J., Molecular biology of GABA$_A$ receptors. *Faseb Journal,* **4** (1990) 1469–80.
2. Schofield, P. R., Darlinson, M. G., Fujita, N., Burt, D. R., Stephenson, F. A., Rodriguez, H., Rhee, L. M., Ramachandran, J., Reale, V., Glencorse, T. A., Seeburg, P. H. & Barnard, E. A., Sequence and functional expression of the GABA-A receptor shows a ligand-gated receptor superfamily. *Nature (Lond.),* **328** (1987) 221–7.
3. Levitan, E. S., Schofield, P. R., Burt, D. R., Rhee, L. M., Wisden, W., Kohler, M., Fujita, N., Rodriguez, H. F., Stephenson, A., Darlison, M. G., Barnard, E. A. & Seeburg, P. H., Structural and functional basis for GABA$_A$ receptor heterogeneity. *Nature (Lond.),* **335** (1988) 76–9.
4. Ymer, S., Schofield, P. R., Draguhn, A., Werner, P., Kohler, M. & Seeburg, P. H., GABA$_A$ receptor β subunit heterogeneity: functional expression of cloned cDNAs. *EMBO J.,* **8** (1989) 1665–70.
5. Pritchett, D. B., Sontheimer, H., Shivers, B. D., Ymer, S., Kettenmann, H., Schofield, P. R. & Seeburg, P. H., Importance of a novel GABA$_A$ receptor subunit for benzodiazepine pharmacology. *Nature,* **338** (1989) 582–5.
6. Shivers, B. D., Killisch, I., Sprengel, R., Sontheimer, H., Kohler, M., Schofield, P. R. & Seeburg, P. H., Two novel GABA$_A$ receptor subunits exist in distinct neuronal subpopulations. *Neuron,* **3** (1989) 327–37.
7. Matsumura, F. & Ghiasuddin, S. M., Evidence for similarities between cyclodiene type insecticides and picrotoxinin in their action mechanisms. *J. Environ. Sci. Health,* **B18** (1983) 1–14.
8. Matsumura, F., Tanaka, K. & Ozoe, Y., GABA-related systems as targets for insecticides. In *Sites of Action of Neurotox Pesticides,* ed. R. M. Hollingworth & M. B. Green. ACS Symposium Series 356, American Chemical Society, Washington, DC, 1987, pp. 44–70.
9. ffrench-Constant, R. H., Roush, R. T., Mortlock, D. & Dively, G. P., Isolation of dieldrin resistance from field populations of *Drosophila melanogaster* (Diptera: Drosophilidae). *J. Econ. Ent.,* **83** (1990) 1733–7.
10. ffrench-Constant, R. H. & Roush, R. T., Gene mapping and cross-resistance in cyclodiene insecticide-resistant *Drosophila melanogaster* (Mg). *Gen. Res.,* **57** (1991) 17–21.
11. ffrench-Constant, R. H. & Roush, R. T., *Drosophila* as a tool for investigating the molecular genetics of insecticide resistance. In *Molecular Approaches to Pure and Applied Entomology,* ed. M. J. Whitten & J. G. Oakeshott. Springer Verlag, Berlin, 1991 (In press).
12. Bender, W., Spierer, P. & Hogness, D. S., Chromosomal walking and jumping to isolate DNA from the *Ace* and *rosy* loci and the bithorax complex in *Drosophila melanogaster. J. Mol. Biol.,* **168** (1983) 17–33.
13. Eldefrawi, A. T. & Eldefrawi, M. E., Receptors for γ-aminobutyric acid and voltage dependent chloride channels as targets for drugs and toxicants. *FASEB J.,* **1** (1987) 262–71.
14. Rauh, J. J., Lummis, S. C. R. & Sattelle D. B., Pharmacological and biochemical properties of insect GABA receptors. *Trends in Pharmacological Sciences,* **11** (1990) 325–9.
15. Georghiou, G. P., The magnitude of the resistance problem. In *Pesticide Resistance: Strategies and Tactics for Management.* National Academy of Sciences, National Academy Press, Washington, DC, 1986, pp. 14–43.
16. Yarbrough, J. D., Roush, R. T., Bonner, J. C. & Wise, D. A., Monogenic inheritance of cyclodiene resistance in the mosquito fish, *Gambusia affinis. Experientia,* **42** (1986) 851–3.

SECTION 6

Neuronal Networks and Behaviour

21

Integration in Neuronal Networks that Control Movement

MALCOLM BURROWS

Department of Zoology, University of Cambridge, Downing Street, Cambridge CB2 3EJ, UK

INTRODUCTION

The objective of this paper is to show how an analysis of the neural mechanisms generating adaptive locomotory behaviour of an insect illuminates the integrative mechanisms of the central nervous system. In particular it seeks to elucidate the essential properties of the neurones involved in these integrative processes and the properties of the circuits that they form. An objective of this meeting is to find ways to control insect pests, and for this a reasonable starting point would be to analyse and understand some of the normal neuronal mechanisms that lead to behaviour. In the course of such an analysis it should then be possible to reveal mechanisms at all levels of organisation which could be disrupted. Studies of insect nervous systems can thus have two interwoven objectives; first, to exploit the relative simplicity of their organisation and use them as models for the elucidation of mechanisms that have relevance in more complex nervous systems; second, to identify features which are specific to them and which would therefore provide targets for control.

The approach that has been adopted here towards understanding the vast topic of integration has been to reduce it to more manageable proportions by concentrating on specific questions; how are sensory signals from mechanoreceptors on a leg processed to produce an appropriate adjustment of posture or locomotion? How are the constituent neurones connected to perform these tasks and what are the integrative properties of the essential local interneurones? Can the networks that are involved be understood, and how are they designed for this processing? The analysis has been carried out on the hind leg and the thoracic nervous system of a locust *Schistocerca gregaria* (Forskål). Intracellular recordings are made from neurones in the segmental ganglia whilst the animal is alert so that the action of a particular neurone can be related directly to the behaviour of the animal. Individual neurones are identified according to their physiological actions and properties, their morphology at the level of the light microscope, and the distribution of their synapses as seen in the electron microscope.

RESULTS

The behaviour

Many thousands of sensory axons from receptors on one hind leg converge onto the segmental, metathoracic ganglion. Some of these afferents are from mechanoreceptors that provide information about touch and about movement of the joints, but many are from chemoreceptors associated with small basiconic sensilla. This paper will consider only the input provided by the singly innervated, trichoid sensilla (hairs) that respond to mechanical stimuli. By contrast, only some 100 motor neurones are responsible for generating the complex patterns of muscular contractions that are required of this leg in locomotion and in adjusting posture. The disparity in numbers between sensory and motor neurones indicates that considerable convergence and integration must occur in the central nervous system. At the same time, however, spatial information provided by the mechanoreceptors must be preserved because specific local reflexes result from the stimulation of restricted arrays of receptors.[1] For example, touching hairs on the ventral tibia causes the trochanter to levate and move the femur forwards, the tibia to extend and the tarsus to depress, whereas touching hairs a few millimetres away on the dorsal tibia causes a different sequence of movements of these joints. These are adaptive compensatory reflexes ensuring that the leg moves to avoid an object. These movements define the integrative problems that must be solved in the central nervous system. How is the signal from a particular hair transformed into a specific movement of the leg and how is this effect integrated with the behaviour in different circumstances?

The neurones

The metathoracic ganglion contains all the motor neurones that move the hind legs. Many of these motor neurones have large cell bodies and stout primary neurites, features that have enabled many to be identified individually. Each muscle is innervated by only a few motor neurones; for example, the large extensor tibiae muscle that generates the main power in jumping is innervated by two excitatory and one inhibitory motor neurone,[2] while the levator tarsi muscle is innervated by just one motor neurone though occasionally a supernumerary neurone may occur.[3] The physiology of these neurones is known in some detail, but information about their pharmacology is limited: we know something of extrajunctional receptors of insect motor neurones[4] but little of their synaptic receptors. This is despite the fact that the putative transmitters of some have been identified by immunocytochemistry,[5,6] and that output synapses of motor neurones have been identified in the central nervous system.[7,8]

The ganglion also contains some 2000 interneurones, of which a few hundred are involved with the control of each leg. A large proportion of these are local interneurones with branches restricted entirely to this ganglion. They either lack an axon completely or have just a short one that links two regions of branches. These local interneurones can be divided into two classes according to their physiological properties; the first normally generate action potentials in response to synaptic inputs or to an experimentally applied depolarisation (*spiking local interneurones*), while the second normally do not spike (*non-spiking local interneurones*).[9,10] The remaining components are intersegmental interneurones with axons that project to adjacent ganglia in the segmental chain, and neurosecretory neurones with axons to the periphery, to adjacent ganglia or with no axon at all. Axons from interneurones that receive sensory stimuli from other legs and other parts of the body also project to this ganglion. These therefore are the neurones that provide the basic framework from which the local networks must be organised.

Integrating the sensory signals

The tactile hairs are usually only stimulated when a leg bumps into an external object but some may also be stimulated when a joint closes completely. Spiking local interneurones with cell bodies in a ventral midline[11] or antero-medial[12] group receive direct excitatory inputs from these afferents.[13,14] The transmitter used by these afferents is unknown but in moths and cockroaches the evidence suggests that hair afferents may use acetylcholine.[15-17] The spiking local interneurones also receive inputs from the other classes of mechanoreceptors; chordotonal organs,[18] strand receptors,[19] campaniform sensilla,[20] and from the mechanoreceptors associated with basiconic sensilla (Burrows, unpublished).

The gain of the first synapse between the hair afferents and the spiking local interneurones can be high so that a single afferent spike can evoke a spike in an interneurone. Each interneurone receives direct inputs from a particular array of receptors that comprise its receptive field. Not all the receptors in a field contribute equally; the afferents from some make high gain synapses while others make synapses of lower gain. These interneurones form a map of the surface of the leg, by virtue of the arrays of direct connections formed by particular afferents.[21] There are two important features of these receptive fields; first, one region of the leg is represented by several interneurones so that there is parallel processing; second, the connections ensure that the spatial information provided by the receptors is preserved in separate channels.

The afferents do not, however, connect only with these interneurones but instead make divergent connections with non-spiking interneurones,[22,23] motor neurones,[17,24] and with intersegmental interneurones so that information is conveyed to other ganglia.[23,25] The afferent signals are therefore distributed widely in the nervous system to several classes of neurone so that their information is processed in a parallel and distributed fashion.

Output connections of the local interneurones

The next stage in the processing depends on the output connections of the local interneurones. Members of the midline population of spiking local interneurones make a series of divergent and inhibitory output connections,[26] whereas members of a second, antero-medial population make excitatory connections.[14] All the effects revealed so far are mediated by spikes and there is no evidence for graded synaptic transmission. This means that each spike in one of these interneurones is followed by either an IPSP or an EPSP in the postsynaptic neurone. Members of the midline group make a few inhibitory connections with motor neurones,[26] but more widespread connections with non-spiking interneurones,[27] with intersegmental interneurones[28] and with antero-medial spiking interneurones.[14] Many of the members of this group of spiking local interneurones stain with an antibody raised against GABA and their inhibitory actions on motor neurones can be blocked by picrotoxin.[29] Their inhibitory actions may thus be mediated by GABA.

Non-spiking interneurones make either excitatory or inhibitory output connections, but they do so without the intervention of spikes. Their effects on postsynaptic neurones are mediated by the graded release of an unknown chemical transmitter(s).[10] They exert marked effects on the patterns of spikes in motor neurones which they organise into sets that are appropriate for the performance of normal movements of a leg.[30] They also make inhibitory connections with other non-spiking interneurones in lateral inhibitory networks.[31] It is the complex web of interconnections of the local interneurones that seems vital for integrative actions and which can be analysed on a neurone by neurone basis in these networks.

Integrative properties of the local interneurones

The information obtained in these detailed analyses of the patterns of connections made by the local interneurones, from studies of their physiological and morphological properties, and from the ultrastructural properties and distribution of their synapses, allows the following conclusions to be drawn.

Midline spiking local interneurones

1. Collate afferent information from arrays of receptors whilst preserving spatial information.
2. Reverse the sign of the sensory signals from excitatory to inhibitory.
3. Limit the receptive fields of non-spiking and intersegmental interneurones by lateral inhibition.

4. Exclude the action of non-spiking interneurones whose motor effects would be inappropriate, and disinhibit those whose action would be appropriate.

Non-spiking interneurones

1. Integrate sensory signals with signals from spiking local and intersegmental interneurones.
2. Release chemical transmitter in a graded manner when activated by small voltage changes generated by individual synaptic potentials.
3. Inhibit local interneurones with inappropriate actions for a particular movement.
4. Excite, inhibit or disinhibit motor neurones by the graded release of transmitter.
5. Control groups of motor neurones in sets appropriate for normal locomotion.

Shaping the motor response

The action called forth by a particular sensory input must be placed in the context of the behaviour of the insect. This implies that the same sensory signal will not always elicit the same motor response. In the pathways described so far what are the key elements that are responsible for adjusting the motor output so that it is appropriate to the prevailing behavioural circumstances? Non-spiking interneurones, by virtue of their inputs from sensory neurones, their interactions with local interneurones, and their connections with pools of motor neurones, appear crucial for the implementation of a local reflex. Manipulating the membrane potential of one of these interneurones alters the effectiveness of a local reflex in a graded manner.[32] Despite the apparent complexity of the pathways and the parallel and distributed processing, a single interneurone therefore plays a substantial role in a reflex pathway. An extrapolation from this observation suggests that a synaptic input to a non-spiking interneurone might be able to change a reflex, and that this might be the way that intersegmental inputs place a local response in the appropriate behavioural context. Some of the intersegmental interneurones responsible for conveying information about events occurring at one leg to the local neuronal circuits controlling the movements of an adjacent leg have recently been identified.[33] These interneurones have cell bodies in the mesothoracic ganglion, receive inputs from mechanoreceptors on a middle leg and have axons that project to the metathoracic ganglion.[33] The receptive fields of these interneurones are shaped by the same series of connections as for the local interneurones: direct excitation from afferents and inhibition by spiking local interneurones.[28] In the ganglion controlling the hind legs these intersegmental interneurones connect with non-spiking interneurones and with some motor neurones.[34] The connections are specific and are related to the receptive field of the intersegmental interneurone and to the output connections of the non-spiking interneurone. The connections are either excitatory or inhibitory depending on the identity of the individual interneurone. Some of these interneurones stain with an antibody raised against GABA.[35] The effect of these connections is to alter the output of a non-spiking interneurone and thus its participation in a local reflex.[32] The final result will be a change in the pattern of spikes in motor neurones that are postsynaptic to this interneurone.

Organisation of the local reflex networks

The analysis that has been described means that the local networks which control the movements of a hind leg can be defined in a way that allows a sensory signal to be followed through its various integrative stages to its emergence as an adaptive motor response. The following characteristics of the networks can be recognised:

1. There is a large convergence from mechanosensory afferents to the local interneurones in the central nervous system in which spatial information is preserved.
2. The afferent signals diverge to make excitatory connections with different classes of neurones. Therefore *parallel distributed processing* of the same signals by neurones with different integrative properties occurs.
3. The afferent signals also diverge to make connections with several interneurones in the same class, resulting in overlapping receptive fields. Thus the position of an afferent on the leg is represented not by one specific central neurone, but by all the interneurones whose receptive fields overlap.
4. An afferent does not connect with all the interneurones in the network, and similarly a single interneurone does not connect with all the output elements. The connections are specific and can be understood in functional terms by their role in organising specific movements.
5. Lateral inhibitory interactions predominate between the local interneurones that are identified as elements of the networks. Disinhibition also contributes to the excitation of the motor neurones.
6. The flow of information is in one direction through the network, in which each interneurone has its own well defined place. Some expected feedback connections between certain elements have not been found. For example, no non-spiking neurones presynaptic to intersegmental interneurones or to spiking local interneurones have yet been found.
7. The complex receptive fields of the interneurones are understood only if their output connections are known by reference to the behaviour.

CONCLUSIONS

The methods of analysing selected neuronal networks in an insect that have been described here offer a precise description of the components and of the way that they interact to produce normal behaviour. An understanding of the operation of these networks necessarily requires knowledge of events at many different levels, from the molecular to the behavioural. Such information is being provided by analyses of various insect networks. First, detailed analyses of visual pathways has led to histamine being identified as the probable transmitter used by retinula cells of the compound eyes[36] and ocelli,[37] a discovery which has led to the characterisation of synaptic receptors for histamine in postsynaptic monopolar cells of the optic neuropil.[38] This is a particularly encouraging development as much of our knowledge of transmitter receptors is based on extrasynaptic ones. Second, the channels which shape the functioning of the component neurones are only just beginning to be examined. Laurent[39] has shown that a voltage sensitive, outward rectifying current in non-spiking interneurones from the circuits described above has dramatic effects on the integration of synaptic potentials by these neurones. Third, the treatment of neurones as individuals reveals not only that many operate without the use of action potentials but also illustrates that their properties must be described if the networks are to be understood. Finally, the method that relates the analysis directly to the behaviour ensures that physiologically relevant properties are always examined. The behavioural observations guide the physiological search and in turn the physiology prompts more behavioural observations. For example, the precision of the receptive fields of the local interneurones led directly to the behavioural descriptions of the local leg reflexes, and conversely the behaviour provided an explanation of the complex arrangements of the receptive fields of these interneurones.

Where information is lacking for these networks is on the chemicals used by many of the interneurones for communication. Moreover, we know little of the role of the neurosecretory neurones that are known to have such profound effects on the performance of the peripheral neural and muscular machinery (for contrast see Chapter 22 in this volume by Harris-Warrick). The effects of some putative neuromodulators when injected into the central nervous system suggest that these neurones could themselves exert profound effects.[40]

The next step in using these networks as models for understanding more complex nervous systems must be to analyse how they operate during the performance of a voluntary movement and to identify the elements and the processes that are responsible for this level of integration.

The next step in using these networks as targets for possible pest related control must be to exploit the profusion of identified neurones, identified synaptic pathways and neurones operating in unconventional ways that have been revealed. A good starting point would be to exploit the connections that are made between the afferents and the interneurones, or between the motor neurones themselves. Moreover, the role of non-spiking interneurones in adjusting the motor output and acting as the summing points for both intra- and intersegmental effects suggests they would be good targets for attempting to modify the behaviour of an insect.

ACKNOWLEDGEMENTS

The experimental work described in this paper is supported by NIH grant NS16058, by a grant from the SERC (UK) and by the Japanese Human Frontier Science Program. Some of the experiments have been undertaken jointly with Drs M. V. S. Siegler, G. J. Laurent and A. H. D. Watson.

REFERENCES

1. Siegler, M. V. S. & Burrows, M., Receptive fields of motor neurones underlying local tactile reflexes in the locust. *J. Neurosci.*, **6** (1986) 507–13.
2. Hoyle, G. & Burrows, M., Neural mechanisms underlying behavior in the locust *Schistocerca gregaria*. I Physiology of identified motorneurons in the metathoracic ganglion. *J. Neurobiol.*, **4** (1973) 3–41.
3. Siegler, M. V. S., Electrical coupling between supernumerary motor neurones in the locust. *J. exp. Biol.*, **101** (1982) 105–19.
4. Davis, J. P. L. & Pitman, R. M., Characterization of receptors mediating the actions of dopamine on an identified inhibitory motoneurone of the cockroach. *J. exp. Biol.*, **155** (1990) 203–17.
5. Bicker, G., Schafer, S., Ottersen, O. P. & Storm-Mathisen, J., Glutamate-like immunoreactivity in identified neuronal populations of insect nervous systems. *J. Neurosci.*, **8** (1988) 2108–22.
6. Watson, A. H. D., The distribution of GABA-like immunoreactivity in the thoracic nervous system of the locust *Schistocerca gregaria*. *Cell Tissue Res.*, **246** (1980) 331–41.
7. Watson, A. H. D. & Burrows, M., Input and output synapses on identified motor neurones of a locust revealed by the intracellular injection of Horseradish Peroxidase. *Cell Tissue Res.*, **215** (1981) 325–32.
8. Burrows, M., Watson, A. H. D. & Brunn, D. E., Physiological and ultrastructural characterization of a central synaptic connection between identified motor neurones in the locust. *European J. Neurosci.*, **1** (1989) 111–26.
9. Pearson, K. G. & Fourtner, C. R., Nonspiking interneurones in walking system of the cockroach. *J. Neurophysiol.*, **38** (1975) 33–52.
10. Burrows, M. & Siegler, M. V. S., Graded synaptic transmission between local interneurones and motoneurones in the metathoracic ganglion of the locust. *J. Physiol.*, **285** (1978) 231–55.
11. Siegler, M. V. S. & Burrows, M., The morphology of two groups of spiking local interneurones in the metathoracic ganglion of the locust. *J. Comp. Neurol.*, **224** (1984) 463–82.
12. Nagayama, T., Morphology of a new population of spiking local interneurones in the locust metathoracic ganglion. *J. Comp. Neurol.*, **283** (1989) 189–211.
13. Siegler, M. V. S. & Burrows, M., Spiking local interneurones as primary integrators of mechanosensory information in the locust. *J. Neurophysiol.*, **50** (1983) 1281–95.
14. Nagayama, T. & Burrows, M., Input and output connections of an antero-medial group of spiking local interneurones in the metathoracic ganglion of the locust. *J. Neurosci.*, **10** (1990) 785–94.
15. Carr, C. E. & Fourtner, C. R., Pharmacological analysis of a monosynaptic reflex in the cockroach, *Periplaneta americana*. *J. exp. Biol.*, **86** (1980) 259–73.

16. Sattelle, D. B., Harrow, I. D., Hue, B., Pelhate, M., Gepner, J. I. & Hall, L. M., α-Bungarotoxin blocks excitatory synaptic transmission between cercal sensory neurones and giant interneurone 2 of the cockroach *Periplaneta americana. J. exp. Biol.*, **107** (1983) 473–89.
17. Weeks, J. C. & Jacobs, G. A., A reflex behavior mediated by monosynaptic connections between hair afferents and motoneurons in the larval tobacco hornworm. *J. Comp. Physiol. A*, **160** (1987) 315–29.
18. Burrows, M., Parallel processing of proprioceptive signals by spiking local interneurons and motor neurons in the locust. *J. Neurosci.*, **7** (1987) 1064–80.
19. Pflüger, H. J. & Burrows, M., A strand receptor with a central cell body synapses upon spiking local interneurones in the locust. *J. Comp. Physiol. A*, **160** (1987) 295–304.
20. Burrows, M. & Pflüger, H. J., Positive feedback loops from proprioceptors involved in leg movements of the locust. *J. Comp. Physiol. A*, **163** (1988) 425–40.
21. Burrows, M. & Siegler, M. V. S., The organization of receptive fields of spiking local interneurones in the locust with inputs from hair afferents. *J. Neurophysiol.*, **53** (1985) 1147–57.
22. Burrows, M., Laurent, G. J. & Field, L. H., Proprioceptive inputs to nonspiking local interneurons contribute to local reflexes of a locust hindleg. *J. Neurosci.*, **8** (1988) 3085–93.
23. Laurent, G. J. & Burrows, M., Direct excitation of nonspiking local interneurons by exteroceptors underlies tactile reflexes in the locust. *J. Comp. Physiol. A*, **162** (1988) 563–72.
24. Laurent, G. J. & Hustert, R., Motor neuronal receptive fields delimit patterns of activity during locomotion of the locust. *J. Neurosci.*, **8** (1988) 4349–66.
25. Laurent, G., Local circuits underlying excitation and inhibition of intersegmental interneurones in the locust. *J. Comp. Physiol. A*, **162** (1988) 145–57.
26. Burrows, M. & Siegler, M. V. S., Spiking local interneurons mediate local reflexes. *Science (N.Y.)*, **217** (1982) 650–52.
27. Burrows, M., Inhibitory interactions between spiking and nonspiking local interneurons in the locust. *J. Neurosci.*, **7** (1987) 3282–92.
28. Laurent, G., The role of spiking local interneurones in shaping the receptive fields of intersegmental interneurones in the locust. *J. Neurosci.*, **7** (1987) 2977–89.
29. Watson, A. H. D. & Burrows, M., Immunocytochemical and pharmacological evidence for GABAergic spiking local interneurones in the locust. *J. Neurosci.*, **7** (1987) 1741–51.
30. Burrows, M., The control of sets of motoneurones by local interneurones in the locust. *J. Physiol.*, **298** (1980) 213–33.
31. Burrows, M., Graded synaptic transmission between local pre-motor interneurons of the locust. *J. Neurophysiol.*, **42** (1979) 1108–23.
32. Laurent, G. & Burrows, M., Intersegmental interneurons can control the gain of reflexes in adjacent segments by their action on nonspiking local interneurons. *J. Neurosci.*, **9** (1989) 3030–39.
33. Laurent, G., The morphology of a population of thoracic intersegmental interneurones in the locust. *J. Comp. Neurol.*, **256** (1987) 412–29.
34. Laurent, G. & Burrows, M., Distribution of intersegmental inputs to nonspiking local interneurons and motor neurons in the locust. *J. Neurosci.*, **9** (1989) 3019–29.
35. Watson, A. H. D. & Laurent, G., GABA-like immunoreactivity in a population of locust intersegmental interneurones and their inputs. *J. Comp. Neurol.*, **302** (1990) 761–7.
36. Hardie, R. C., Is histamine a neurotransmitter in insect photoreceptors? *J. Comp. Physiol. A*, **161** (1987) 201–13.

37. Simmons, P. J. & Hardie, R. C., Evidence that histamine is a neurotransmitter of photoreceptors in the locust ocellus. *J. exp. Biol.*, **138** (1988) 205–19.
38. Hardie, R. C., Effects of antagonists on putative histamine receptors in the first visual neuropile of the housefly (*Musca domestica*). *J. exp. Biol.*, **138** (1988) 221–41.
39. Laurent, G., Voltage-dependent nonlinearities in the membrane of locust nonspiking local interneurons, and their significance for synaptic integration. *J. Neurosci.*, **10** (1990) 2268–80.
40. Stevenson, P. A. & Kutsch, W., Demonstration of functional connectivity of the flight motor system in all stages of the locust. *J. Comp. Physiol. A*, **162** (1988) 247–59.

22

Neuromodulation of Small Neural Networks in Crustacea

Ronald M. Harris-Warrick, Robert E. Flamm, Bruce R. Johnson, Paul S. Katz, Ole Kiehn & Bing Zhang

Section of Neurobiology and Behavior, Cornell University, Seeley G. Mudd Hall, Ithaca, New York 14850, USA

INTRODUCTION

One of the important goals of motor systems research is to elucidate the neural networks that underlie simple rhythmic movements.[1] In some cases, the component neurons and synaptic connectivity of such networks (sometimes called Central Pattern Generators, or CPGs) have been partially or even fully elucidated.[2] Despite a relatively fixed neuroanatomical structure, the motor patterns that these networks generate are extremely flexible and plastic, allowing the animal to adapt its behavior to changing demands of the environment. This flexibility arises from sensory feedback and from modulatory inputs to the motor networks from other neural centers. Modulatory inputs can change the intrinsic electrophysiological properties of the neurons in the network and alter the strength of the synaptic connections within the network.[3] This causes a reconfiguration of the network's motor output, leading to an altered behavior.

The crustacean stomatogastric ganglion (STG) is an excellent model system for studying neuromodulation of network output.[4] In this paper, we describe experiments showing how application of identified transmitters or stimulation of identified neuromodulatory neurons can alter the properties of a small network in the STG, leading to the production of a number of variants on a basic rhythmic motor pattern. The network can thus be thought of as a library of potential components, each with variable properties which can be selected by modulatory input to build a functional circuit for each variant motor pattern.

THE STOMATOGASTRIC GANGLION

The STG of decapod crustaceans is a small ganglion of about 30 neurons that controls rhythmic movements of the foregut.[4] Two motor networks are localized within the ganglion: one controls a slow (0·05–0·1 Hz) movement of the gastric mill teeth, while the other controls a rapid (1–2 Hz) rhythmic contraction of the pylorus, a filter for food particles. In the spiny lobster, *Panulirus interruptus,* the pyloric network operates independently of the other network under cetain conditions, and can thus be studied in isolation.

The STG, its motor nerves, and several anterior ganglia sending input to the STG can be dissected free from the foregut, and the activity of the pyloric network neurons monitored with a combination of extracellular recordings

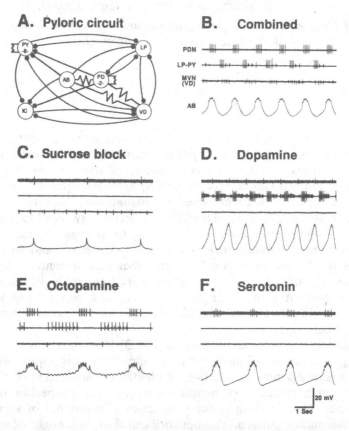

FIG. 1. Multiple motor patterns generated by amines in the pyloric circuit. (A) Summary of the pyloric circuit. The synaptic connections are either electrical (resistor symbols) or chemical inhibitory (filled circles). Some weak connections are not included (see Fig. 4). (B) Typical motor pattern from combined preparation, with intact modulatory input from the commissural and esophageal ganglia. Extracellular recordings from nerves carrying axons of the PD, LP, PY and VD motoneurons are shown, along with intracellular recordings from the AB interneuron. (C) Loss of activity upon blockade of modulatory inputs from other ganglia by placing a pool of isotonic sucrose on the input nerve. (D), (E), (F) Effects of bath-applied dopamine (10^{-4} M), octopamine (10^{-4} M) and serotonin (10^{-5} M) on the pyloric rhythm in the isolated STG. Reprinted from Ref. 26.

from the appropriate nerves and intracellular recordings from the cell bodies in the STG. The pyloric network generates a strong rhythmic motor pattern when the STG is receiving modulatory inputs from the anterior ganglia: intracellular recordings show that each cell undergoes rhythmic oscillations in membrane potential, with bursts of action potentials at the peak of the oscillations [Fig. 1(B)]. The cells fire in a stereotyped order with characteristic phasing of the activity of each cell within the cycle.

The neural network underlying this motor pattern has been completely elucidated in *Panulirus*[4] [Figs 1(A), 4(A)]. It contains 14 neurons in 6 major classes. Thirteen of the component neurons are also motoneurons sending the signals that evoke muscle contraction; the final cell is an interneuron (AB) and is the major pacemaker in the network. The synaptic interconnectivity is well understood, and is composed of electrotonic connections and chemical inhibitory synapses using either acetylcholine or glutamate as transmitters.

The pyloric motor rhythm is active only when modulatory inputs from other ganglia are functional.[5] These inputs can be reversibly eliminated by placing a pool of isotonic sucrose on the sole input nerve to block action potential conduction from the other ganglia. Under these conditions, the pyloric rhythm stops, and cells are either silent or fire tonically at low frequencies [Fig. 1(C)]. Bal *et al.*[6] showed that this arises from the loss of intrinsic bursting or plateau properties in all the pyloric cells: each cell is a conditional oscillator which fires in rhythmic bursts as long as modulatory input is intact to elicit these properties.

MODULATION OF THE PYLORIC RHYTHM BY BATH-APPLIED MONOAMINES

We are interested in seeing how different modulatory inputs can alter the properties of the pyloric network to produce multiple motor programs. Our first approach was to remove all modulatory input by blocking action potential conduction in the input nerve, then to add back the modulatory transmitters one by one, in order to see what each modulator does in isolation. Figure 1 shows the pyloric motor patterns evoked in *Panulirus* upon bath application of the monoamines dopamine (DA), octopamine (Oct) and serotonin (5HT), which are endogenous transmitters or neurohormones in the lobster.[7] Each amine reactivates a unique pyloric motor pattern. The most obvious difference between these patterns is that different subgroups of neurons are active; the remaining neurons, though anatomically part of the circuit, are not functionally active. Thus, the neuronal components of the functional circuit vary with different modulatory agents. Other major differences in the pattern include the cycle frequency, which is fastest with DA and slowest with Oct; the intensity of firing of the cells, which is highest with DA and lower with Oct or 5HT; and the phasing and duty cycle of activity of the cells.

One major conclusion is apparent from these studies: a single anatomically defined neural network can generate many different motor patterns by the selective actions of neuromodulators: the anatomy of the circuit alone does not determine the resulting motor pattern.

CELLULAR MECHANISMS OF MODULATION OF NEURAL NETWORK OUTPUT

We were naturally interested in determining the cellular mechanisms by which each of these amines produces a unique motor pattern. Our research has so far uncovered two mechanisms.

Neuromodulators alter the intrinsic electrophysiological properties of the component neurons

To demonstrate the actions of each amine directly on each cell, we studied their actions on *Panulirus* pyloric neurons that had been totally isolated from all detectable synaptic input.[8] This was done with three steps: (1) remove inputs from other ganglia with a sucrose block on the sole input nerve; (2) eliminate electrically coupled or cholinergic pre-synaptic neurons by photoinactivation;[9] (3) block the remaining glutamatergic synapses with the antagonist picrotoxin.[10]

After synaptic isolation, pyloric cells are either silent or fire tonically at low frequency. Bath application of DA, 5HT or Oct has characteristic and reproducible effects on the excitability of nearly all the pyloric neurons[8] (Fig. 2). A number of conclusions can be drawn from these results. First, each amine has a unique constellation of effects on the neurons of the pyloric circuit. This underlies the unique variants of the pyloric motor pattern generated by the circuit. Second, nearly every neuron in the circuit is directly modulated by each amine; there is no single or major target for amine action in this circuit. Third, each amine can have multiple physiological effects on different cells, even within this small network. These effects alter the intrinsic firing properties of the cells. For example, DA induces endogenous oscillatory properties in one cell, strongly inhibits two others, and evokes a prolonged depolarization with tonic spike activity in the remaining cells (Fig. 2). Similar variations in cell activity were seen with Oct and 5HT. These changes in the electrophysiological properties of the pyloric cells have pronounced effects on the output from the entire pyloric network. For example, the induction of rhythmic bursting in the AB cell is essential for the amines to activate a rhythmic motor pattern in the isolated STG. Amine-induced activation or inhibition of cells causes them to be recruited or removed from the motor pattern, and imparts their characteristic intensity of firing.

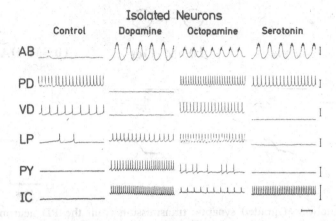

FIG. 2. Effects of amines on synaptically isolated cells in the STG. Each neuron was isolated from all detectable synaptic input by a combination of pharmacological blockade and 5,6-carboxyfluorescein photoinactivation of pre-synaptic cells. Bath-applied amines induced reproducible changes in cellular activity that were different for each cell type. All effects were reversible. Voltage scale: 10 mV; time scale: 1 s.
Reprinted from Ref. 26.

Neuromodulators alter the strength of synaptic interactions within the pyloric network

By altering the strength of the synaptic connections within the network, neuromodulators can functionally 'rewire' the network. In the pyloric circuit of the STG, the component neurons communicate by both spike-evoked and graded chemical and electrical synaptic transmission.[11] In the active pyloric rhythm, the cells release transmitter both as discrete spike-evoked events and as a continuous function of the oscillating membrane potential. Graded transmission can be monitored in the presence of TTX; depolarization of the pre-synaptic cell causes a graded release of transmitter which hyperpolarizes the post-synaptic cell, with an initial peak and a delayed plateau (Fig. 3). Bath-applied monoamines alter the synaptic efficacy: the amount and sign of the alteration varies greatly between amines and at different synapses.[12] For example, at the cholinergic PD:LP synapse, Oct enhances the synaptic efficacy while DA completely eliminates this synapse (Fig. 3). The cellular mechanisms of these effects are unclear, due to the fact that the synaptic contacts are located in the neuropil, far from the recording site in the cell body. Some evidence suggests that octopamine specifically enhances the release of trans-mitter from the PD terminals. In contrast, dopamine appears to reduce the input resistance of pre- and/or post-synaptic cells. This shunts and reduces the effective flow of current between input and output sites within the neuropil.[12]

We have studied the effects of DA, 5HT and Oct on all of the synapses within the pyloric network (Johnson, B., Peck, J. & Harris-Warrick, R., unpublished). As with the amines' modulation of the intrinsic properties of

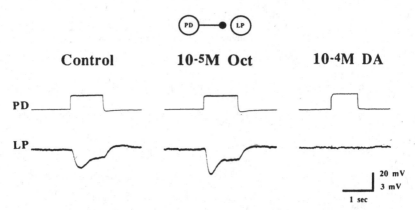

FIG. 3. Modulation of graded synaptic transmission from the PD neuron to the LP neuron by octopamine and dopamine. Experiment done in the presence of 10^{-7} M TTX to abolish action potentials. Other synaptic inputs to these cells have been eliminated.

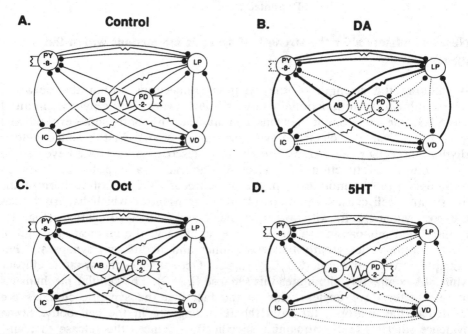

FIG. 4. Effects of monoamines on synaptic efficacy in the pyloric network of *Panulirus interruptus*. (A) Complete synaptic diagram of the pyloric network. Strength of the synapses is not indicated. (B), (C), (D) Effects of dopamine, serotonin and octopamine on the synaptic strengths in the network. Synapses that are functionally strengthened show a thick line; those that are unaffected show a thin line, while those that are weakened show a dashed line. (From Johnson, B., Peck, J. & Harris-Warrick, R., unpublished.)

pyloric cells, their effects on synaptic efficacy are complex, with varying effects at different synapses. Each amine can increase the efficacy at some synapses, leave others unchanged, and decrease efficacy at others (Fig. 4). Oct enhances transmission at most of the chemical synapses while leaving the electrical synapses unaffected. DA mainly has either strong excitatory or inhibitory effects on graded transmission, while 5HT has only weak excitatory or inhibitory effects. These effects correlate somewhat with the effect of an amine on the intrinsic firing pattern of the cell, but there are many exceptions: for example, dopamine inhibits and hyperpolarizes the VD neuron (Fig. 2), but strengthens the VD:LP synapse [Fig. 4(B)].

Changes in the functional rewiring of the network by these amines is dramatized by changes in the sign of synaptic interactions or in the appearance of new connections between neurons. For example, under control conditions, PD depolarization weakly inhibits the IC cell. In DA, this chemical inhibition is blocked, revealing an electrical connection that causes the IC to depolarize in phase with the PD cell. In another case, only a chemical synaptic interaction is normally observed between the AB and LP cells; in DA, an electrical connection between these two cells 'appears' (Johnson, B., Peck, J. & Harris-Warrick, R., unpublished observations).

THE GASTRO-PYLORIC RECEPTOR CELLS: SENSORY/MODULATORY NEURONS THAT USE ACETYLCHOLINE AND SEROTONIN AS CO-TRANSMITTERS

The results described above were obtained using bath application of monoamines. The next step is to identify the cells that release these transmitters and to compare the effects of cell-released to bath-applied transmitter. In addition, one would like to know what activates these modulatory neurons, in order to place the release of known neuromodulators in some behavioral context.

We have begun this approach by studying a set of serotonin-containing neurons in the Jonah crab, *Cancer borealis*. In 1984, Beltz *et al.*[13] detected a set of peripheral 5HT-immunoreactive neurons in motor nerves of the STG. We have demonstrated that these peripheral cells are the sole source of serotonin (5HT) to the STG in the crab.[14] There are two bilaterally symmetrical pairs of these cells; each sends processes extending over a set of muscles at the gastro-pyloric border of the foregut. The cells send their axons to the STG, where they branch profusely throughout the neuropil, and then continue and terminate in the anterior ganglia.

Unlike other peripheral aminergic neurons,[15,16] the GPR cells had no detectable effect on muscle tension or neuromuscular transmission. Instead, the GPR cells are primary mechanoreceptors monitoring tension of the muscles they innervate.[14] We have verified in semi-intact foregut preparations that the GPR cells are activated by stretch induced by contraction of a set of gastric mill muscles[14] (Richardson, F. & Kiehn, O., unpublished); thus, they typically fire slow bursts in phase with the slow gastric mill. In addition, the GPR cells are sometimes rhythmically active even when the gastric mill network is silent, and even when the GPR cell is physically isolated. Thus, the GPR cells appear to be endogenous bursters which can be coupled to the gastric mill rhythm by periodic changes in muscle tension.[14,17]

HPLC measurements proved that the immunoreactive material in the GPR cells is authentic 5HT. Since crustacean mechanoreceptors are generally cholinergic, we performed choline acetyltransferase assays and showed that these cells also synthesize acetylcholine (ACh). Combining these results with the pharmacological studies we will describe below, we conclude that the GPR cells in crabs use both ACh and 5HT as co-transmitters.[14,17,18]

EFFECTS OF GPR STIMULATION ON THE PYLORIC RHYTHM

When the GPR cells are stimulated in an in-vitro preparation in bursts that approximate those seen in the semi-intact foregut preparation, they evoke dramatic changes in the pyloric rhythm.[18] If the pyloric network is silent, GPR stimulation can activate rhythmic cycling. If it is already cycling weakly, GPR stimulation can evoke both transient and prolonged changes in pyloric activity (Fig. 5). *During* the GPR stimulation, the pyloric rhythm pauses, due primarily to inhibition of the PY cells and excitation of the LP neuron, which in turn inhibits the pacemaker cells. *After* the stimulation ends, the pyloric rhythm speeds up and many of the cells become more active. These changes can last up to several minutes after GPR stimulation. If short trains of GPR stimuli are repeated at a frequency mimicking the slow gastric mill rhythm in the semi-intact foregut preparation, the pyloric rhythm pauses during each stimulation and slowly accelerates between GPR trains. With continued stimulation, GPR stimulation has less and less of an effect, as if there were a ceiling to their effects on the pyloric neurons.[18] These results suggest that one possible function of the GPR cells is to coordinate the pyloric rhythm to the gastric mill rhythm and to enhance pyloric cycling if it is weak. This modulation of one CPG by another CPG through a peripheral feedback loop has not been previously described.

CELLULAR TARGETS OF GPR ACTIVITY

We have studied the effects of GPR stimulation on all the neurons in the crab STG.[17–20] As with the effects of bath-applied amines, the synaptic responses to GPR stimulation vary widely between different post-synaptic cells. These effects fell into two classes of synaptic responses: (1) rapid epsps that are

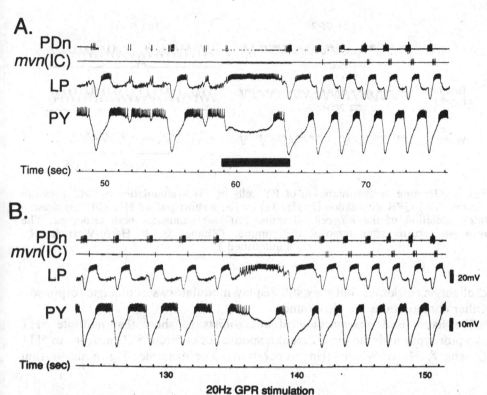

FIG. 5. Effects of GPR stimulation on the pyloric motor pattern in the crab, *Cancer borealis*. (A) The pyloric pattern is initially slow and irregular. One 5-s, 20-Hz GPR train (bar) causes a pause during which LP is excited and PY is inhibited. Following the stimulation, cycle frequency and regularity increase. (B) One minute following the first GPR train, the cycle frequency and cell activity are still elevated. A second stimulus train interrupts the ongoing cycling, then enhances it. The time line for (A) and (B) shows the temporal relationship between the traces. Reprinted from Ref. 17.

blocked by *d*-tubocurarine and other nicotinic cholinergic antagonists; (2) slow modulatory responses that alter the intrinsic electrophysiological properties of the target cell by a non-cholinergic mechanism. The modulatory responses vary markedly among the different neurons and can include inhibition, prolonged excitation, enhancement or initiation of rhythmic bursting, and induction of bistable properties. Exogenous serotonin (applied by bath application or pressure ejection from a micropipette) mimics these modulatory responses, and selective antagonists block the responses to both GPR stimulation and exogenously applied serotonin.

The rapid and slow responses to GPR stimulation are found in varying combination in different cells. Most of the gastric mill cells have a dual response to GPR stimulation: a rapid nicotinic epsp and a slow serotonergic modulatory response. Most of the pyloric neurons have no detectable

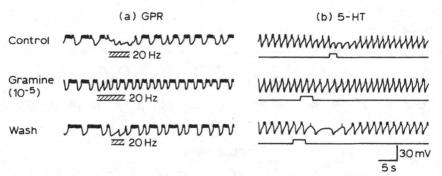

FIG. 6. Gramine blocks inhibition of PY cells by GPR stimulation or 5HT pressure ejection. (A) GPR stimulation (20 Hz, 3 s) or (B) a short puff of 5HT (10^{-3} M) cause a brief inhibition of the PY cell. Gramine (10^{-5} M) eliminates both responses. The response returns after removal of gramine. (Zhang, B. & Harris-Warrick, R., unpublished.)

cholinergic responses, but they still display modulatory serotonergic responses. Other combinations are also found.

We have used pharmacological antagonists to show that multiple 5HT receptor types underlie the different responses of different STG neurons to 5HT (Zhang & Harris-Warrick, in preparation). For example, Fig. 6 shows that

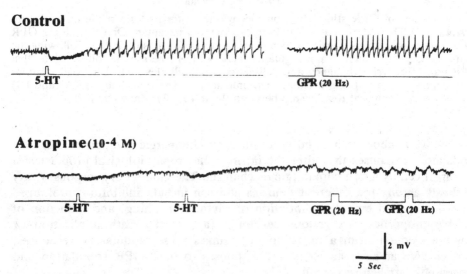

FIG. 7. Atropine blocks induction of bursting in AB cell by GPR stimulation and 5HT application. (A) GPR stimulation (20 Hz) or (B) a short puff of 5HT (10^{-3} M) induce bursting in the AB neuron. 5HT induces a transient inhibition that appears to be an artifact of the puff. Atropine (10^{-4} M) eliminates the response. The response recovers after removal of atropine (not shown). (Zhang, B. & Harris-Warrick, R., unpublished.)

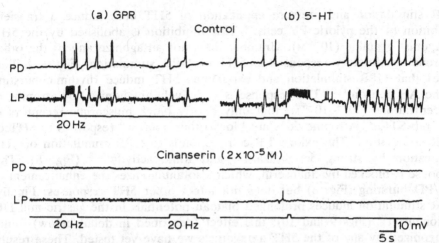

FIG. 8. Cinanserin blocks tonic activation of LP and PD bursting by GPR stimulation and 5HT application. (A) GPR stimulation (20 Hz) or (B) a short puff of serotonin (10^{-3} M) activate tonic spiking in the LP cell and bursting (after a delay) in the PD cell. Cinanserin (2×10^{-5} M) eliminates the response. The response recovers after removal of cinanserin (not shown). (Zhang, B. & Harris-Warrick, R., unpublished.)

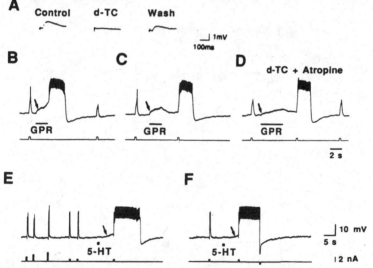

FIG. 9. GPR stimulation induces plateau potentials in the gastric mill DG motoneuron. (A) A GPR-evoked epsp is eliminated by 0·1 mM *d*-tubocurarine (*d*-TC). (B) A 10-Hz GPR train produces epsps which trigger a plateau potential. (C) A 10 Hz GPR train does not elicit a plateau, but a subsequent small current pulse (which is ineffective before GPR stimulation) elicits a plateau potential. (D) 0·1 mM *d*-TC eliminates the GPR-evoked rapid epsps but does not block a slow depolarization and a plateau potential upon current injection. (E) Pressure ejection of 5HT (10^{-3} M) causes a small depolarization; a 1 nA pulse (which was ineffective before the 5HT puff) evokes a plateau potential. Adapted from Ref. 17 and unpublished results of Kiehn & Harris-Warrick.

GPR stimulation and pressure application of 5HT both induce a transient inhibition of the pyloric PY cells.[18] This inhibition is abolished by the 5HT antagonist gramine (10^{-5} M). Gramine does not antagonize any of the other neurons' modulatory responses to 5HT or GPR stimulation. Figure 7 illustrates that GPR stimulation and exogenous 5HT induce rhythmic bursting in the AB neuron.[17,18] This response is selectively antagonized by atropine at concentrations (10^{-4}–10^{-5} M) that only poorly block muscarinic receptors in the crab STG.[20] Atropine does not block other neurons' responses to 5HT or GPR stimulation. The pyloric LP cell responds to GPR stimulation or 5HT application by strong depolarization and tonic activity[17,18] (Fig. 8). This response is blocked by cinanserin, which also antagonizes the enhancement of AB/PD bursting (Fig. 8) but does not affect other 5HT responses. Finally, GPR stimulation induces prolonged plateau potentials in the gastric mill DG motoneuron[17] (Figs 9 and 10); this effect (described in detail below) is not antagonized by any of the 5HT antagonists we have yet tested. These results show that four different responses to 5HT (inhibition, tonic excitation, burst induction, plateau potential initiation) are all mediated by different 5HT receptors with different pharmacological profiles. These profiles are not similar to those described for various subclasses of 5HT receptors in the vertebrates.[21]

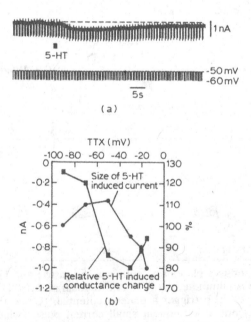

FIG. 10. Voltage clamp analysis of DG responses to 5HT puff. (A) The cell is held at −50 mV in 10^{-7} M TTX, and brief steps to −60 mV are given. A puff of 5HT (10^{-3} M) evokes a slow inward current and a conductance decrease. (B) Voltage dependence of the 5HT-induced current amplitude and of the conductance change before and after 5HT (Kiehn, O. & Harris-Warrick, R., unpublished.)

SYNERGISTIC ACTION OF ACh AND 5HT TRIGGERS PLATEAU POTENTIALS IN THE DG MOTONEURON

We have carried out a more detailed study of the cellular and ionic mechanisms by which the GPR cells induce plateau potentials in the gastric mill DG motoneuron[17,22,23] (Kiehn & Harris-Warrick, in preparation). GPR stimulation at low frequency (1 Hz) evokes only a rapid small epsp with a typical nicotinic pharmacological profile [Fig. 9(A)]. When the GPR cell is fired in a brief train at 10 Hz, the epsps summate and then suddenly trigger a plateau potential with high frequency spiking lasting several seconds [Fig. 9(B)]. This plateau potential far outlasts the time of GPR stimulation, and can be prematurely terminated by a brief hyperpolarizing pulse or by synaptic inhibition.

The ability of DG to fire a plateau potential results from 5HT release from the GPR cells.[17,22,23] Sometimes when the GPR cell is stimulated, the DG does not fire a plateau, but merely shows summating epsps superimposed on a low amplitude depolarization. After the epsps have decayed, a brief depolarizing current pulse (which was ineffective before GPR stimulation) can elicit a full plateau potential [Fig. 9(C)]. If the nicotinic epsps are completely suppressed with *d*-tubocurarine, the GPR train elicits only a small prolonged depolarization, during which the cell still fires a plateau potential in response to a short depolarizing current pulse [Fig. 9(D)]. This latter response can be mimicked by a puff of serotonin from a micropipette, which induces a small, prolonged depolarization, during which a brief depolarizing pulse can induce a prolonged plateau potential [Fig. 9(E)].

Thus, it appears that the two transmitters used by GPR in the crab act synergistically to induce plateau potentials in the DG neuron.[17] Serotonin induces the capability to fire a plateau, provided the cell is sufficiently depolarized above a critical threshold value. The co-released ACh provides a rapid depolarizing drive to carry the cell above threshold to evoke the plateau potential.

IONIC MECHANISMS OF PLATEAU POTENTIAL INDUCTION BY 5HT IN THE DG NEURON

When the DG cell is a voltage-clamped, a puff of serotonin induces an inward current of prolonged duration[22,23] [Fig. 10(A); Kiehn & Harris-Warrick, in preparation]. The I/V relation of this 5HT-induced current is inward over the entire voltage range and has an inverted U-shape with a minimum near the normal resting potential [Fig. 10(B)]. 5HT-induced changes in input conductance show a complex voltage dependence [Fig. 10(B)], with a conductance increase from hyperpolarized voltages and a conductance decrease from moderately depolarized voltages. Clearly, serotonin's effects on the DG cell are not mediated by a single current.

We have so far identified two ionic currents that are modulated by serotonin in the DG cell. A conductance decrease in a calcium-activated potassium current, $I_{K(Ca)}$, accounts for at least part of the decreased conductance seen with 5HT puffs from depolarized voltage levels.[23] In normal saline, this current is significantly decreased by 5HT, while in low Ca^{2+} saline serotonin has no detectable effect on the remaining outward currents. $I_{K(Ca)}$ appears to contribute significantly to the DG resting potential, since reduction of extracellular Ca^{2+} causes a depolarization of 10–15 mV.

Serotonin increases the input conductance at hyperpolarized potentials at least in part by an enhancement of the hyperpolarization-activated inward current, I_h.[23] This very slow inward current is activated by hyperpolarization from rest and closes only slowly upon repolarization; it could be activated during the rhythmic hyperpolarization of DG in the gastric mill rhythm. Serotonin enhances this current by increasing the rate constant for activation and shifting the voltage dependence of activation to more depolarized voltages, thus enhancing the amount of I_h at physiologically relevant membrane potentials. This current is very sensitive to low concentrations of Cs^+, which cause the DG cell to hyperpolarize. This suggests that, like $I_{K(Ca)}$, I_h contributes to the normal resting potential.

This combined conductance increase and decrease response to 5HT gives at least a partial mechanism for the induction of plateau potentials in the DG neuron. Since both currents normally contribute to the resting potential, serotonin's actions will lead to a depolarization which may reach threshold for plateau potential generation. At this time, we cannot exclude the possibility that 5HT modulates additional currents. However, these experiments provide a first insight in the ionic mechanisms by which 5HT evokes bistability in a motoneuron.

CONCLUSION

Our work and that of others has demonstrated that a single anatomically defined neural network can generate an enormous variety of related motor patterns in the presence of different neuromodulators.[24] The cellular mechanisms of this modulation are being studied, with particular emphasis on the modulation of the intrinsic electrophysiological properties of the component neurons and modulation of synaptic efficacy within the network. Thus, a motor network should be considered as a library of potential components and potential connections whose properties can be changed, enhanced or decreased by different neuromodulators. These mechanisms of motor pattern generation and modulation are not just found in 'simpler' invertebrate preparations: they are also central to the generation and control of vertebrate behaviors such as locomotion[27] and respiration.[28]

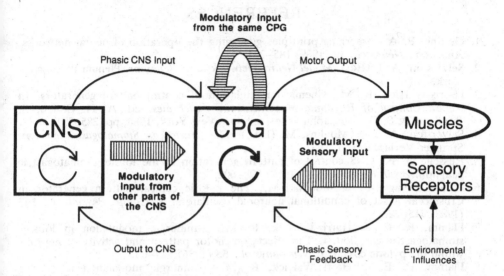

FIG. 11. Sources of modulation to central pattern generator networks. Both central sources and sensory neurons can provide slow modulatory inputs as well as rapid input. Reprinted from Ref. 25.

A number of modulatory neurons have been identified that use these transmitters to alter the motor rhythms in the STG.[25] Most of them use multiple transmitters just as the GPR does, and exert both rapid and slow modulatory effects on their neuronal targets. In the future we must understand the actions of these cells *in vivo*, in order to understand the behavioral context for the different modulated motor patterns.

Finally, our results have shown that in addition to modulatory neurons descending from other neural centers, primary sensory neurons can also evoke slow modulatory effects on motor networks (Fig. 11). Sensory neurons have long been known to evoke rapid feedback on a cycle-by-cycle basis to correct the motor pattern. The GPR cells serve this function by the release of acetylcholine onto rapid nicotinic receptors. However, in addition, they release 5HT which alters the intrinsic properties of the target cells and can markedly change the motor pattern generated by the network. This action lasts for many cycles, and need not be phase-locked to the rhythm. Thus, sensory neurons can be *instructive* as well as *corrective* in their input to motor networks.

ACKNOWLEDGMENTS

The work reported here was supported by NIH grant #NS17323 and USDA Hatch Act Grant NYC-191410 (to R. H-W.). B.R.J. was supported on NRSA #NS07859. R.E.F. and P.S.K. were supported on NIH Training Grant 2 T32 MG-07469. O.K. was supported by SLF #128776.

REFERENCES

1. Getting, P. A., Emerging principles governing the operation of neural networks. *Ann. Rev. Neurosci.*, **12** (1989) 185–204.
2. Selverston, A. I. (Ed.), *Model Neural Networks and Behavior*, Plenum Press, New York, 1985.
3. Harris-Warrick, R. M., Chemical modulation of central pattern generators. In *Neural Control of Rhythmic Movements in Vertebrates*, ed. A. H. Cohen, S. Rossignol & S. Grillner, John Wiley & Sons, New York, 1988, pp. 285–331.
4. Selverston, A. I. & Moulins, M. (Eds.), *The Crustacean Stomatogastric System*. Springer-Verlag, Berlin, 1987.
5. Russell, D. F., CNS control of pattern generation in the lobster stomatogastric ganglion. *Brain Research*, **177** (1979) 598–602.
6. Bal, T., Nagy, F. & Moulins, M., The pyloric central pattern generator in Crustacea: a set of conditional neuronal oscillators. *J. Comp. Physiol.* A, **163** (1988) 715–27.
7. Flamm, R. E. & Harris-Warrick, R. M., Aminergic modulation in lobster stomatogastric ganglion. I. The effects on motor pattern and activity of neurons within the pyloric circuit. *J. Neurophysiol.*, **55** (1986) 847–65.
8. Flamm, R. E. & Harris-Warrick, R. M., Aminergic modulation in lobster stomatogastric ganglion. II. Target neurons of dopamine, octopamine, and serotonin within the pyloric circuit. *J. Neurophysiol.*, **55** (1986) 866–81.
9. Miller, J. P. & Selverston, A. I., Rapid killing of single neurons by irradiation of intracellular injected dye. *Science*, **206** (1979) 702–4.
10. Bidaut, M., Pharmacological dissection of pyloric network of the lobster stomatogastric ganglion using picrotoxin. *J. Neurophysiol.*, **44** (1980) 1089–101.
11. Graubard, K., Raper, J. A. & Hartline, D. K., Graded synaptic transmission between identified spiking neurons. *J. Neurophysiol.*, **50** (1983) 508–21.
12. Jonnson, B. R. & Harris-Warrick, R. M., Aminergic modulation of graded synaptic transmission in the lobster stomatogastric ganglion. *J. Neurosci.*, **10** (1990) 2066–76.
13. Beltz, B., Eisen, J. S., Flamm, R., Harris-Warrick, R. M., Hooper, S. & Marder, E., Serotonergic innervation and modulation of the stomatogastric ganglion of three decapod crustaceans. *J. Exp. Biol.*, **109** (1984) 35–54.
14. Katz, P. S., Eigg, M. H. & Harris-Warrick, R. M., Serotonergic/cholinergic muscle receptor cells in the crab stomatogastric nervous system. I. Identification and characterization of the gastropyloric receptor cells. *J. Neurophysiol.*, **62** (1989) 558–70.
15. Kupfermann, I. & Weiss, K. R., The role of serotonin in arousal of feeding behavior of *Aplysia*. In *Serotonin Neurotransmission and Behavior*, ed. A. Gelperin & B. Jacobs, MIT Press, Cambridge, 1981, pp. 255–87.
16. Evans, P. D. & O'Shea, M., An octopaminergic neurone modulates neuromuscular transmission in the locust. *Nature, Lond.*, **270** (1977) 257–9.
17. Katz, P. S. & Harris-Warrick, R. M., Serotonergic/cholinergic muscle receptor cells in the crab stomatogastric nervous system. II. Rapid nicotinic and prolonged modulatory effects on neurons in the stomatogastric ganglion. *J. Neurophysiol.*, **62** (1989) 571–81.
18. Katz, P. S. & Harris-Warrick, R. M., Neuromodulation of the crab pyloric central pattern generator by serotonergic/cholinergic proprioceptive afferents. *J. Neurosci.*, **10** (1990) 1495–512.
19. Katz, P. S. & Harris-Warrick, R. M., Recruitment of crab gastric mill neurons into the pyloric motor pattern by mechanosensory afferent stimulation. *J. Neurophysiol.*, **65** (1991) 1442–51.

20. Katz, P. S., Motor pattern modulation by serotonergic sensory cells in the stomatogastric nervous system. PhD thesis, Cornell University, Ithaca, New York, 1989.
21. Peroutka, S. J. (Ed.), *Serotonin Receptor Subtypes: Basic and Clinical Aspects.* Wiley–Liss, New York, 1991.
22. Kiehn, O. & Harris-Warrick, R. M., Serotonin induces plateau potentials in a stomatogastric motoneuron by a mixed conductance decrease and increase mechanism. *Soc. Neurosci. Abst.,* **16** (1990) 633.
23. Kiehn, O. & Harris-Warrick, R. M., Serotonergic cells induce plateau properties by enhancing a hyperpolarization activated inward current (I_h) and decreasing a calcium-dependent outward current. *European J. Neurosci.* Supplement No. 4, 3224P.
24. Harris-Warrick, R. M. & Marder, E., Modulation of neural networks for behavior. *Ann. Rev. Neurosci.,* **14** (1991) 39–57.
25. Katz, P. S. & Harris-Warrick, R. M., Identified neuromodulatory neurons in a simple motor system. *Trends in Neurosci.,* **13** (1990) 367–73.
26. Harris-Warrick, R. M. & Flamm, R. E., Chemical modulation of a small central pattern generator circuit. *Trends in Neurosci.,* **9** (1986) 432–7.
27. Kiehn, O., Plateau potentials and active integration in the 'final common pathway' for motor behavior. *Trends in Neurosci.,* **14** (1991) 68–73.
28. Feldman, J. L. *et al.,* Neurogenesis of respiratory rhythm and pattern: emerging concepts. *Am. J. Physiol.,* **259** (1990) R879–86.

23

Olfaction in *Manduca sexta*: Cellular Mechanisms of Responses to Sex Pheromone

JOHN G. HILDEBRAND, THOMAS A. CHRISTENSEN, EDMUND A. ARBAS, JON H. HAYASHI, UWE HOMBERG,† RYOHEI KANZAKI* & MONIKA STENGL†

Arizona Research Laboratories, Division of Neurobiology, University of Arizona, 611 Gould-Simpson Building, Tucson, Arizona 85721, USA

INTRODUCTION

In insects, olfaction plays a major role in the control of many kinds of behavior. Orientation and movement toward, and interactions with, receptive mating partners, appropriate sites for oviposition, sources of food, and hosts for parasitism usually involve olfactory signals that initiate, sustain, and guide the behaviors. Because of their prominence in the zoosphere, their economic and medical importance, and their usefulness as models for both behavioral and neurobiological research, insects have been extensively studied by investigators interested in mechanisms of olfactory control of behavior. Insects respond to a variety of semiochemicals, including pheromones (chemical messengers within a species, such as sex attractants) and kairomones (chemical messengers between species and adaptively favorable to the recipient, such as attractants and stimulants for oviposition and feeding emitted by a host plant). Studies of insect responses to such biologically significant odors have shown that the quality and quantity of odorants in complex mixtures present in the environment are encoded in patterns of activity in multiple olfactory receptor cells (ORCs) in the antennae. These 'messages' are decoded and integrated in the olfactory centers of the central nervous system (CNS), and it is there that olfactorily induced changes in the behavior or physiology of the insect are initiated.

Of paramount interest, both historically and currently, is the attraction of a mating partner by means of a chemical signal—the sex pheromone—released by a receptive individual of one sex and detected by conspecifics of the opposite sex. In moths, these chemical signals are the primary means by which

* Present address: Institute of Biological Sciences, The University of Tsukuba, Tsukuba City, Ibaraki, 305 Japan.
† Present address: Fakultät für Biologie, Universität Konstanz, Universitätsstrasse 10, Postfach 5560, D-7750 Konstanz 1, FRG.

females broadcast their sexual receptiveness over relatively long distances to conspecific males. The male moths respond to the sex-pheromonal stimulus with well-characterized mate-seeking behaviors involving arousal, patterned anemotactic flight, short-range orientation to the calling female, and mating.[1]

Behaviorally relevant olfactory mechanisms have been probed in depth in several insect taxa, including cockroaches,[2] honey bees,[3,4] and fruit flies,[5] but the most extensive explorations of the detection and processing of, and behavioral responses to, sex pheromones have been carried out in moths. Building upon earlier and recent work of others[6-11] and paralleling the current research of others using different insect species, we study the olfactory system of the experimentally favorable giant sphinx moth *Manduca sexta*. We and our co-workers have investigated this model olfactory system extensively by means of anatomical, neurophysiological, biochemical, cytochemical, developmental, and cell-culture methods.[12-14] This brief review focuses on some of our findings and speculations about the functional organization and physiology of the olfactory system in *M. sexta*; more detailed reviews of this and other aspects of our work have been presented elsewhere.[12-15]

A principal long-term goal of our research is to understand the neurobiological mechanisms through which information about a specific olfactory signal, the female's sex pheromone, is detected and integrated with inputs of other modalities in the male moth's brain and how the message ultimately initiates and controls the characteristic behavioral response of a conspecific male moth. Pursuit of that goal promises to teach us much about how the brain processes chemosensory information and uses it to shape behavior. Our studies along this line to date have persuaded us that the male's olfactory system consists of two parallel subsystems, one a sexually dimorphic 'labeled-line' pathway specialized to detect and process information about the sex pheromone, and the other a more complex pathway that processes information about plant (and probably other environmental) odors encoded in 'across-fiber' patterns of physiological activity.

THE SEX-PHEROMONAL STIMULUS

The sex pheromones of moths generally are mixtures of two or more chemical components, typically aldehydes, acetates, alcohols, or hydrocarbons produced in specialized glands by metabolism of fatty acids.[1] Often, the species-specific blend of components is the message, and males of many moth species, including *M. sexta*, give their characteristic, qualitatively and quantitatively normal behavioral responses only when stimulated by the correct blend of sex-pheromone components and not to individual components or partial blends lacking key components.[16,17]

Solvent washes of the pheromone gland of female *M. sexta* yield 12 substances: $8 C_{16}$ and $4 C_{18}$ aldehydes.[17] A synthetic mixture of these components elicits the same behavioral responses in males as does the signal released by a calling female.[17,18] A blend of two of the components, the dienal (E,Z)-10,12-hexadecadienal (E10,Z12-16:AL or bombykal[18]) and the trienal (E,E,Z)-10,12,14-hexadecatrienal (E10,E12,Z14-16:AL), elicits the normal sequence of male behaviors in a wind tunnel, but the individual components are ineffective.[17] Field trapping studies show that a blend of all 8 of the C_{16} aldehydes (including the E,E-isomer of bombykal and the E,E,E-isomer of the trienal) present in the gland washes is significantly more effective in attracting males than are blends of fewer components, suggesting that all of the C_{16} aldehydes have a role in the communication system of *M. sexta*, i.e. that the sex pheromone of this species comprises those 8 C_{16} aldehydes (Ref. 17 and Tumlinson, pers. comm., 1990). Our neurophysiological studies have focused on three important properties of the sex-pheromonal signal: its quality (chemical composition of the blend), quantity (concentrations of components), and intermittency (owing to the fact that the pheromone in the plume downwind from the source exists in filaments and blobs of odor-bearing air interspersed with clean air[19,20]). Each of these properties of the pheromonal message is important, as the male moth gives his characteristic behavioral responses only when the necessary and sufficient pheromone components are present in the blend[17] (qualitative criterion), when the concentrations and blend proportions of the components fall within acceptable ranges[21] (quantitative criterion), and when the pheromone blend stimulates his antennae discontinuously[9,22] (intermittency criterion). We examine how each of these important aspects of the odor stimulus affects the activity of neurons at various levels in the olfactory pathway.

DETECTION OF THE SEX PHEROMONE

Each antenna of adult *M. sexta* comprises two basal segments (scape and pedicel), containing mechanosensory organs, and a long sexually dimorphic flagellum divided into c. 80 annuli bearing numerous sensilla of several types, the great majority of which are olfactory.[23,24] Antennal flagella of both male and female *M. sexta* (whose larvae feed, and adult females oviposit, exclusively on plants of the family *Solanaceae*) have many ORCs that respond to volatiles given off by plants[25] and presumably are involved in host-plant recognition and discrimination. In addition, the antennae of the male moth possess ORCs specialized to detect individual components of the female's sex pheromone.[25,26]

A male flagellum has c. 3×10^5 sensory neurons associated with c. 10^5 recognized sensilla, of which c. 40% are male-specific sensory hairs or *sensilla trichodea*.[23,25,26–28] Type-I trichoid hairs ($\leq 600 \mu m$ long) are typical olfactory sensilla, having single walls with pores, and are innervated by two ORCs that send their unbranched dendrites through the lumena of the fluid-filled hairs to their tips.[23,27–29] One of these two male-specific ORCs is highly sensitive and specific to E10,Z12-16:AL, while the second ORC is 'tuned' to E10,E12,Z14-16:AL in about 85% of these sensilla or to E10,E12,E14-16:AL in the remaining 15% of the sensilla.[26] Pheromone-specific ORCs in moth antennae thus represent information about stimulus quality by means of their specialization as narrowly tuned input channels. Groups of these cells can 'follow' intermittent pheromonal stimuli at naturally occurring frequencies (up to c. 10 Hz).[30]

FUNCTIONAL ORGANIZATION OF CENTRAL OLFACTORY PATHWAYS

Axons of antennal ORCs project through the antennal nerve to enter the brain at the level of the ipsilateral antennal lobe (AL) of the deutocerebrum.[24] ORC axons and possibly also fibers of other modalities (such as hygroreceptors) project from the flagellum to targets in the AL, but axons from mechanosensory neurons in the basal segments of the antenna bypass the AL and project instead to a deutocerebral 'antennal mechanosensory and motor center' in the deutocerebrum posteroventral (with respect to the body axis of the animal) to the AL[24,30,31]. In moths and some other insect taxa, sex-pheromonal information is processed in a prominent male-specific area of neuropil in each AL called the macroglomerular complex (MGC).[24,31,32]

The AL has a central zone of coarse neuropil (largely neurites of AL neurons) surrounded by an orderly array of glomeruli, including 64 ± 1 spheroidal 'ordinary' glomeruli and, in the male, the sexually dimorphic MGC near the entrance of the antennal nerve into the AL (Ref. 31 and Rospars, J. P. & Hildebrand, J. G. (submitted for publication)). Bordering this neuropil are three groups of AL neuronal somata—lateral, medial, and anterior—totalling about 1200 cells.[12,32,33] Each olfactory axon projecting from the antennal flagellum into the brain terminates within a single glomerulus in the ipsilateral AL (Refs 31 and 34 and Christensen, T. A., Cuzzocrea, C., Harrow, I. D., Kent, K. S., Schneiderman, A. M., & Hildebrand, J. G., unpublished observations), where it makes chemical synapses with dendritic processes of AL neurons (Refs 32 and 34 and Christensen, T. A., Waldrop, B., Harrow, I. D. & Hildebrand, J. G., in preparation). These inputs are believed to be cholinergic[35] and the postsynaptic elements, to be largely or exclusively LNs (Christensen, T. A., Waldrop, B., Harrow, I. D. & Hildebrand, J. G., in preparation). The ordinary glomeruli, which are condensed neuropil structures 50–100 μm in diameter, contain terminals of sensory axons, dendritic arborizations of AL neurons, and all of the primary-afferent synapses and synaptic connections among AL neurons, and they are nearly surrounded by an investment of glial processes.[32,34,36]

With very few exceptions, the neurons in the medial, lateral, and anterior cell groups of the AL fall into two main classes.[12,32,33,37,38] Projection neurons (PNs or output neurons) have dendritic arborizations in the AL neuropil and axons that project out of the AL, and local interneurons (LNs) lack axons and have more or less extensive arborizations confined to the AL neuropil. The

PNs relay information about odors, synaptically processed and integrated in the AL by neural circuitry involving sensory axons, LNs, and PNs, to higher-order centers in the brain.[33] Many PNs have dendritic arborizations confined to single AL glomeruli and axons that project via the inner antenno-cerebral tract through the ipsilateral protocerebrum, sending branches into the calyces of the ipsilateral mushroom body and terminating in characteristic olfactory foci in the lateral protocerebrum. Other PNs have arborizations in one or more AL glomeruli and send axons via the middle, outer, dorsal, and dorso-medial antenno-cerebral tracts to characteristic regions of the protocerebrum, including lateral and inferior regions.[12,33]

Axons of male-specific antennal ORCs specialized to detect components of the species-specific sex pheromone project exclusively to the MGC (Ref. 31 and Harrow, I. D., Christensen, T. A., and Hildebrand, J. G. unpublished), and all AL neurons that respond to antennal stimulation with sex-pheromone components have arborizations in the MGC.[32,37,38] The MGC in *M. sexta* has two major, easily distinguishable divisions: a doughnut-shaped neuropil structure (the 'toroid') and a globular structure (the 'cumulus') adjacent to the toroid and closer to the entrance of the antennal nerve into the AL.[39] AL PNs that respond to antennal stimulation with E10,Z12-16:AL have arborizations in the toroid and PNs responsive to E10,E12,Z14-16:AL, in the cumulus.[39] Thus first-order synaptic processing of sensory information about these important sex-pheromone components, which are necessary and sufficient to elicit and sustain the male's behavioral response, apparently is confined to different, distinctive neuropil regions of the MGC.

PROCESSING OF SEX-PHEROMONAL INFORMATION IN THE MALE AL

In our studies of central processing of sex-pheromonal information in *M. sexta,* we again focus on the qualitative, quantitative, and intermittency properties of the pheromonal signal. We examine how each of these aspects of the pheromonal stimulus affects the activity of the two main classes of AL neurons associated with the MGC: the male-specific LNs and PNs.

Stimulus quality

By means of intracellular recording and staining methods, we have examined the activity of AL neurons in response to stimulation of the ipsilateral antenna with each of the sex-pheromone components as well as partial and complete blends.[40] In accordance with results of behavioral and sensory-receptor studies, we have found that E10,Z12-16:AL and E10,E12,Z14-16:AL are the most effective and potent components for eliciting physiological responses in the AL neurons. On the basis of these responses, we classified the neurons into two broad categories: pheromone generalists and pheromone specialists.[41] Pheromone generalists are neurons that respond similarly to stimulation of either the E10,Z12-16:AL input channel or the E10,E12,Z14-16:AL input channel and do not respond differently when the complete, natural blend is presented to the antenna. These cells may therefore be involved in mediating general arousal in response to sex pheromone but apparently do not account for species recognition. In contrast, we refer to neurons that can discriminate between antennal stimulation with E10,Z12-16:AL from stimulation with E10,E12,Z14-16:AL as pheromone specialists. There are several types of pheromone specialists. Some receive input only from the dienal channel or the trienal channel. These cells therefore preserve information about individual components of the species-specific blend. Similar specialist responses have been reported even for higher-order neurons descending from the brain in the silkmoth *Bombyx mori*.[42] These observations suggest that information about specific components of the blend, and not only the complete blend, is relevant to chemical communication in these animals.

An important subset of pheromone-specialist PNs found in male *M. sexta* receives input from both E10,Z12-16:AL and E10,E12,Z14-16:AL channels, but the physiological effects of the two inputs are opposite.[37] That is, if antennal stimulation with the dienal leads to excitation, then stimulation with the trienal inhibits the interneuron, and vice versa. Simultaneous stimulation of the antenna with both E10,Z12-16:AL and E10,E12,Z14-16:AL elicits a mixed inhibitory and excitatory response in these special PNs. Thus these neurons can discriminate between the two inputs based upon how each affects the spiking activity of the cell. These cells also respond uniquely to the natural pheromone blend released by the female: only this subset of pheromone-specialist neurons can follow intermittent pheromonal stimuli occurring at natural frequencies of up to 10 Hz[43] (see below).

While most of the olfactory interneurons encountered in male *M. sexta* ALs respond preferentially to E10,Z12-16:AL and E10,E12,Z14-16:AL, some neurons also respond to other C_{16} components in the female's pheromone blend, including the isomeric trienal E10,E12,E14-16:AL.[39] Taken together with the aforementioned field-trapping evidence that all of the C_{16} components in the solvent washes of female sex-pheromone glands have behavioral effects, these findings suggest that 'minor' components of the blend may play subtle but important roles in pheromonal communication. In addition, sex-pheromonal components from another species can sometimes stimulate ORCs in a male moth's antennae and thereby affect his behavior, as is the case of ORCs in the antennae of male *Helicoverpa zea*.[41]

Stimulus quantity

Numerous studies in the field and in wind tunnels have shown that pheromone-mediated orientation is a dose-dependent phenomenon.[44] We therefore have examined the ability of AL neurons to encode changes in the concentration of a pheromonal stimulus.[41] When a male's antenna is stimulated with a series of pheromonal stimuli of graded concentrations, MGC PNs exhibit various dose–response relationships. In some of these PNs, the dynamic range of the cell extends up to the highest concentration tested [0·1 female equivalents (FE) of sex pheromone], but in other MGC PNs, inactivation of spiking occurs between 0·01 and 0·05 FE. Some PNs that have this ability to encode quantitative information about the pheromone yield a dose–response relationship, as measured by number of spikes elicited, that is quite linear up to 0·05 FE but falls off above this concentration. The response as measured by maximum instantaneous frequency of spiking, however, continues to increase up to the maximum concentration tested (0·1 FE). A corresponding increase in the amplitude of the membrane depolarization can also be seen. In contrast, some other concentration-sensitive PNs give a spiking response that falls off between 0·01 and 0·05 FE. Membrane depolarization continues to increase with concentration, suggesting that attenuation and gradual disappearance of the spikes are due to sodium inactivation at these high levels of membrane depolarization.

Stimulus intermittency

One of the most important characteristics of a female moth's sex-pheromone plume is its non-uniformity. Simulations of odor plumes using ionized air have shown clearly that a plume is not a simple concentration gradient but instead possesses a distinctly discontinuous and filamentous structure.[19,20] Furthermore, abundant behavioral evidence shows that a male moth's ability to locate a pheromone source is greatly improved if the odor plume is discontinuous.[44] Because spatial discontinuity of the pheromonal signal in the environment is detected by a flying male moth as temporally intermittent stimuli, we have sought to learn how intermittent pheromonal stimuli delivered to a male's antenna are registered by MGC PNs in his ALs. We discovered that only certain pheromone-specialist PNs can follow rapidly pulsed pheromonal stimuli with corresponding bursts of impulses. These are the cells cited above that can discriminate between E10,Z12-16:AL and E10,E12,Z14-16:AL because one of these key components excites the cell while the other inhibits it. This inhibitory input to such PNs enhances their ability to follow brief pulses of pheromone blends delivered at frequencies up to 10 Hz by controlling the duration of excitatory responses and preparing the PN for the next bout of excitation.[43]

SYNAPTIC MECHANISMS IN THE AL

Having characterized many AL LNs and PNs both morphologically and physiologically, we have begun to try to explain how their characteristic patterns of responses to olfactory stimuli are generated. To accomplish this mechanistic goal, we must analyze the synaptic 'wiring' of the AL, test physiologically for synaptic interactions between known types of AL neurons, identify neurotransmitters and synaptic mechanisms employed by AL neurons for intercellular communication, and seek evidence for and mechanisms of integration of other modalities with olfactory inputs in the ALs.

Cholinergic excitation in the ALs

Biochemical assays have shown that the most prominent 'classical' neurotransmitter candidate in the ALs is acetylcholine (ACh).[45,46] ACh appears to be the transmitter employed by axons of ORCs at their synapses in the glomeruli. To a variety of biochemical evidence,[45–49] we have added physiological and pharmacological findings[35] in support of this role of ACh in the AL. The primary-afferent synapses are excitatory and appear to be nicotinic.[35]

GABAergic inhibition in the ALs

Another prominent neurotransmitter in the ALs is γ-aminobutyric acid (GABA).[46] GABA immunocytochemistry has revealed that all of the GABA-immunoreactive neurons in the AL have somata in the large lateral cell group of the AL.[50] There are approximately 350 GABA-immunoreactive LNs and 110 GABA-positive PNs (i.e. about 30% of the neurons in the lateral cell group appear to be GABAergic).[12,50] Most (and possibly all) of the LNs are GABA-immunoreactive, and most (or all) of these cells are probably inhibitory interneurons.

The important inhibitory postsynaptic potential (IPSP) that enables certain PNs to follow intermittent pheromonal stimuli (see above) appears to be due to chemical-synaptic transmission mediated by GABA.[51] This IPSP reverses below the PNs' resting potential and is mediated by an increased Cl^- conductance. This IPSP can be inhibited reversibly by picrotoxin, which blocks GABA-receptor-gated Cl^- channels, and by bicuculline, a blocker of vertebrate $GABA_A$ receptors. Furthermore, applied GABA also hyperpolarizes the postsynaptic neuron, and this response can be blocked reversibly by bicuculline, indicating that bicuculline directly blocks the GABA receptors. Such GABAergic synaptic transmission is essential to the critical ability of the specialized AL PNs to follow intermittent pheromonal stimuli.[43]

Synaptic interactions between AL neurons

We have tested the idea that this inhibition of PNs is mediated through LNs by directly examining synaptic interactions between pairs of AL neurons (Christensen, T. A., Waldrop, B., Harrow, I. D. & Hildebrand, J. G., in preparation). Using intracellular techniques, we could pass current into one neuron while monitoring postsynaptic activity in the other. None of the PN–PN pairs examined showed any current-induced interactions, but a significant proportion of the LN–PN pairs studied exhibited such interactions, all of which were unidirectional. That is, LN activity could influence PN activity, but not vice versa. Depolarizing current injected into an LN, causing it to produce spikes, was associated with a cessation of firing in the PN. Spike-triggered averaging revealed a weak, prolonged IPSP in the PN. Cross-correlation analysis also revealed a weak inhibitory interaction, polarized in the direction from LN to PN. When the simultaneous responses of the two neurons to olfactory stimulation of the ipsilateral antenna with E10,Z12-16:AL were recorded, a brief period of inhibition was observed in the LN, and this was followed shortly thereafter by a transient increase in the firing frequency in the PN. This suggests that the excitation of the PN is due to disinhibition. The LN that synaptically inhibits the PN thus may itself be inhibited by olfactory inputs, probably through another interneuron.

Other putative neurotransmitters and neuropeptides in the AL

Immunocytochemistry using a specific antiserum against chemically fixed 5-hydroxytryptamine (5HT or serotonin) has shown that the AL possesses a single 5HT-immunoreactive neuron.[52] This remarkable cell has aborizations in the ipsilateral and contralateral protocerebrum and the contralateral AL and appears to be a centrifugal neuron with heterolateral feedback characteristics.[53] Other monoamines detected immunocytochemically in the AL include histamine[54] and dopamine (U. Homberg & J. Hildebrand, unpublished).

The AL also expresses diverse neuropeptide-like immunoreactivities. For example, about 80 neurons in the lateral cell group of the AL exhibit FMRFamide-like immunoreactivity (FLI).[55] Neuronal processes exhibiting FLI appear throughout the AL neuropil, including all of the glomeruli. Extracts of brain tissue of *M. sexta* contain two peptides that are recognized by anti-FMRFamide antibodies, a major lower-molecular weight substance and a minor higher-molecular weight substance.[56] The smaller, more abundant compound has been characterized and shown to be a new member of the FLRFamide subfamily of the FMRFamide family of neuropeptides, a decapeptide we have named *Manduca*FLRFamide.[56] Work still under way on this line of inquiry focuses on several other neuropeptide-like immunoreactivities expressed in the AL.

IN-VITRO STUDIES

Because the brain and other ganglia of the insect CNS are very compact and surrounded by a strong sheath that serves as a 'blood–brain barrier', neurons and synapses in the AL, as in other regions of the insect CNS, are relatively inaccessible to pharmacological manipulation, patch-clamp experiments, and manipulations of the ionic composition of their extracellular environment. Experiments aimed at revealing the biophysical properties of neuronal membranes and eludicating mechanisms of synaptic transmission and other forms of cell–cell interaction therefore benefit greatly from the use of cell-culture methodology.

Olfactory receptor cells in primary cell culture

To investigate the mechanisms of sensory transduction in pheromone-specific ORCs, we dispersed developing antennae from male *M. sexta* pupae into single cells and kept these cells in long-term primary cultures *in vitro*.[57] Among the diverse cells in our cultures, the ORCs could be identified with the aid of two monoclonal antibodies, one of which recognizes all ORCs and the other is specific to mature pheromone-specific ORCs *in situ*.[57,58]

Patch-clamp methods have been used to characterize the membrane ion channels of these neurons and to probe the mechanisms underlying their responses to sex-pheromone components. These cells express at least three different kinds of K^+ channels and at least one type of TTX-blockable Na^+ channel after 3 weeks *in vitro*.[59] A subset of the ORCs in culture respond to DMSO extracts of female pheromone glands applied by puffer pipette. In most cases an inwardly rectifying unspecific cation channel (I_{cat}) opens in cell-attached recordings.[60] Current through this non-specific cation channel is the only inward current observed in ORCs *in vitro* at potentials more negative than the resting potential (-60 to -80 mV). We have begun to investigate the possible involvement of G-proteins and second messengers in the gating of the I_{cat}.[61] Patch-excision opened cation channels (possibly via exposure to high Ca^{2+}) that resembled the pheromone-dependent I_{cat} in reversal potential, amplitude substates, and ion selectivity and were blocked by tetraethylammonium (TEA) or cGMP + ATP. After perfusion of the cells with GTPγS + ATP, IP$_3$, or 10^{-6} M Ca^{2+}, cation conductances were activated which resembled the pheromone-sensitive cation channel by several criteria. These and other observations raise the hypotheses, which should be further tested experimentally, that the pheromone-dependent I_{cat} is controlled by Ca^{2+} and that second-messenger-controlled Ca^{2+}-permeable ion channels play roles in the response of male-specific ORCs to components of the female's sex pheromone.

AL neurons in primary cell culture

In parallel studies, we have found conditions under which neurons derived from the AL thrive and develop in long-term primary cultures *in vitro*.[62] Cultures prepared from the medial group of AL neurons (which includes only PNs[12,33]) consistently exhibit six different morphological types of neurons *in vitro*. Cultures derived from the large lateral group of AL neurons, which includes both PNs and LNs, contain neurons resembling those six morphological types of PNs as well as three additional, distinct types of neurons. Cells of two of these types clearly lack axons and are therefore presumably LNs. These neurons exhibit vigorous process outgrowth, but our patch-clamp studies have revealed that, although they produce large outward currents, these neurons apparently cannot develop voltage-gated inward currents after 2 weeks *in vitro*. In contrast, all PNs develop inward currents after about a week in culture. A subset of PNs expresses TTX-sensitive Na^+ channels. The current through these channels activates at about $-40\,mV$ and is associated with a delayed-rectifier type of K^+ channel. We also find slowly activating, slowly inactivating Ca^{2+} channels that are permeable to Ba^{2+} and blocked by Cd^{2+}. Each type of PN exhibits a characteristic set of currents, and some give an outward current in response to a puff of GABA solution applied externally. These studies have shown that dissociated AL neurons develop Na^+ and K^+ channels and possibly two types of Ca^{2+} channels in the absence of direct contact with other cells *in vitro*.[63] In the future, cultures of AL neurons and co-cultures of antennal and AL neurons should help us greatly to study synaptic and modulatory mechanisms involved in AL function.

HIGHER-ORDER PROCESSING OF PHEROMONAL INFORMATION IN THE CNS

After synaptic processing in the AL, information about sex pheromone and other odors is relayed to higher centers in the protocerebrum by way of the axons AL PNs. Toward the goal of understanding how pheromonal information controls the behavior of male moths, we have begun to explore the physiological and morphological properties of neurons in the protocerebrum that respond to stimulation of the antennae with sex pheromone or its components.[64,65]

We find that many pheromone-responsive protocerebral neurons have arborizations in a particular neuropil region, the lateral accessory lobe (LAL), which appears to be important for processing of olfactory information.[64] The LALs are situated lateral to the central body on each side of the protocerebrum. Each LAL is linked by neurons that innervate it to the ipsilateral superior protocerebrum as well as the lateral protocerebrum, where axons of AL PNs terminate.[33,37,38] The LALs are also linked to each other by bilateral neurons with arborizations in each LAL. Neuropil adjacent to the LAL contains branches of many neurons that descend in the ventral nerve cord. Local neurons link the LAL to this adjacent neuropil. Some descending neurons also have arborizations in the LAL. Thus, the LAL is interposed in the pathway of olfactory information flow from the AL through the lateral protocerebrum to descending neurons.

All protocerebral neurons observed to date to respond to antennal stimulation with pheromone were excited. Although brief IPSPs were sometimes elicited in mixed inhibitory/excitatory responses, sustained inhibition was not observed. Certain protocerebral neurons show long-lasting excitation (LLE) that sometimes outlasts the olfactory stimuli by up to 30 s. In some other protocerebral neurons, pheromonal stimuli elicit brief excitations that recover to background firing rates <1 s after stimulation. LLE is more frequently elicited by the sex-pheromone blend than by E10,Z12-16:AL or E10,E12,Z14-16:AL. LLE responses to pheromonal stimuli were obserbed in >50% of the bilateral protocerebral neurons sampled that had arborizations in the LALs. Fewer than 10% of the protocerebral local neurons examined exhibited LLE in response to similar stimuli. AL PNs responding to pheromone components do not show LLE.[37,38] Thus, LLE appears not to be produced at early stages of olfactory processing in the AL, but first occurs at the level of the protocerebrum.

These findings suggest that the LAL is an important region of convergence of olfactory neurons from other regions of the protocerebrum. Synaptic interactions in the LAL may mediate integration of both ipsilateral and bilateral olfactory information prior to its transmission to the bilateral pool of descending neurons. LLE appears to be one important kind of physiological response that is transmitted to thoracic motor centers. How this LLE might participate in the generation of the male moth's characteristic behavioral response is a subject of ongoing research.

ACKNOWLEDGEMENTS

We thank our many present and former co-workers and collaborators who contributed to the work reviewed in this chapter. Our research in this area is currently supported by NIH grants AI-23253, DC-00348, and NS-28495 (Proj. 2).

REFERENCES

1. Birch, M. C. & Haynes, K. F., *Insect Pheromones.* Arnold, London, 1982.
2. Boeckh, J. & Ernst, K.-D., Contribution of single unit analysis in insects to an understanding of olfactory function. *J. Comp. Physiol. A,* **161** (1987) 549–65.
3. Arnold, G., Masson, C. & Budharugsa, S., Comparative study of the antennal lobes and their afferent pathway in the worker bee and the drone (*Apis mellifera*). *Cell Tiss. Res.,* **242** (1985) 593–605.
4. Flanagan, D. & Mercer, A. R., Morphology and response characteristics of neurones in the deutocerebrum of the brain in the honeybee *Apis mellifera. J. Comp. Physiol. A,* **164** (1989) 483–94.
5. Rodriguez, V., Spatial coding of olfactory information in the antennal lobe of *Drosophila melanogaster. Brain Res.,* **453** (1988) 299–307.
6. Boeckh, J. & Boeckh, V., Threshold and odor specificity of pheromone-sensitive neurons in the deutocerebrum of *Antheraea pernyi* and *A. polyphemus* (Saturnidae). *J. Comp. Physiol.,* **132** (1979) 235–42.
7. Boeckh, J., Ernst, K.-D., Sass, H. & Waldow, U., Anatomical and physiological characteristics of individual neurones in the central antennal pathway of insects. *J. Insect Physiol.,* **30** (1984) 15–26.
8. Kaissling, K.-E., *R. H. Wright Lectures on Insect Olfaction.* Simon Fraser University, Burnaby, BC, 1987.
9. Kaissling, K.-E., Sensory basis of pheromone-mediated orientation in moths. *Verh. Dtsch. Zool. Ges.,* **83** (1990) 109–31.
10. Masson, C. & Mustaparta, H., Chemical information processing in the olfactory system of insects. *Physiol. Revs.,* **70** (1990) 199–245.

11. Rospars, J. P., Structure and development of the insect antennodeutocerebral system. *Int. J. Insect Morphol. Embryol.*, **17** (1988) 243–94.
12. Homberg, U., Christensen, T. A. & Hildebrand, J. G., Structure and function of the deutocerebrum in insects. *Ann. Rev. Entomol.*, **34** (1989) 477–501.
13. Christensen, T. A. & Hildebrand, J. G., Functions, organization, and physiology of the olfactory pathways in the lepidopteran brain. In *Arthropod Brain: Its Evolution, Development, Structure and Functions*, ed. A. P. Gupta. John Wiley, New York, 1987, pp. 457–84.
14. Hildebrand, J. G., Metamorphosis of the insect nervous system: Influences of the periphery on the postembryonic development of the antennal sensory pathway in the brain of *Manduca sexta*. In *Model Neural Networks and Behavior*, ed. A. I. Selverston. Plenum, New York, 1985, pp. 129–48.
15. Hildebrand, J. G., Homberg, U., Kingan, T. G., Christensen, T. A. & Waldrop, B. R., Neurotransmitters and neuropeptides in the olfactory pathway of the sphinx moth *Manduca sexta*. In *Insect Neurochemistry and Neurophysiology 1986*, ed. A. B. Borkovec & D. B. Gelman. The Humana Press, Clifton, NJ, 1986, pp. 255–8.
16. Linn, C. E. Jr & Roelofs, W. L., Response specificity of male moths to multicomponent pheromones. *Chemical Senses*, **14** (1989) 421–37.
17. Tumlinson, J. H., Brennan, M. M., Doolittle, R. E., Mitchell, E. R., Brabham, A., Mazomenos, B. E., Baumhover, A. H. & Jackson, D. M., Identification of a pheromone blend attractive to *Manduca sexta* (L.) males in a wind tunnel. *Arch. Insect Biochem. Physiol.*, **10** (1989) 255–71.
18. Starratt, A. N., Dahm, K. H., Allen, N., Hildebrand, J. G., Payne, T. L. & Röller, H., Bombykal, a sex pheromone of the sphinx moth *Manduca sexta*. *Z. Naturforsch.*, **34c** (1979) 9–12.
19. Murlis, J. & Jones, C. D., Fine scale structure of odour plumes. *Physiol. Entomol.*, **6** (1981) 71–86.
20. Murlis, J., Willis, M. A. & Cardé, R. T., Odour signals: Patterns in time and space. In *ISOT X. Proceedings of the Tenth International Symposium on Olfaction and Taste*, ed. K. B. Døving. University of Oslo, Norway, 1990, pp. 6–17.
21. Linn, C. E. Jr, Campbell, M. G. & Roelofs, W. L., Male moth sensitivity to multicomponent pheromones: The critical role of the female-released blend in determining the functional role of components and the active space of the pheromone. *J. Chem. Ecol.*, **12** (1985) 659–68.
22. Baker, T. C., Upwind flight and casting flight: Complimentary phasic and tonic systems used for location of sex pheromone sources by male moths. In *ISOT X. Proceedings of the Tenth International Symposium on Olfaction and Taste*, ed. K. B. Døving. University of Oslo, Norway, 1990, pp. 18–25.
23. Sanes, J. R. & Hildebrand, J. G., Structure and development of antennae in a moth, *Manduca sexta* (Lepidoptera: Sphingidae). *Devel. Biol.*, **51** (1976) 282–99.
24. Hildebrand, J. G., Matsumoto, S. G., Camazine, S. M., Tolbert, L. P., Blank, S., Ferguson, H. & Ecker, V., Organisation and physiology of antennal centres in the brain of the moth *Manduca sexta*. In *Insect Neurobiology and Pesticide Action (Neurotox 79)*. Soc. Chem. Ind., London, 1980, pp. 375–82.
25. Schweitzer, E. S., Sanes, J. R. & Hildebrand, J. G., Ontogeny of electroantennogram responses in the moth *Manduca sexta*. *J. Insect Physiol.*, **22** (1976) 955–60.
26. Kaissling, K.-E., Hildebrand, J. G. & Tumlinson, J. H., Pheromone receptor cells in the male moth *Manduca sexta*. *Arch. Insect Biochem. Physiol.*, **10** (1989) 273–9.
27. Keil, T. A., Fine structure of the pheromone-sensitive sensilla on the antenna of the hawkmoth, *Manduca sexta*. *Tiss. Cell*, **21** (1989) 139–51.
28. Lee, J.-K. & Strausfeld, N. J., Structure, distribution and number of surface sensilla and their receptor cells on the olfactory appendage of the male moth *Manduca sexta*. *J. Neurocytol.*, **19** (1990) 519–38.

29. Sanes, J. R. & Hildebrand, J. G., Origin and morphogenesis of sensory neurons in an insect antenna. *Devel. Biol.*, **51** (1976) 300–19.
30. Marion-Poll, F. & Tobin, T. R., Olfactory responses of receptor neurons to pulsed stimulation with sex pheromone. In *Neural Mechanisms of Behavior*, ed. J. Erber, R. Menzel, H.-J. Pflüger & D. Todt, Thieme Verlag, Stuttgart, 1989, pp. 247–8.
31. Camazine, S. M. & Hildebrand, J. G., Central projections of antennal sensory neurons in mature and developing *Manduca sexta*. *Soc. Neurosci. Abstr.*, **5** (1979) 155.
32. Matsumoto, S. G. & Hildebrand, J. G., Olfactory mechanisms in the moth *Manduca sexta*: Response characteristics and morphology of central neurons in the antennal lobes. *Proc. Roy. Soc. London. B*, **213** (1981) 249–77.
33. Homberg, U., Montague, R. A. & Hildebrand, J. G. Anatomy of antenno-cerebral pathways in the brain of the sphinx moth *Manduca sexta*. *Cell Tiss. Res.*, **254** (1988) 255–81.
34. Tolbert, L. P. & Hildebrand, J. G., Organization and synaptic ultrastructure of glomeruli in the antennal lobes of the moth *Manduca sexta*: A study using thin sections and freeze-fracture. *Proc. Roy. Soc. London. B*, **213** (1981) 279–301.
35. Waldrop, B. & Hildebrand, J. G., Physiology and pharmacology of acetylcholinergic responses of interneurons in the antennal lobe of the moth *Manduca sexta*. *J. Comp. Physiol. A*, **164** (1989) 433–41.
36. Oland, L. A. & Tolbert, L. P., Glial patterns during early development of antennal lobes in *Manduca sexta*: A comparison between normal lobes and lobes deprived of antennal axons. *J. Comp. Neurol.*, **255** (1987) 196–207.
37. Christensen, T. A. & Hildebrand, J. G., Male-specific, sex pheromone-selective projection neurons in the antennal lobes of the moth *Manduca sexta*. *J. Comp. Physiol. A*, **160** (1987) 553–69.
38. Kanzaki, R., Arbas, E. A., Strausfeld, N. J. & Hildebrand, J. G., Physiology and morphology of projection neurons in the antennal lobe of the male moth *Manduca sexta*. *J. Comp. Physiol. A*, **165** (1989) 427–53.
39. Hansson, B., Christensen, T. A. & Hildebrand, J. G., Functionally distinct subdivisions of the macroglomerular complex in the antennal lobes of the sphinx moth *Manduca sexta*. *J. Comp. Neurol.*, **312** (1991) 264–78.
40. Christensen, T. A., Hildebrand, J. G., Tumlinson, J. H. & Doolittle, R. E., The sex-pheromone blend of *Manduca sexta*: Responses of central olfactory interneurons to antennal stimulation in male moths. *Arch. Insect Biochem. Physiol.*, **10** (1989) 281–91.
41. Christensen, T. A. & Hildebrand, J. G., Representation of sex-pheromonal information in the insect brain. In *ISOT X. Proceedings of the Tenth International Symposium on Olfaction and Taste*, ed. K. B. Døving. University of Oslo, Norway, 1990, pp. 142–50.
42. Olberg, R. M., Pheromone-triggered flip-flopping interneurons in the ventral nerve cord of the silkworm moth, *Bombyx mori*. *J. Comp. Physiol. A*, **152** (1983) 297–307.
43. Christensen, T. A. & Hildebrand, J. G., Frequency coding by central olfactory neurons in the sphinx moth *Manduca sexta*. *Chemical Senses*, **13** (1988) 123–30.
44. Baker, T. C., Sex pheromone communication in the Lepidoptera: New research progress. *Experientia*, **45** (1989) 248–62.
45. Sanes, J. R., Prescott, D. J. & Hildebrand, J. G., Cholinergic neurochemical development of normal and deafferented antennal lobes in the brain of the moth, *Manduca sexta*. *Brain Res.*, **119** (1977) 389–402.
46. Maxwell, G. D., Tait, J. F. & Hildebrand, J. G., Regional synthesis of neurotransmitter candidates in the CNS of the moth *Manduca sexta*. *Comp Biochem. Physiol.*, **61C** (1978) 109–19.

47. Sanes, J. R. & Hildebrand, J. G., Acetylcholine and its metabolic enzymes in developing antennae of the moth, *Manduca sexta. Devel. Biol.*, **52** (1976) 105–20.
48. Hildebrand, J. G., Hall, L. M. & Osmond, B. C., Distribution of binding sites for ^{125}I-labeled α-bungarotoxin in normal and deafferented antennal lobes of *Manduca sexta. Proc. Natl Acad. Sci. USA*, **76** (1979) 499–503.
49. Stengl, M., Homberg, U. & Hildebrand, J. G., Acetylcholinesterase activity in antennal receptor neurons of the sphinx moth *Manduca sexta. Cell Tiss. Res.*, **262** (1990) 245–52.
50. Hoskins, S. G., Homberg, U., Kingan, T. G., Christensen, T. A. & Hildebrand, J. G., Immunocytochemistry of GABA in the antennal lobes of the sphinx moth *Manduca sexta. Cell Tiss. Res.*, **244** (1986) 243–52.
51. Waldrop, B., Christensen, T. A. & Hildebrand, J. G., GABA-mediated synaptic inhibition of projection neurons in the antennal lobes of the sphinx moth *Manduca sexta. J. Comp. Physiol. A*, **161** (1987) 23–32.
52. Kent, K. S., Hoskins, S. G. & Hildebrand, J. G., A novel serotonin-immunoreactive neuron in the antennal lobe of the sphinx moth *Manduca sexta* persists throughout postembryonic life. *J. Neurobiol.*, **18** (1987) 451–65.
53. Homberg, U. & Hildebrand, J. G., Serotonin-immunoreactive neurons in the median protocerebrum and suboesophageal ganglion of the sphinx moth *Manduca sexta. Cell Tiss. Res.*, **258** (1989) 1–24.
54. Homberg, U. & Hildebrand, J. G., Histamine-immunoreactive neurons in the midbrain and suboesophageal ganglion of the sphinx moth *Manduca sexta. J. Comp. Neurol.*, **307** (1991) 647–57.
55. Homberg, U., Kingan, T. G. & Hildebrand, J. G., Distribution of FMRFamide-like immunoreactivity in the brain and suboesophageal ganglion of the sphinx moth *Manduca sexta* and colocalization with SCP$_B$-, BPP-, and GABA-like immunoreactivity. *Cell. Tiss. Res.*, **259** (1990) 401–19.
56. Kingan, T. G., Teplow, D. B., Phillips, J. M., Riehm, J. P., Rao, K. R., Hildebrand, J. G., Homberg, U., Kammer, A. E., Jardine, I., Griffin, P. R. & Hunt, D. F., A new peptide in the FMRFamide family isolated from the CNS of the hawkmoth, *Manduca sexta. Peptides*, **11** (1990) 849–56.
57. Stengl, M. & Hildebrand, J. G., Insect olfactory neurons *in vitro*: Morphological and immunocytochemical characterization of male-specific antennal receptor cells from developing antennae of male *Manduca sexta. J. Neurosci.*, **10** (1990) 837–47.
58. Hishinuma, A., Hockfield, S., McKay, R. & Hildebrand, J. G., Monoclonal antibodies reveal cell-type-specific antigens in the sexually dimorphic olfactory system of *Manduca sexta*. I. Generation of monoclonal antibodies and partial characterization of the antigens. *J. Neurosci.*, **8** (1988) 296–307.
59. Zufall, F., Stengal, M., Franke, C., Hildebrand, J. G. & Hatt, H., Ionic currents of cultured olfactory receptor neurons from antennae of male *Manduca sexta. J. Neurosci.*, **11** (1991) 956–65.
60. Stengl, M., Levine, R. B. & Hildebrand, J. G., Second messengers modulate ion channels in cultured olfactory receptor neurons from male *Manduca sexta* pupae. *ECRO IX Abstr* (1990) p. 96.
61. Stengl, M. & Hildebrand, J., Patch clamp studies on second-messenger mediated calcium regulation in cultured insect olfactory receptor neurons. In *Synapse– Transmission–Modulation. Proceedings of the 19th Göttingen Neurobiology Conference*, ed. N. Elsner & H. Penzlin. Thieme Verlag, Stuttgart, 1991, abstr. 23.
62. Hayashi, J. H. & Hildebrand, J. G., Insect olfactory neurons *in vitro*: Morphological and physiological characterization of cells from the developing antennal lobe of *Manduca sexta. J. Neurosci.*, **10** (1990) 848–59.
63. Hayashi, J. H. & Hildebrand, J. G., Calcium channels in insect central olfactory neurons developing *in vitro. Soc. Neurosci. Abstr.*, **16** (1990) 177.

64. Kanzaki, R., Arbas, E. A. & Hildebrand, J. G., Physiology and morphology of protocerebral olfactory neurons in the male moth *Manduca sexta*. *J. Comp. Physiol. A,* **168** (1991) 281–98.
65. Kanzaki, R., Arbas, E. A. & Hildebrand, J. G., Physiology and morphology of descending neurons in pheromone-processing olfactory pathways in the male moth *Manduca sexta*. *J. Comp. Physiol. A,* **169** (1991) 1–14.

24

Molecular Determinants of Pheromone Activity

JOHN A. PICKETT, LESTER J. WADHAMS & CHRISTINE M. WOODCOCK

Department of Insecticides & Fungicides, AFRC Institute of Arable Crops Research, Rothamsted Experimental Station, Harpenden, Hertfordshire, AL5 2JQ, UK

INTRODUCTION

Some pheromones derive specificity from their molecular structures. However, many contain compounds that are common to pheromones from different species, or that are components of other semiochemical systems. In these cases, specificity can be obtained by employment of mixtures having unique relative proportions, with ecological or temporal isolation of the organism often adding further specificity. Initially, it was expected that pheromones would be single-component, but it is now widely accepted that most are multi-component systems. It is also clear that in many pheromonal systems, behavioural effects do not take place through interactions with the pheromone alone, but require the presence of other semiochemicals, for example, compounds from the host plant.[1,2]

Examples are provided, particularly from aphid chemical ecology, of molecular determinants of pheromonal activity that allow discrimination from other semiochemicals containing similar structures. In addition, a regulated release of individual components and the evolution of highly tuned receptors allow the pheromone to embody an extremely specific message. The roles of other semiochemicals in determining pheromonal activity are also discussed. The very restricted possibilities for creating pheromone mimics and analogues are emphasised.

MOLECULAR STRUCTURE OF PHEROMONES

Structures as simple as that of ethanol can be employed as pheromone components.[3,4] However, within the constraints of volatility required for aerial transport (the structural requirements in aquatic systems are surprisingly similar), considerable specificity can be achieved within one empirical formula by a combination of structural isomerism and stereoisomerism encompassing both optical and geometrical isomerism. Thus, for the lepidopterous sex pheromones, where the basic structures usually comprise only C_{12}—C_{16} straight carbon chains with functionalities of alcohol, aldehyde and acetate, tremendous diversity is obtained by a combination of the position, degree and geometrical isomerism of unsaturated double bonds.[5,6]

Insect pheromones can employ a range of structures, from the fatty acid-based compounds to polyketides and polyisoprenoids. Polyketides can be simple compounds, for example 4-methyl-3,5-heptanedione (I), the aggregation pheromone of the pea and bean weevil, *Sitona lineatus*,[1] or structures with several chiral centres, such as *exo*-brevicomin (II), a component of the

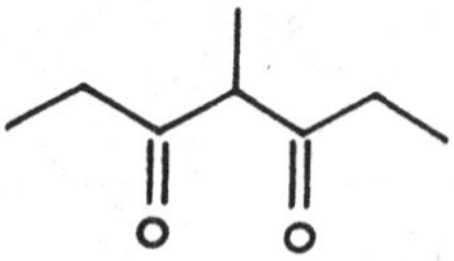

I

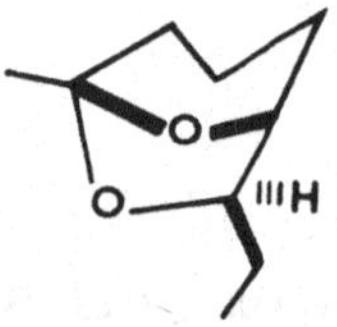

II

aggregation pheromone of bark beetles, including the mountain pine beetle, *Dendroctonus ponderosae*.[7,8] Simple terpenoids can also be employed, for example (*E*)-citral (III) in the Nasonov pheromone of the honeybee[9,10] or the

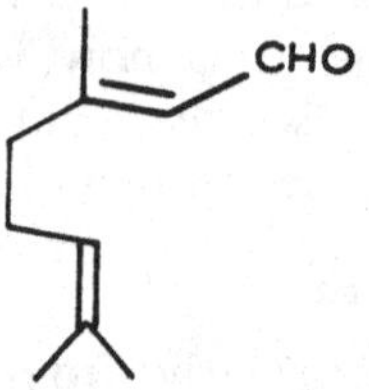

III

highly chiral cyclopentanoid structures comprising the sex pheromones produced by female aphids. For many aphid species, two biosynthetically related compounds are involved, a nepetalactol and a nepetalactone. There are 16 and 8 possible isomers respectively, but only the nepetalactol isomer IV and the nepetalactone V are active.[11] However, the lactone V and some isomers are abundant components of certain plants from the family Labiatae, particularly

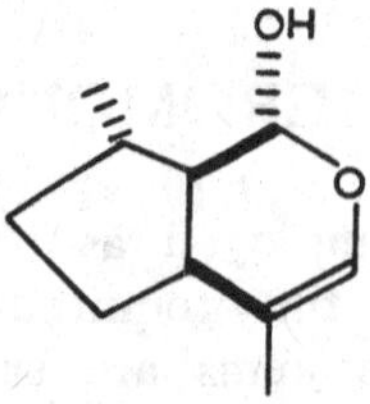

IV

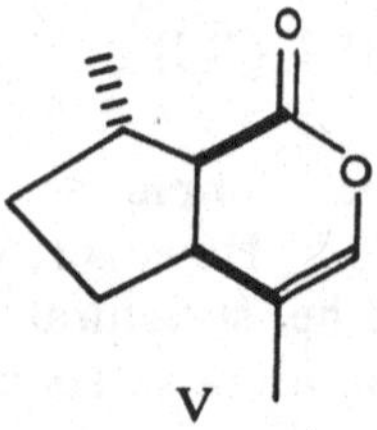

V

those in the *Nepeta* genus, and aphids have developed additional ways to avoid interference with their own pheromonal signalling processes by plant chemicals. This has partly been achieved, together with some species specificity, by employment of particular ratios between components, regulated during release. The regulation mechanism involves enzymic oxidation of IV–V. Thus, pheromone samples entrained from air above scenting females contain a larger proportion of lactone than those extracted directly from the scent gland.[11,12] At the receptor, further specificity is possible and can be studied by electrophysiological recording techniques, particularly single cell recording (SCR).[13] For aphids, the sex pheromone receptors are principally the secondary rhinaria (placoid sensilla) on the male antenna. These are innervated by a number of cells, including two having the dendritic receptors tuned either to the lactol IV or to the lactone V, e.g. for the vetch aphid, *Megoura viciae* (Fig. 1). Dose–response studies on these two cell types clearly show the discrimination each has for its specific compound. For most aphids, the two compounds need to be present in the correct ratio to give a full behavioural response.[11]

OH   OH

VI   VII

The sex pheromone for the damson-hop aphid, *Phorodon humuli*, comprises the diastereoisomeric pair of lactols VI and VII. These compounds alone, placed in slow-release vials above water traps, catch thousands of males, compared to control catches of only tens.[14] The male aphids can be seen struggling to fly upwind, only to be caught in the trap by the synthetic pheromone lure. Single ion monitoring in coupled gas chromatography–mass

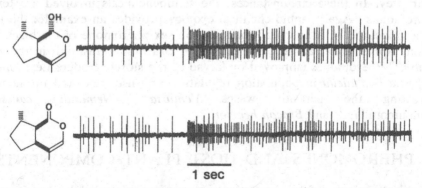

FIG. 1. *Megoura viciae* olfactory cells, secondary rhinarium: response to the lactol IV and the lactone V. (Stimulus concentration = 10 μg/ml, 10 μl applied.)

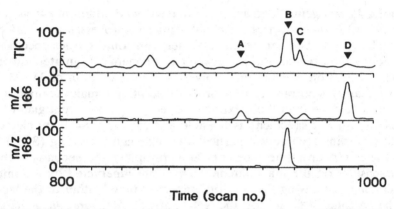

Fɪɢ. 2. GC–MS of volatiles entrained from female *Phorodon humuli*. Total ion current (TIC) and single ion monitoring at *m/z* 166 and 168. A = unknown. B = lactols VI and VII. C and D = lactones isomeric with V.

spectrometry (GC–MS)[15] of molecular ions, from cyclopentanoid lactones at *m/z* 166 and lactols at *m/z* 168, showed the presence of at least two lactones in the volatiles entrained from scenting females (Fig. 2). The two so far characterised showed no activity. Nonetheless, high selection pressure through use of the pheromonal lactols for control purposes could result in *P. humuli* evolving to employ the other compounds, thereby resisting the control strategy. However, the great power of GC–MS in allowing identification of candidate structures that could be the basis of resistance is demonstrated.

PHEROMONES AS KAIROMONES

Use of pheromone signalling systems in herbivores presents an opportunity for interactions at the third trophic level. There are a number of examples where parasitoids and predators have evolved the ability to detect the pheromones of their prey. In these circumstances, the semiochemicals involved are termed 'kairomones'. Again, aphid chemical ecology provides an example: the lactol IV and lactone V, at the ratio found in the sex pheromone of the black bean aphid, *Aphis fabae*, attracts two species of Braconid parasitoids in the genus *Praon*.[16] Pheromones employed by larvae of the stored product Lepidopteran *Ephestia kuehniella* in population regulation are also used as kairomones in attracting the parasitic wasps *Venturia* (=*Nemeritis*) *canescens*, Ichneumonidae,[17] and *Bracon hebetor*.[18]

PHEROMONES AND HOST PLANT COMPONENTS

Chemicals released from host plants are commonly used by herbivores to locate the plant and are also classed as kairomones. The activity of the pea and bean weevil aggregation pheromone (I) is strongly synergised by such host plant-derived kairomones.[1,19] This type of interaction between semiochemicals is also found with many bark beetles[20] and other insects.[2] A number of interesting observations have recently been made in connection with *P. humuli*. Sexual female aphids attract the males on the primary (winter) hosts and for *P. humuli*, these are trees in the genus *Prunus*, Rosaceae. Behavioural studies have demonstrated that volatiles from *Prunus* bark can synergise the activity of the synthetic sex pheromone, and electrophysiological studies are being made on sensilla on the fifth and sixth antennal segments, the primary rhinaria, where cells are situated that respond to plant components. In the case of the migrants which recolonise the rapidly growing hops, *Humulus lupulus* (Cannabaceae), in the spring, host plant attractants have already been identified by coupled gas chromatography–single cell recording (GC–SCR). However, in the same sensilla, it has been possible to detect cells that respond to non-host plant compounds such as the isothiocyanates (C. M. Woodcock, unpublished), which are typical of plants in the Cruciferae. Precursors that slowly release these materials have been used in the field to repel spring migrants from hops and from attractant hop plant components (C. A. M. Campbell, unpublished).

When aphids are attacked, they produce an alarm pheromone, which for most species comprises the sesquiterpene hydrocarbon (E)-β-farnesene (VIII). This compound is detected principally by cells in the primary rhinaria, in this case predominantly on the sixth antennal segment. (E)-β-Farnesene is also commonly found in plants, but usually in association with larger amounts of $(-)$-β-caryophyllene (IX). Other cell types in the fifth antennal segment

VIII   IX

respond to β-caryophyllene and, to a lesser extent, to the molecularly related humulene, e.g. for the peach-potato aphid, *Myzus persicae* (Fig. 3), and these compounds act as pheromone inhibitors: down to a ratio of 0·3:1 of caryophyllene to (E)-β-farnesene, a significant reduction in response was detected.[21] This mechanism is believed to allow aphids to distinguish between (E)-β-farnesene released by plants, and the same molecule released by aphids, when it would be beneficial to respond in typical alarm dispersal behaviour. Aphid alarm pheromones can comprise other components, but even for species where (E)-β-farnesene is the only pheromonal component, other semiochemicals can be required for full activity. The turnip or mustard aphid, *Lipaphis* (=*Hyadaphis*) *erysimi*, employs (E)-β-farnesene as its alarm pheromone, but

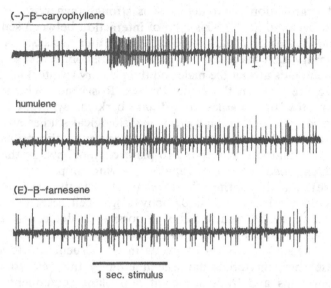

(−)-β-caryophyllene

humulene

(E)-β-farnesene

1 sec. stimulus

FIG. 3. *Myzus persicae* olfactory cell, proximal primary rhinarium: response to plant sesquiterpenes. (Stimulus concentration = 1 mg/ml, 10 μl applied.)

responds only weakly to pure synthetic samples. GC–SCR on the primary rhinaria, using volatiles from crushed aphids, allowed identification of host plant-derived components, the isothiocyanates,which synergised the activity of (E)-β-farnesene. Dose–response analysis in terms of action potential frequency showed allyl isothiocyanate to be the most active and this compound, although behaviourally inactive alone, in combination with synthetic (E)-β-farnesene gave an alarm response equal to that produced by volatiles from the crushed aphids.[22,23]

MOLECULAR DETERMINANTS IN PHEROMONE ANALOGUE ACTIVITY

Pheromone receptors are continuously bombarded by numerous extraneous olfactory signals. Although there are other processes that intervene before the molecular interaction with the receptor cell dendrites, the receptors are considerably more specific than those for internal signals such as neurotransmitters and hormones. It is therefore much more difficult to make analogues with similar activity to the parent pheromone. For lepidopterous sex pheromones, it has been established that the length of the terminal alkyl groups is a critical molecular determinant of activity in these compounds.[24] Normal isosteric replacement, for example replacing the α-methylene of an aldehyde with oxygen to give the formate, which may work in other systems, usually results in an order of magnitude reduction in activity.[25] However, some interesting results have been obtained by replacing hydrogen with fluorine in pheromone molecules. In the case of the aphid alarm pheromone, three hydrogens have been replaced to produce trifluoro-(E)-β-farnesene, and this compound retains high activity.[26] Through merit of the fluorine replacement, it is also more volatile. Indeed, the norfarnesene, with two fluorines at carbon-1 (X), is even more volatile and active. Field trials on *M. persicae*, using

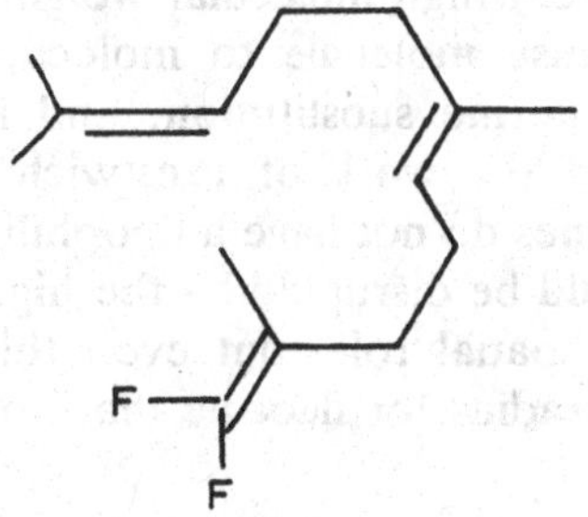

X 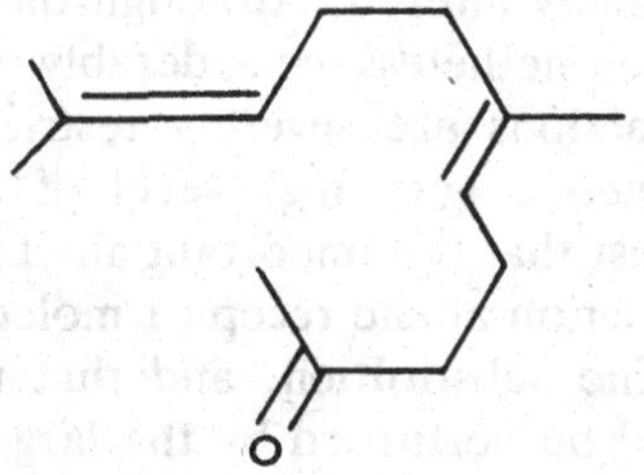   XI

(E)-β-farnesene to increase aphid mobility and thereby improve kill by contact insecticides such as pyrethroids, were more successful in China than in the UK. The possibility that higher field temperatures increased the vapour pressure of the (E)-β-farnesene was investigated by using the more volatile difluoronor-farnesene X, but there was no significant improvement. It had been observed that the keto compound geranylacetone (XI) was also active and since the difluoromethylene of X could be considered to have replaced the carbonyl group in XI, this substitution was also investigated for the sex pheromone of the diamondback moth, *Plutella xylostella*. However, the product had no activity at the levels tested.[26] Nonetheless, a number of studies describe fluorine substitution in lepidopterous pheromones, with either retention or modification of activity through this molecular transformation.[27-29] Recently, Prestwich[30] has produced high fluorine substitution in the terminal alkyl part of lepidopterous pheromones, with interesting modifications to biological activity. However, greater fluorine substitution has been obtained in the mosquito oviposition pheromone. This pheromone comprises a long chain, fatty acid-type acetoxylacetone with two chiral centres (XIIa).[31] The compound itself has

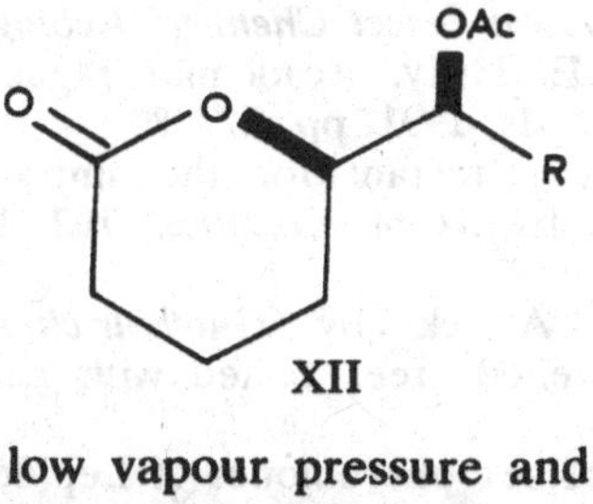

XIIa R = $(CH_2)_9CH_3$

XIIb R = $(CH_2)_2(CF_2)_7CF_3$

XII

a relatively low vapour pressure and attracts mosquitoes to oviposit over a relatively short range. To improve the volatility, the acetoxy group was replaced by trifluoroacetoxy, with surprising retention of activity, although shortening the alkyl chain removed activity,[32] and in spite of the fact that the trifluoroacetoxy group would quite dramatically affect the electron density at the functionality of this compound. Therefore, as an extreme example, all the hydrogen atoms in the alkyl chain, except for the two methylene groups adjacent to the acetoxy function, were replaced with fluorine (XIIb). The two methylene bridging groups were left so that the functionality would be minimally altered. Although the compound has a very high molecular weight, it is nonetheless considerably more volatile because molecule to molecule interactions are severely restricted by the high fluorine substitution, and it retained a very high level of activity.[33] This and the work of Prestwich[30] suggest that the important alkyl groups of pheromones do not have a lipophilic interaction at the receptor molecule, since this would be disrupted by the high fluorine substitution, and that they must have a spatial role, but even this would be perturbed by the larger Van der Waal's radius for fluorine than for hydrogen.

CONCLUSIONS

In describing molecular determinants of pheromone activity, the use of these agents in crop protection has not been discussed, but although this prospect is increasingly promising, and is having greater demands placed on it, there remain many difficulties.[34] From this paper, it is clear that pheromones cannot be developed for agricultural use solely by analogue preparation. The electrophysiological methods and behavioural considerations described here are essential, and must be employed in concert with studies on interactions at the molecular receptors[35] and the further processing of olfactory information within the insect nervous system.[36]

REFERENCES

1. Blight, M. M., Pickett, J. A., Smith, M. C. & Wadhams, L. J., An aggregation pheromone of *Sitona lineatus*. *Naturwissenschaften*, **71** (1984) 480–81.
2. Dickens, J., Green leaf volatiles: a ubiquitous chemical signal modifies insect pheromone responses. In *Proc. Int. Congr. on Insect Chemical Ecology*, Tábor, Czechoslovakia, 12–18 August 1990, ed. E. Hrdy. Academia Prague and SPB Acad. Publishers, The Hague, The Netherlands, 1991, pp. 277–80.
3. Moeck, H. A., Ethanol as the primary attractant for the ambrosia beetle *Trypodendron lineatum* (Coleoptera:Scolytidae). *Can. Entomol.*, **102** (1970) 985–95.
4. McLean, J. A. & Borden, J. H., Attack by *Gnathotrichus sulcatus* (Coleoptera:Scolytidae) on stumps and felled trees baited with sulcatol and ethanol. *Can. Entomol.*, **109** (1977) 675–86.
5. Arn, H., Tóth, M. & Priesner, E., List of sex pheromones of Lepidoptera and related attractants. OILB-SROP Working Group: use of pheromones and other semiochemicals in integrated control, 1986.

6. Mayer, M. S. & McLaughlin, J. R., *Handbook of Insect Pheromones and Sex Attractants.* CRC Press, Florida, 1991.
7. Rudinsky, J. A., Morgan, M. E., Libbey, L. M. & Putnam, T. B., Antiaggregative rivalry pheromone of the mountain pine beetle, and a new arrestant of the southern pine beetle. *Environ. Entomol.,* **3** (1974) 90–98.
8. Libbey, L. M., Ryker, L. C. & Yandell, K. L., Laboratory and field studies of volatiles released by *Dendroctonus ponderosae* Hopkins (Coleoptera: Scolytidae). *Z. Angew. Entomol.,* **100** (1985) 381–92.
9. Pickett, J. A., Williams, I. H., Martin, A. P. & Smith, M. C., Nasonov pheromone of the honey bee, *Apis mellifera* L. (Hymenoptera: Apidae). Part I. Chemical characterization. *J. Chem. Ecol.,* **6** (1980) 425–34.
10. Pickett, J. A., Williams, I. H., Smith, M. C. & Martin, A. P., Nasonov pheromone of the honey bee, *Apis mellifera* L. (Hymenoptera, Apidae). Part III. Regulation of pheromone composition and production. *J. Chem. Ecol.,* **7** (1981) 543–54.
11. Dawson, G. W., Griffiths, D. C., Merritt, L. A., Mudd, A., Pickett, J. A., Wadhams, L. J. & Woodcock, C. M., Aphid semiochemicals—a review, and recent advances on the sex pheromone. *J. Chem. Ecol.,* **16** (1990) 3019–30.
12. Hardie, J., Holyoak, M., Nicholas, J. Nottingham, S. F., Pickett, J. A., Wadhams, L. J. & Woodcock, C. M., Aphid sex pheromone components: age-dependent release by females and species-specific male response. *Chemoecology,* **1** (1990) 63–8.
13. Wadhams, L. J., The use of coupled gas chromatography: electrophysiological techniques in the identification of insect pheromones. In *Chromatography and Isolation of Insect Hormones and Pheromones,* ed. A. R. McCaffery & I. D. Wilson, Plenum, New York/London, 1990, pp. 289–98.
14. Campbell, C. A. M., Dawson, G. W., Griffiths, D. C., Pettersson, J., Pickett, J. A., Wadhams, L. J. & Woodcock, C. M., Sex attractant pheromone of damson-hop aphid *Phorodon humuli* (Homoptera, Aphididae). *J. Chem. Ecol.,* **16** (1990) 3455–65.
15. Pickett, J. A., Gas chromatography–mass spectrometry in insect pheromone identification: three extreme case histories. In *Chromatography and Isolation of Insect Hormones and Pheromones,* ed. A. R. McCaffery & I. D. Wilson. Plenum, New York/London, 1990, pp. 299–309.
16. Hardie, J., Nottingham, S. F., Powell, W. & Wadhams, L. J., Synthetic aphid sex pheromone lures female parasitoids. *Entomol. Exp. Appl.,* **61** (1991) 97–9.
17. Mudd, A. & Corbet, S. A., Response of the ichneumonid parasite *Nemeritis canescens* (Grav.) to kairomones from the flour moth *Ephestia kuehniella* Zeller. *J. Chem. Ecol.,* **8** (1982) 843–50.
18. Strand, M. R., Williams, H. J., Vinson, S. B. & Mudd, A., Arrestment and trail following response of *Bracon hebetor* (Say) to kairomones from *Ephestia kuehniella* Zeller. *J. Chem. Ecol.,* **15** (1989) 1491–1500.
19. Blight, M. M. & Wadhams, L. J., Male-produced aggregation pheromone in pea and bean weevil, *Sitona lineatus* (L.). *J. Chem. Ecol.,* **13** (1987) 733–9.
20. Borden, J. H., Aggregation pheromones in the Scolytidae. In *Pheromones,* ed. M. C. Birch. North-Holland, Amsterdam/London, 1974, pp. 135–60.
21. Dawson, G. W., Griffiths, D. C., Pickett, J. A., Smith, M. C. & Woodcock, C. M., Natural inhibition of the aphid alarm pheromone. *Entomol. exp. appl.,* **36** (1984) 197–9.
22. Dawson, G. W., Griffiths, D. C., Pickett, J. A., Wadhams, L. J. & Woodcock, C. M., Plant compounds that synergise activity of the aphid alarm pheromone. *1986 Brit. Crop Prot. Conf.—Pests and Diseases,* (1986) 829–33.
23. Dawson, G. W., Griffiths, D. C., Pickett, J. A., Wadhams, L. J. & Woodcock, C. M., Plant-derived synergists of alarm pheromone from turnip aphid, *Lipaphis* (*Hyadaphis*) *erysimi* (Homoptera, Aphididae). *J. Chem. Ecol.,* **13** (1987) 1663–71.

24. Bestmann, H. J., Cai-Hong, W., Döhla, B., Li-Kedong & Kaissling, K. E., Functional group recognition of pheromone molecules by sensory cells of *Antheraea polyphemus* and *Antheraea pernyi* (Lepidoptera: Saturniidae). *Pheromones,* **56** (1986) 435–41.
25. Beevor, P. S., Hall, D. R., Nesbitt, B. F., Dyck, V. A., Arida, G., Lippold, P. C. & Oloumi-Sadeghi, H., Field trials of the synthetic sex pheromones of the striped rice borer, *Chilo suppressalis* (Walker) (Lepidoptera: Pyralidae), and of related compounds. *Bull. ent. Res.,* **67** (1977) 439–47.
26. Briggs, G. G., Cayley, G. R., Dawson, G. W., Griffiths, D. C., Macaulay, E. D. M., Pickett, J. A., Pile, M. M., Wadhams, L. J. & Woodcock, C. M., Some fluorine-containing pheromone analogues. *Pestic. Sci.,* **17** (1986) 441–8.
27. Camps, F., Coll, J., Fabrias, G. & Guerrero, A., Synthesis of dienic fluorinated analogs of insect sex pheromones. *Tetrahedron,* **40** (1984) 2871–8.
28. Sun, W.-C. & Prestwich, G. D., Partially fluorinated analogs of (Z)-9-dodecenyl acetate: probes for pheromone hydrophobicity requirements. *Tetrahedron Lett.,* **31** (1990) 801–4.
29. Bengtsson, M., Rauscher, St., Arn, H., Sun, W.-C. & Prestwich, G. D., Fluorine-substituted pheromone components affect the behavior of the grape berry moth. *Experientia,* **46** (1990) 1211–13.
30. Prestwich, G. D., Sun, W.-C., Mayer, M. S. & Dickens, J. C., Perfluorinated moth pheromones: synthesis and electrophysiological activity. *J. Chem. Ecol.,* **16** (1990) 1761–78.
31. (a) Laurence, B. R. & Pickett, J. A., *erythro*-6-Acetoxy-5-hexadecanolide, the major component of a mosquito oviposition attractant pheromone. *J. Chem. Soc., Chem. Commun.* (1982) 59–60; (b) Laurence, B. R., Mori, K., Otsuka, T., Pickett, J. A. & Wadhams, L. J., Absolute configuration of mosquito oviposition attractant pheromone, 6-acetoxy-5-hexadecanolide. *J. Chem. Ecol.,* **11** (1985) 643–8.
32. Laurence, B. R. & Pickett, J. A., An oviposition attractant pheromone in *Culex quinquefasciatus* Say (Diptera: Culicidae). *Bull. ent. Res.,* **75** (1985) 283–90.
33. Dawson, G. W., Mudd, A., Pickett, J. A., Pile, M. M. & Wadhams, L. J., Convenient synthesis of mosquito oviposition pheromone and a highly fluorinated analog retaining biological activity. *J. Chem. Ecol.,* **16** (1990) 1779–89.
34. Pickett, J. A., Wadhams, L. J. & Woodcock, C. M., New approaches to the development of semiochemicals for insect control. In *Proc. Int. Congr. on Insect Chemical Ecology,* Tábor, Czechoslovakia, 12–18 August 1990, ed. E. Hrdy. Academia Prague and SPB Publishers, The Hague, The Netherlands, 1991, pp. 333–45.
35. Prestwich, G. D., Adventures in photoaffinity labeling, hormone carriers and receptors in arthropods. In *Proc. Int. Congr. on Insect Chemical Ecology,* Tábor, Czechoslovakia, 12–18 August 1990, ed. E. Hrdy. Academia Prague and SPB Publishers, The Hague, The Netherlands, 1991, pp. 21–33.
36. Hildebrand, J. G., *et al.* Olfaction in *Manduca sexta:* cellular mechanisms of responses to sex pheromone. In *Neurotox '91: Molecular Basis of Drug and Pesticide Action,* ed. I. R. Duce. Elsevier Science Publishers, London, 1992, pp. 323–38.

NEUROTOX '91

Epilogue

The science presented at NEUROTOX '91 was unquestionably first class and clearly illustrates the progress that has been made in insect neuroscience since York 1979. It is normally the duty of the author of the Epilogue to summarise the science, to identify the highlights and to present a view of the future. I will part from this tradition because I believe there is value in responding to points raised in the Prologue to this volume. In other words, I wish to dwell upon the *raison d'être* for the NEUROTOX symposia, which has been brought into sharp focus by Dr Gunther Voss and Dr Rainer Neumann.

It was the very recognition that academic or basic insect neuroscientists and industrial pesticide scientists populate contrasting worlds, yet have some common goals, that led the late Dr Charles Potter and I to propose to the Society of Chemical Industry the establishment of the NEUROTOX symposia. What objectives did we have in mind at that time? In contemplating this, it is important to bear in mind that the forces driving pesticide research were very different from their contemporary counterparts and academic science, at least in the United Kingdom, had not yet even conceived the idea of a Research Defence Society. The first essential was to provide a meeting place and talk-shop where industrialists and academics could exhibit their research wares. It was hoped that this market would engender collaborative and cooperative programmes of research which would bridge the industrial/academic divide. Perhaps naively, we believed that this would encourage some academics to reorientate their research towards industry, and catalyse within the chemical industry a greater sense of responsibility for the promotion of basic insect neuroscience. A further, major objective was the need to foster younger neuroscientists, to provide them with the experience of scientific discourse in a critical but supportive environment and to expose them to prospective employers, both academic and industrial. So far, I have purposely avoided mentioning rational discovery of toxicants, since this was not an explicit feature of the NEUROTOX philosophy. It is true that at that time Dr Potter and I strongly adhered to the view that rational discovery and design of pesticides should be given greater consideration by chemical companies; a view which was discussed at NEUROTOX '79, but also a view that had been promulgated extensively many years before that meeting.

NEUROTOX is no longer the sole international meeting of its type and it will survive in a competitive environment only if it continues to meet with the approval of its core membership. NEUROTOX '79 was criticised for its failure to cater sufficiently for pesticide chemists. NEUROTOX '85 was criticised for failing adequately to reflect developments in molecular biology and to address the problems of resistance. NEUROTOX '88 escaped major criticisms, but perhaps that was a criticism itself! Maybe it is a sign of the growing maturity of the NEUROTOX series that the presenter of the Prologue for NEUROTOX '91 felt able to launch a strong criticism of the concept of rational pesticide design and discovery which indirectly questioned the usefulness of meetings like NEUROTOX. The response of participants was thoughtfully receptive. However, to ensure that this response is not misconstrued, let me attempt to answer some of the questions raised in their Prologue by our two colleagues from CIBA-GEIGY Ltd.

I am convinced that the academic insect neuroscience community has a clear view of the difficulties facing the pesticide industry in its attempts to discover new, 'green' insecticides. I remain less convinced that industry understands the difficulties facing the academic insect neuroscience community. Even implicit demands that the research of academic insect neuroscientists should be subservient to the needs of industry are unlikely to be welcomed by peer groups assessing research grant proposals, since these are mainly comprised of academics who have no contact with, and little interest in, the pesticide industry.

The irritation and disappointment felt by Drs Voss and Neumann over the failure of rational science, as opposed, presumably, to random screening, to produce new insecticides is understandable, and their dissatisfaction is shared by many academics. However, those of us who who have always looked for success in the longer term would probably say that we remain impatient rather than irritated and disappointed. Is it really appropriate for the industrial scientist to even expect success in terms of new products to emanate directly from academia? I think not. Any suggestions that academic scientists have reneged on promises to make such discoveries proves that NEUROTOX has not yet attained its prime objective of bridging the gulf in understanding between its two communities.

What then is the perceived role of the academic insect neuroscientist? An overt industry-orientated strategy is a recipe for disaster because of peer review, unless industry is willing to fund most of the insect neuroscience research in universities and polytechnics which is directed towards its interests. Should academic insect neuroscientists retreat into their ivory towers, thereby competing on more equal terms for government funding, to randomly sally forth with tidings of new advances in basis insect neuroscience for the consumption of their peers and for consideration by their industrial counterparts? This dichotomy of basic and applied interests would take us full circle back to pre-NEUROTOX days and really identify responsibility for pesticide discovery and development, where it has always been, i.e. in the province of the industrial scientist. Should academic insect neuroscientists interested in solving basic research problems work with commercially important insects? Yes, but only if in doing so they do not prejudice the outcome of their research and its relevance to basic neuroscience. Research goals, feasibility and compatibility factors must define which insect is the most appropriate subject for a basic research project. Until industry can offer career-long research support matched by institutional recognition of industry-orientated research, academic insect neuroscientists will have to continue to accept the disciplines of the world of basic science in which they spend a large proportion of their time. Basic insect neuroscience research, driven solely by individual and corporate curiosity will continue for as long as funds are made available to support it.

Does the academic neuroscientist have the responsibility of supporting industry in its public defence of neuroactive pesticides? I think not. Academic freedom is a privilege which should be prized. There is much fertile, common ground between academic and industrial neuroscientists, but academics would be wise to maintain their independence of thought and action. At times, this may lead them into conflict with industry, but on other occasions it may enable them to give uncompromised support to the aims and aspirations of their industrial counterparts.

So what does the academic scientist have to offer the pesticide industry? Information, advice and expertise are the three areas which immediately spring to mind.

Information generated by basic research can provide industrial scientists with an invaluable database for objective decision making. Whether this process leads to rationally-based pesticide research either sponsored or in-house will be a measure of the quality of industrial science. Of course, some academics will continue to prevail upon the pesticide industry for investment in their specific goal-orientated research programmes. If there is to be any criticism of the manner in which industry has assessed past proposals from academia, it has been a failure to seek independent expert advice. This lack of peer review, understandably due to the need for confidentiality, has probably contributed to some of the disillusionment over the outcome of industry/academia collaborative research.

Advice from the academic neuroscientist could be of various types, ranging from assistance with establishing new *in-vitro* screens and the identification of novel targets, through to help with the evaluation of site and mode of action of neurotoxicants. Provision of expertise in the form of secondment of established scientists and of a pool of trained, younger personnel is the third contribution that can be made by academic neuroscientists in their support of the pesticide industry. In fact, a survey of contemporary establishments of industrial pesticide biologists gives clear testimony to the past success of academia in meeting industry's demands for well-trained insect neuroscientists.

It is not possible, of course, without resort to anecdote, to quantify the role played by NEUROTOX in meeting these three objectives, but the continuing support of industry, both upfront financially, and in terms of attendance suggests that the NEUROTOX symposia have made important contributions.

I do not wish at this time to engage in the debate on the so-called biorational pesticide discovery versus random screening. This is a problem exclusively for the pesticide industry; as much a matter of internal politics as a matter of survival. It serves little purpose for the academic to draw parallels with the pharmaceutical industry, in which the biorational approach has had a much longer lifespan and is a proven success, since the objectives, traditions and funding of pharmaceutical and agrochemical research differ in many important respects.

What of the future? The development of our knowledge of insect neuro-science will continue apace and opportunities for exploitation of ideas arising from this development will provide opportunities for the pesticide industry. My first hope is that my industrial colleagues will continue to have the inclination and resources to avail themselves of these opportunities. My second hope is that industrial and academic scientists will take very seriously the comments made by Drs Voss and Neumann and that they will be constructively and extensively debated.

A prosperous future for the pesticide industry seems secure. Can the same be said about the future of academic insect neuroscience?

NEUROTOX '94 is on the horizon and will soon be taking shape. It should be an interesting meeting!

Professor Peter N. R. Usherwood
Department of Life Science
University of Nottingham
Nottingham
NG7 2RD
UK

Author Index

353

Subject Index

Printed in the United States
By Bookmasters